D1228344

DYNAMICS

Thomas R. Kane

STANFORD UNIVERSITY

HOLT, RINEHART AND WINSTON, INC.

NEW YORK • CHICAGO • SAN FRANCISCO • ATLANTA • DALLAS

MONTREAL • TORONTO • LONDON

The author thanks Academic Press, Inc., for permission to use material from *Analytical Elements of Mechanics*, Vols. I and II.

To ANN, LINDA, and JEFFREY

Library of Congress Catalog Card Number: 68–26826
2711604
Printed in the United States of America
1 2 3 4 5 6 7 8 9

Preface

TO EXPLAIN the underlying rationale for this book, it may be helpful to describe how the material is presently being used by the author.

The Department of Applied Mechanics of Stanford University offers a sequence of three graduate courses entitled Dynamics. Each of these lasts for one academic quarter, a period of about ten weeks, with three fifty-minute class meetings per week. Classes are composed of students holding B.S. or M.S. degrees in various branches of engineering or related fields.

Normally, two of the three class meetings each week are devoted to the discussion of material treated in the text, whereas the third deals primarily with questions related to the Problem Sets at the end of each part of the book. The sections of the text that form the theoretical basis for problem solutions are listed at the beginning of each Problem Set, and students are encouraged to attack problems as soon as the requisite theory has been covered in class. By the end of the third quarter, the entire text has usually been studied, and most students have solved nearly every problem in the Problem Sets.

Two basic decisions were made in selecting and arranging the contents of the book. First, when it became clear that limitations on students' time would not permit both thorough exposition of general methods for solving problems and extensive study of classical solutions of specific problems, it was decided to stress the former, bringing in the latter only to the extent that it might enhance understanding of fundamental principles. In other words, primary emphasis was placed on presenting material that students could use to solve *new* problems, rather than on attempting to create an encyclopedia of solutions obtained in the past. Consequently, such famous subjects as the motion of a symmetrical top are touched upon in connection with specific points of theory (for example, see the examples in Sections 4.10 and 6.9), but are not necessarily treated in a unified or formal manner. The second major decision concerned the arrangement of topics. The need for such a decision arises from the fact that the two objectives of stimulating interest by providing strong motivation and facilitating understanding by arranging topics in logically sound order cannot always be accomplished simultaneously: The attempt to explain why a subject must be studied tends to disrupt the logical exposition of theory, and preoccupation with purely rational aspects of a question can stifle interest. To resolve this dilemma, it was decided to strive for a logically sound arrangement of topics in the book, but to depart from this when necessary in the classroom in order to provide adequate motivation. For example, the formal statement of a law of motion is deferred to Section 5.1, because it involves quantities that have not been studied fully until this point has been reached; and the special form of this law that applies to the solution of equilibrium problems appears even later, in Section 5.9. But this need not prevent a teacher from using Equation (5.8) to solve an illustrative problem immediately after discussing Section 3.2, where generalized forces are introduced; and the law of motion expressed in Equations (5.1) can be employed for the solution of interesting particle dynamics problems as soon as generalized inertia forces have been defined, as in Section 3.9. Furthermore, one can jump to Sections 3.2 and 3.9 directly from Section 2.13; that is, a preliminary look at generalized forces requires only prior consideration of partial rates of change of position. Hence, a teacher using the text with discretion can quite easily let students know right from the outset and throughout the course why each topic is being studied; and the endeavor to do this can be stimulating for teachers as well as students.

Related questions arise in connection with reviewing material that, presumably, has been covered in earlier courses. Such review is generally necessary, for one cannot assume that all students have reached the same level of proficiency in dealing with every topic. For example, experience indicates that some students can be expected to be thoroughly familiar with three-dimensional kinematics, whereas others possess only rudimentary notions of such concepts as, for example, angular velocity and angular acceleration. Review material is, therefore, included in the text, usually immediately *following* the point at which it is first needed, this arrangement being chosen in order to show as clearly as possible why the topic under consideration is being discussed. For the purpose of classroom presentation, one can, of course, reverse the order of presentation, omit certain sections, or add items as required for a particular group of students.

Finally, still on the subject of the use of the book, a few words about proofs are in order. Although they are presented in the text in considerable detail, they do not necessarily need to receive extensive attention in the classroom. On the contrary, if one believes that sound proofs ultimately form the only firm basis for the methods employed in the solution of problems, but that students gain more from leisurely perusal of proofs than from a presentation conducted at a pace selected by the instructor, one can make sparing use of this material in lectures, but encourage students to study it outside of class. The attempt may then be made in the classroom to clarify particularly difficult points or to focus attention on the subtler facets of a derivation.

The book contains a number of departures from conventional presentations of the subject of dynamics. These were made solely in order to facilitate specific tasks. For example, the introduction of *partial rates of change of orientation* and *partial rates of change of position* (see Sections 2.4, 2.12, and 2.23) makes it possible to define two classes of generalized forces in a unified way, both for holonomic and nonholonomic systems (see Sections 3.2 and 2.9); it leads to formulas that are helpful in the actual evaluation of generalized forces (see Sections 3.4, 3.5, and 3.10); it provides a means for dealing with virtual velocities, virtual displacements, and virtual work in simple, analytic terms (see Sections 2.25–2.30 and 4.14–4.16); and it brings the discussion of generalized impulse and generalized momentum into close harmony with topics treated earlier [see Sections 5.18 and Problems 13(a) and 13(b)]. Similarly, the use of *second moments* permits the introduction of products and moments of inertia in coordinate-independent form (see Section 3.12); it simplifies the derivation of transformation formulas (see Section 3.13); it sheds light on the subject of principal moments of inertia (see Section 3.16); and it provides a convenient approach to the discussion of inertia dyadics (see Section 3.21). As a final example, the *matrix notation* introduced in Section 6.2 opens the door to a presentation of Hamilton-Jacobi theory (see Sections 6.11–6.14) without use of the calculus of variations, thus making it unnecessary either to teach this subject during a course in dynamics or to require that it be studied prior to such a course.

Some comments on the subject of notation may prove helpful. In dynamics, as in other areas of applied mathematics, one is faced with a dilemma arising from the fact that, frequently, two attributes of a good notation are mutually exclusive, namely, maximum simplicity and the capability of conveying a maximum amount of information. For example, the equation

$$^{R}\mathbf{v}^{P} = {}^{R}\mathbf{v}^{\bar{B}} + {}^{B}\mathbf{v}^{P}$$

which appears in Section 2.16, is a velocity relationship involving four entities called B, P, R, and $\bar{B}$, and each of these comes into evidence explicitly in one or more superscripts. If one adopts the convention that a right superscript refers to an object in motion and a left superscript to the reference frame under consideration, then each term in the equation becomes essentially self-explanatory; that is, the equation is seen to state that the velocity of P in reference frame R is equal to the sum of the velocity of $\bar{B}$ in R and the velocity of P in reference frame B. However, a simpler representation of this relationship, such as, for example,

$$\mathbf{u} = \mathbf{v} + \mathbf{w}$$

can of course, be devised, and this is easier to read, to write, and to remember, but not to comprehend, than the more elaborate form of the equation. Where conflicts of this sort arose, the attempt was made to resolve them in favor of clarity, but with minimum complexity. Thus, the relationship between the velocities in a reference frame R of two points P and Q which are fixed on a rigid body B is stated in Section 2.15 as

$$\mathbf{v}^P = \mathbf{v}^Q + \boldsymbol{\omega} \times \mathbf{r}$$

and the reader must remember without the aid of clues the $\boldsymbol{\omega}$ represents the angular velocity of body B in reference R and that $\mathbf{r}$ designates the position vector of point P relative to point Q. The more elaborate form

$${}^R\mathbf{v}^P = {}^R\mathbf{v}^Q + {}^R\boldsymbol{\omega}^B \times \mathbf{r}^{P/Q}$$

which is explicit on these points, was rejected because of its possibly excessive complexity.

To simplify the solution of certain frequently encountered notation problems, a number of practices were adopted arbitrarily. These are described in the next paragraph.

The phrase "in a reference frame R" sometimes conveys essential information about angular velocities, velocities of points, accelerations of points, angular momenta, and so forth. In connection with some of these, a phrase such as "relative to point P" must be added to identify a particular reference point. This is the case, for example, when one speaks of the angular momentum *in inertial space* I of a satellite S *relative to the mass center* C of S. Under these circumstances, a letter appearing as a left superscript designates the reference frame, a right superscript refers to the body under consideration, and a slash followed by a letter in a right superscript indicates the reference point. Thus, if $\mathbf{H}$ stands for angular momentum, then ${}^I\mathbf{H}^{S/C}$ denotes the angular momentum in reference frame I of body S relative to point C. Omission of left superscripts throughout an equation implies that *any* reference frame may be used, provided that the same one be used throughout the equation.

In conclusion, I wish to acknowledge my indebtedness to all authors on whose work I have drawn and to express my gratitude both to Stanford University and to the many students who have generously provided assistance in the preparation of the manuscript for this book.

Stanford, California
June 1968

THOMAS R. KANE

Contents

PART I

KINEMATICS

THE CONCEPTS OF VELOCITY OF a point and angular velocity of a rigid body are familiar from elementary mechanics, where each is generally regarded as a function of a single scalar variable, the time t, and is defined in terms of derivatives of certain vectors with respect to this variable. The theory of ordinary differentiation of vector functions of a single variable thus lies at the roots of elementary kinematics. When using various advanced methods of mechanics, one must regard velocities and angular velocities as functions of several scalar variables—such as, for example, the so-called generalized coordinates and their time derivatives. A theory of differentiation of these functions is then required.

Purely mathematical aspects of this topic are considered in the first part of this chapter, and the theory there set forth is then used, in the second part, to define and discuss the quantities later needed for the solution of physical problems.

Chapter **1**

Differentiation of Vector Functions

1.1 Vector Functions. If either the magnitude of a vector **v** and/or the direction of **v** in a reference frame R depends on a scalar variable q, **v** is called a *vector function of* q *in* R. Otherwise, **v** is said to be *independent of* q *in* R.

▪ EXAMPLE

In Figure 1.1, P represents a point that is free to move on the surface

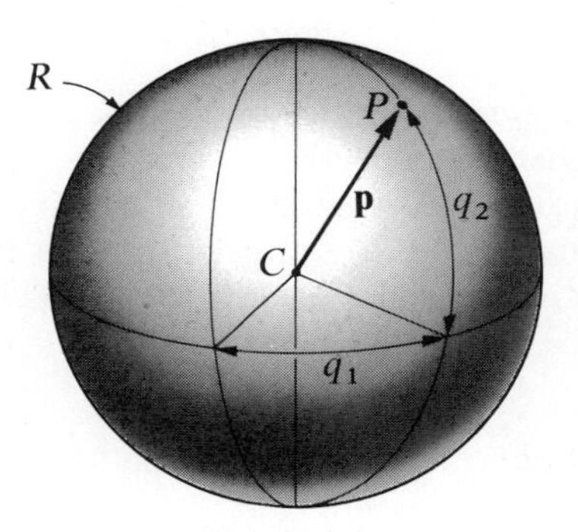

Figure 1.1

of a rigid sphere. The sphere may be regarded as a reference frame R, and, if **p** is the position vector of P relative to the center C of the sphere, and

q_1 and q_2 are the angles shown, then **p** is a vector function of q_1 and q_2 in R, because the direction of **p** in R depends on q_1 and q_2.

1.2 Reference Frames. A vector **v** may be a function of a variable q in one reference frame, but be independent of this variable in another reference frame.

▪ EXAMPLE

Figure 1.2 is a schematic representation of a gyroscope. If A, B, and C

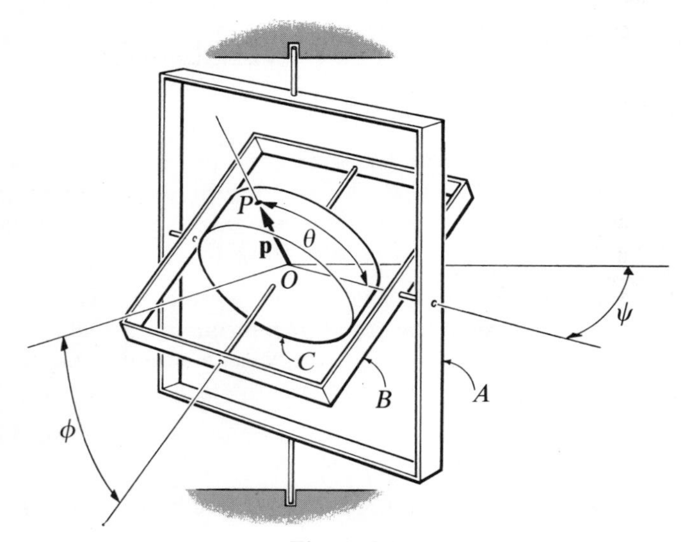

Figure 1.2

designate reference frames comprised of the outer gimbal ring, the inner gimbal ring, and the rotor, respectively, and P is a point on the periphery of the rotor, then the position vector **p** of P relative to the center O of the rotor is a vector function of the angles θ and ϕ in reference frame A; depends on θ, but is independent of ϕ, in B; and is independent of both θ and ϕ in C.

1.3 Scalar Functions. Given a reference frame R and a vector function **v** of n scalar variables $q_1, \cdots, q_n$ in R (see Section 1.1), let $\mathbf{n}_1, \mathbf{n}_2, \mathbf{n}_3$ be a set of nonparallel, noncoplanar (but not necessarily mutually perpendicular) unit vectors fixed in R. Then there exists a unique set of scalar functions v_1, v_2, v_3 of $q_1, \cdots, q_n$ such that

$$\mathbf{v} = v_1\mathbf{n}_1 + v_2\mathbf{n}_2 + v_3\mathbf{n}_3 \tag{1.1}$$

The functions v_1, v_2, and v_3 characterize the behavior of **v** in R completely. However, their specification does not provide sufficient information for a description of the behavior of **v** in any reference frame other than R. For example, if v_1, v_2, and v_3 are all independent of, say, q_r, then **v** is independent of q_r in R, but **v** may nevertheless be a function of q_r in some other reference frame.

Use of Equation (1.1) can be both a help and hindrance. On the one hand, it is restrictive, because it necessitates selection of a particular set of unit vectors. On the other hand, it facilitates many theoretical developments by providing a bridge that connects scalar and vector analysis and thus makes it possible to extend concepts familiar from scalar analysis, such as continuity, differentiability, and so forth, to vector analysis.

▪ EXAMPLE

In Figure 1.3 (which shows the gyroscope previously discussed in the example in Section 1.2), $\mathbf{a}_1$, $\mathbf{a}_2$, $\mathbf{a}_3$ and $\mathbf{b}_1$, $\mathbf{b}_2$, $\mathbf{b}_3$ designate mutually perpendicular unit vectors fixed in reference frames A and B, respectively. The vector **p** can be referred to each of these sets of unit vectors; that is, **p** can be expressed both as

$$\mathbf{p} = {}^{A}p_1\mathbf{a}_1 + {}^{A}p_2\mathbf{a}_2 + {}^{A}p_3\mathbf{a}_3 \tag{a}$$

and as

$$\mathbf{p} = {}^{B}p_1\mathbf{b}_1 + {}^{B}p_2\mathbf{b}_2 + {}^{B}p_3\mathbf{b}_3 \tag{b}$$

Specifically, if r is the radius of the rotor, it can be seen from Figure 1.3

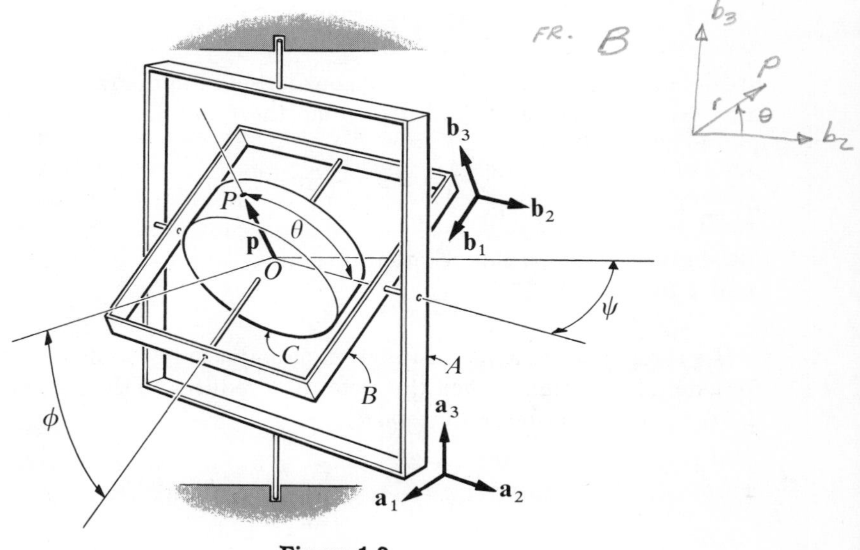

Figure 1.3

that

$$\mathbf{p} = r(\cos\theta\mathbf{b}_2 + \sin\theta\mathbf{b}_3) \tag{c}$$

Comparison with (b) then shows that

$$^{B}p_1 = 0 \qquad ^{B}p_2 = r\cos\theta \qquad ^{B}p_3 = r\sin\theta$$

Again from Figure 1.3,

$$\mathbf{b}_2 = \mathbf{a}_2$$

and

$$\mathbf{b}_3 = \sin\phi\mathbf{a}_1 + \cos\phi\mathbf{a}_3$$

Substitution into (c) thus gives

$$\mathbf{p} = r(\sin\theta\sin\phi\mathbf{a}_1 + \cos\theta\mathbf{a}_2 + \sin\theta\cos\phi\mathbf{a}_3) \tag{d}$$

and comparison with (a) shows that

$$^{A}p_1 = r\sin\theta\sin\phi \qquad ^{A}p_2 = r\cos\theta$$

$$^{A}p_3 = r\sin\theta\cos\phi$$

1.4 First Partial Derivatives, Ordinary Derivative. If $\mathbf{v}$ is a vector function of n scalar variables $q_1, \cdots, q_n$ in a reference frame R (see Section 1.1), then n vectors, called *first partial derivatives of* $\mathbf{v}$ *in* R and denoted by the symbols

$$\frac{^{R}\partial\mathbf{v}}{\partial q_r} \quad \text{or} \quad \frac{^{R}\partial}{\partial q_r}(\mathbf{v}) \quad \text{or} \quad {^{R}\partial\mathbf{v}}/\partial q_r \qquad r = 1, \cdots, n$$

are defined as follows: Let $\mathbf{n}_1$, $\mathbf{n}_2$, $\mathbf{n}_3$ be any set of nonparallel, noncoplanar unit vectors fixed in R, and determine the functions v_1, v_2, v_3 such that (see Section 1.3)

$$\mathbf{v} = v_1\mathbf{n}_1 + v_2\mathbf{n}_2 + v_3\mathbf{n}_3 \tag{1.2}$$

Then

$$\frac{^{R}\partial\mathbf{v}}{\partial q_r} = \sum_{i=1}^{3} \frac{\partial v_i}{\partial q_r}\mathbf{n}_i \qquad r = 1, \cdots, n \tag{1.3}$$

If $\mathbf{v}$ is regarded as a vector function of only a single scalar variable in R, for example the time t, then this definition reduces to that of the *ordinary derivative of* $\mathbf{v}$ *with respect to* t *in* R,

$$\frac{^{R}d\mathbf{v}}{dt} = \sum_{i=1}^{3} \frac{dv_i}{dt}\mathbf{n}_i \tag{1.4}$$

▪ EXAMPLE

The first partial derivatives of $\mathbf{p}$ with respect to θ and ϕ in reference frames A and B of the example in Section 1.3 are found as follows:

To evaluate partial derivatives of $\mathbf{p}$ in A, refer $\mathbf{p}$ to the unit vectors $\mathbf{a}_1$, $\mathbf{a}_2$, $\mathbf{a}_3$ fixed in A [see Equation (d) in the foregoing example]:

$$\mathbf{p} = r(\sin\theta \sin\phi \mathbf{a}_1 + \cos\theta \mathbf{a}_2 + \sin\theta\cos\phi \mathbf{a}_3)$$

Then

$$\begin{aligned}\frac{{}^A\partial}{\partial\theta}(\mathbf{p}) &= \left[\frac{\partial}{\partial\theta}(r\sin\theta\sin\phi)\right]\mathbf{a}_1 \\ &\quad + \left[\frac{\partial}{\partial\theta}(r\cos\theta)\right]\mathbf{a}_2 + \left[\frac{\partial}{\partial\theta}(r\sin\theta\cos\phi)\right]\mathbf{a}_3 \\ &= r(\cos\theta\sin\phi\mathbf{a}_1 - \sin\theta\mathbf{a}_2 + \cos\theta\cos\phi\mathbf{a}_3)\end{aligned}$$

and

$$\begin{aligned}\frac{{}^A\partial}{\partial\phi}(\mathbf{p}) &= \left[\frac{\partial}{\partial\phi}(r\sin\theta\sin\phi)\right]\mathbf{a}_1 \\ &\quad + \left[\frac{\partial}{\partial\phi}(r\cos\theta)\right]\mathbf{a}_2 + \left[\frac{\partial}{\partial\phi}(r\sin\theta\cos\phi)\right]\mathbf{a}_3 \\ &= r(\sin\theta\cos\phi\mathbf{a}_1 - \sin\theta\sin\phi\mathbf{a}_3)\end{aligned}$$

${}^B\partial\mathbf{p}/\partial\theta$ and ${}^B\partial\mathbf{p}/\partial\phi$ are found by using $\mathbf{p}$ as given in Equation (c) of the example in Section 1.3:

$$\begin{aligned}\frac{{}^B\partial}{\partial\theta}(\mathbf{p}) &= \left[\frac{\partial}{\partial\theta}(0)\right]\mathbf{b}_1 + \left[\frac{\partial}{\partial\theta}(r\cos\theta)\right]\mathbf{b}_2 + \left[\frac{\partial}{\partial\theta}(r\sin\theta)\right]\mathbf{b}_3 \\ &= r(-\sin\theta\mathbf{b}_2 + \cos\theta\mathbf{b}_3)\end{aligned}$$

and

$$\begin{aligned}\frac{{}^B\partial}{\partial\phi}(\mathbf{p}) &= \left[\frac{\partial}{\partial\phi}(0)\right]\mathbf{b}_1 + \left[\frac{\partial}{\partial\phi}(r\cos\theta)\right]\mathbf{b}_2 + \left[\frac{\partial}{\partial\phi}(r\sin\theta)\right]\mathbf{b}_3 \\ &= 0\end{aligned}$$

The last result agrees with the statement in the example in Section 1.2 that $\mathbf{p}$ is independent of ϕ in reference frame B.

1.5 Notation. From the fact that partial derivatives of vector functions depend on the reference frame in which the differentiations are performed, it follows that a notation such as $\partial\mathbf{v}/\partial q_r$—that is, one that contains no mention of a reference frame—is meaningful only when either the context clearly implies use of a particular reference frame or the reference frame can be chosen arbitrarily. In the sequel, it is to be under-

stood that, wherever no reference frame is mentioned explicitly, any reference frame may be used, but all partial differentiations in any one equation must be performed in the same reference frame.

1.6 Partial Differentiation of a Sum. If $\mathbf{v}_1, \cdots, \mathbf{v}_N$ are vector functions of n scalar variables $q_1, \cdots, q_n$ (see Section 1.1), then

$$\frac{\partial}{\partial q_r} \sum_{i=1}^{N} \mathbf{v}_i = \sum_{i=1}^{N} \frac{\partial}{\partial q_r} (\mathbf{v}_i) \qquad r = 1, \cdots, n \tag{1.5}$$

Proof: Considering any reference frame, let $\mathbf{n}_1, \mathbf{n}_2, \mathbf{n}_3$ be a set of non-coplanar unit vectors fixed in this reference frame, and express $\mathbf{v}_i$ in the form[1]

$$\mathbf{v}_i \underset{(1.1)}{=} \sum_{j=1}^{3} v_{ij}\mathbf{n}_j \tag{a}$$

Then

$$\sum_{i=1}^{N} \mathbf{v}_i \underset{(a)}{=} \sum_{i=1}^{N} \left(\sum_{j=1}^{3} v_{ij}\mathbf{n}_j \right) = \sum_{j=1}^{3} \left(\sum_{i=1}^{N} v_{ij} \right) \mathbf{n}_j \tag{b}$$

and

$$\frac{\partial}{\partial q_r} \sum_{i=1}^{N} \mathbf{v}_i \underset{(1.3)}{=} \sum_{j=1}^{3} \left(\frac{\partial}{\partial q_r} \sum_{i=1}^{N} v_{ij} \right) \mathbf{n}_j \tag{c}$$

Now, from scalar calculus,

$$\frac{\partial}{\partial q_r} \sum_{i=1}^{N} v_{ij} = \sum_{i=1}^{N} \frac{\partial v_{ij}}{\partial q_r} \tag{d}$$

Hence

$$\begin{aligned} \frac{\partial}{\partial q_r} \sum_{i=1}^{N} \mathbf{v}_i &\underset{(c),(d)}{=} \sum_{j=1}^{3} \left(\sum_{i=1}^{N} \frac{\partial v_{ij}}{\partial q_r} \right) \mathbf{n}_j \\ &= \sum_{i=1}^{N} \left(\sum_{j=1}^{3} \frac{\partial v_{ij}}{\partial q_r} \mathbf{n}_j \right) \\ &\underset{(1.3)}{=} \sum_{i=1}^{N} \frac{\partial}{\partial q_r} (\mathbf{v}_i) \end{aligned}$$

[1] Numbers or letters beneath equal signs are intended to direct the reader's attention to corresponding equations.

1.7 Partial Differentiation of Products. The rules governing partial differentiation of products involving vector functions are analogous to those for ordinary differentiation of such products. Thus, if s is a scalar function of n scalar variables $q_1, \cdots, q_n$, and $\mathbf{v}$ and $\mathbf{w}$ are vector functions of these variables (see Section 1.1), then

$$\frac{\partial}{\partial q_r}(s\mathbf{v}) = \frac{\partial s}{\partial q_r}\mathbf{v} + s\frac{\partial \mathbf{v}}{\partial q_r} \tag{1.6}$$

$$\frac{\partial}{\partial q_r}(\mathbf{v}\cdot\mathbf{w}) = \frac{\partial \mathbf{v}}{\partial q_r}\cdot\mathbf{w} + \mathbf{v}\cdot\frac{\partial \mathbf{w}}{\partial q_r} \tag{1.7}$$

$$\frac{\partial}{\partial q_r}(\mathbf{v}\times\mathbf{w}) = \frac{\partial \mathbf{v}}{\partial q_r}\times\mathbf{w} + \mathbf{v}\times\frac{\partial \mathbf{w}}{\partial q_r} \tag{1.8}$$

More generally, if P is the product of N scalar and/or vector functions F_i, $i = 1, \cdots, N$, that is,

$$P = F_1 F_2 \cdots F_N$$

then, if all symbols of operation—such as dots, crosses, parentheses—are kept in place,

$$\frac{\partial P}{\partial q_r} = \frac{\partial F_1}{\partial q_r}F_2F_3 \cdots F_N + F_1\frac{\partial F_2}{\partial q_r}F_3 \cdots F_N + \cdots + F_1F_2 \cdots F_{N-1}\frac{\partial F_N}{\partial q_r} \tag{1.9}$$

▪ EXAMPLE

$$\frac{\partial}{\partial q_r}[s\mathbf{u}\times(\mathbf{v}\times\mathbf{w})] = \frac{\partial s}{\partial q_r}\mathbf{u}\times(\mathbf{v}\times\mathbf{w}) + s\frac{\partial \mathbf{u}}{\partial q_r}\times(\mathbf{v}\times\mathbf{w}) + s\mathbf{u}\times\left(\frac{\partial \mathbf{v}}{\partial q_r}\times w\right) + s\mathbf{u}\times\left(\mathbf{v}\times\frac{\partial \mathbf{w}}{\partial q_r}\right)$$

1.8 Second Partial Derivatives. In general, ${}^R\partial\mathbf{v}/\partial q_r$ (see Section 1.4) is a vector function of $q_1, \cdots, q_n$ both in R and in any other reference frame R', and can, therefore, be differentiated with respect to any one of these variables in either R or R'. The result is called a second partial derivative and, if the two differentiations are performed in different reference frames, the order in which the differentiations are performed *cannot*, in general, be interchanged without affecting the result. If both differentiations are performed in the same reference frame, the order is immaterial.

1.9 Total Derivative. If $\mathbf{v}$ is a vector function of n scalar variables $q_1, \cdots, q_n$ in a reference frame R (see Section 1.1), and $q_1, \cdots, q_n$ are functions of a single scalar variable t, then $\mathbf{v}$ may be regarded as a vector function of t in R, and the ordinary or *total derivative* of $\mathbf{v}$ with respect to t in R [see Equation (1.3)] can be expressed in terms of partial derivatives as

$$\frac{{}^R d\mathbf{v}}{dt} = \sum_{r=1}^{n} \frac{{}^R \partial \mathbf{v}}{\partial q_r} \dot{q}_r \tag{1.10}$$

where $\dot{q}_r$ denotes the first derivative of q_r with respect to t.

Proof: With v_1, v_2, v_3 and $\mathbf{n}_1, \mathbf{n}_2, \mathbf{n}_3$ defined as in Section 1.4, v_i may be regarded as a function of $q_1, \cdots, q_n$, for which, from scalar calculus,

$$\frac{dv_i}{dt} = \sum_{r=1}^{n} \frac{\partial v_i}{\partial q_r} \dot{q}_r$$

Hence,

$$\frac{{}^R d\mathbf{v}}{dt} \underset{(1.4)}{=} \sum_{i=1}^{3} \sum_{r=1}^{n} \frac{\partial v_i}{\partial q_r} \dot{q}_r \mathbf{n}_i$$

$$= \sum_{r=1}^{n} \left(\sum_{i=1}^{3} \frac{\partial v_i}{\partial q_r} \mathbf{n}_i \right) \dot{q}_r$$

$$\underset{(1.3)}{=} \sum_{r=1}^{n} \frac{{}^R \partial \mathbf{v}}{\partial q_r} \dot{q}_r$$

▪ EXAMPLE

Referring to the example in Section 1.3, and assuming that the gyroscope is made to move in such a way that

$$\theta = 0.275\omega t \qquad \phi = 0.2 \sin 0.5\omega t$$

where ω is a constant, one may use the results obtained in the example in Section 1.4 to determine the velocity of point P in reference frame A at time $t = 20\pi/\omega$. To this end, recall that the velocity ${}^A\mathbf{v}^P$ of P in A (see Figure 1.3 for the vector $\mathbf{p}$) is defined as

$${}^A\mathbf{v}^P = \frac{{}^A d\mathbf{p}}{dt}$$

Hence

$${}^A\mathbf{v}^P \underset{(1.10)}{=} \frac{{}^A \partial \mathbf{p}}{\partial \theta} \dot{\theta} + \frac{{}^A \partial \mathbf{p}}{\partial \phi} \dot{\phi}$$

$$= r(\cos\theta \sin\phi \mathbf{a}_1 - \sin\theta \mathbf{a}_2 + \cos\theta \cos\phi \mathbf{a}_3)\dot{\theta}$$
$$+ r(\sin\theta \cos\phi \mathbf{a}_1 - \sin\theta \sin\phi \mathbf{a}_3)\dot{\phi}$$

From the given expressions for θ and ϕ,

$$\dot{\theta} = 0.275\omega \qquad \dot{\phi} = 0.1\omega \cos (0.5\omega t)$$

and, at $t = 20\pi/\omega$,

$$\theta = \frac{11\pi}{2} \qquad \dot{\theta} = 0.275\omega$$

$$\phi = 0 \qquad \dot{\phi} = 0.1\omega$$

Hence, at $t = 20\pi/\omega$,

$$^{A}\mathbf{v}^{P} = r\omega(-0.1\mathbf{a}_1 + 0.275\mathbf{a}_2)$$

1.10 Relationship between the Total Derivative and Partial Derivatives. If $q_1, \cdots, q_n$ are n scalar functions of a single variable t, and $\mathbf{v}$ is a vector function of the $n + 1$ variables $q_1, \cdots, q_n$, and t in a reference frame R (see Section 1.1), then $\mathbf{v}$ may be regarded as a vector function of t in R, and the ordinary or *total derivative of* $\mathbf{v}$ with respect to t in R [see Equation (1.3)] can be expressed in terms of partial derivatives as

$$\frac{^{R}d\mathbf{v}}{dt} = \sum_{r=1}^{n} \frac{^{R}\partial \mathbf{v}}{\partial q_r} \dot{q}_r + \frac{^{R}\partial \mathbf{v}}{\partial t} \tag{1.11}$$

where $\dot{q}_r$ denotes the first derivative of q_r with respect to t.

Proof: Referring to Section 1.9, replace the symbol n with $\bar{n}$. Then

$$\frac{^{R}d\mathbf{v}}{dt} \underset{(1.10)}{=} \sum_{r=1}^{\bar{n}} \frac{^{R}\partial \mathbf{v}}{\partial q_r} \dot{q}_r$$

$$= \sum_{r=1}^{\bar{n}-1} \frac{^{R}\partial \mathbf{v}}{\partial q_r} \dot{q}_r + \frac{^{R}\partial \mathbf{v}}{\partial q_{\bar{n}}} \dot{q}_{\bar{n}}$$

Now let

$$q_{\bar{n}} = t$$

so that

$$\dot{q}_{\bar{n}} = 1$$

and take

$$\bar{n} = n + 1$$

Then Equation (1.11) follows immediately.

▪ EXAMPLE

Referring to the example in Section 1.3, suppose that the rotor is driven in such a way that θ is given by

$$\theta = \Omega t$$

where Ω is a constant. Then [see Equation (d) in the example of Section 1.3] $\mathbf{p}$ can be expressed as

$$\mathbf{p} = r(\sin \Omega t \sin \phi \mathbf{a}_1 + \cos \Omega t \mathbf{a}_2 + \sin \Omega t \cos \phi \mathbf{a}_3)$$

and $\mathbf{p}$ can be regarded as a vector function of the two variables ϕ and t in reference frame A, where ϕ is, itself, a function of t. The velocity ${}^A\mathbf{v}^P$ of P in A can then be found as follows:

$$\begin{aligned}
{}^A\mathbf{v}^P = \frac{{}^A d}{dt}(\mathbf{p}) &\underset{(1.11)}{=} \frac{{}^A\partial}{\partial \phi}(\mathbf{p})\dot{\phi} + \frac{{}^A\partial}{\partial t}(\mathbf{p}) \\
&\underset{(1.3)}{=} r(\sin \Omega t \cos \phi \mathbf{a}_1 - \sin \Omega t \sin \phi \mathbf{a}_3)\dot{\phi} \\
&\qquad + r(\Omega \cos \Omega t \sin \phi \mathbf{a}_1 - \Omega \sin \Omega t \mathbf{a}_2 \\
&\qquad + \Omega \cos \Omega t \cos \phi \mathbf{a}_3)
\end{aligned}$$

(It may be verified that, for $\Omega = 0.275\omega$, $\phi = 0.2 \sin 0.5\omega t$, and $t = 20\pi/\omega$, this expression gives the same result as that obtained in the example in Section 1.9.)

1.11 Total and Partial Differentiation. If $\mathbf{v}$ is a vector function of the $n + 1$ scalar variables $q_1, \cdots, q_n$, and t in a reference frame R (see Section 1.1), and $q_1, \cdots, q_n$ are, themselves, functions of t, the order of the operations of total differentiation with respect to t and partial differentiation with respect to q_r may be interchanged:

$$\frac{d}{dt}\frac{\partial \mathbf{v}}{\partial q_r} = \frac{\partial}{\partial q_r}\frac{d\mathbf{v}}{dt} \qquad r = 1, \cdots, n \tag{1.12}$$

Proof:

$$\begin{aligned}
\frac{d}{dt}\left(\frac{\partial \mathbf{v}}{\partial q_r}\right) &\underset{(1.11)}{=} \sum_{s=1}^{n} \frac{\partial}{\partial q_s}\left(\frac{\partial \mathbf{v}}{\partial q_r}\right)\dot{q}_s + \frac{\partial}{\partial t}\left(\frac{\partial \mathbf{v}}{\partial q_r}\right) \\
&\underset{(\text{Sec. } 1.8)}{=} \sum_{s=1}^{n} \frac{\partial}{\partial q_r}\left(\frac{\partial \mathbf{v}}{\partial q_s}\right)\dot{q}_s + \frac{\partial}{\partial q_r}\left(\frac{\partial \mathbf{v}}{\partial t}\right) \\
&\underset{(\text{Sec. } 1.6)}{=} \frac{\partial}{\partial q_r}\left(\sum_{s=1}^{n} \frac{\partial \mathbf{v}}{\partial q_s}\dot{q}_s + \frac{\partial \mathbf{v}}{\partial t}\right) \\
&\underset{(1.11)}{=} \frac{\partial}{\partial q_r}\left(\frac{d\mathbf{v}}{dt}\right)
\end{aligned}$$

Chapter **2**

Rates of Change of Position and Orientation

2.1 Configuration, Constraints, Constraint Equations. The *configuration* of a set S of N particles $P_1, \cdots, P_N$ in a reference frame R is known whenever the position vector of each particle relative to a point fixed in R is known. Thus N vector quantities or, equivalently, $3N$ scalar quantities are required for the specification of the configuration of S in R.

If the motion of S is affected by the presence of bodies that come into contact with one or more of the particles of S, restrictions may be imposed on the positions that the particles may occupy or on the manner in which these positions are permitted to change. S is then said to be *subject to constraints*. For example, when a sphere rolls on a plane fixed in a reference frame R, the particle at the center of the sphere must remain at a fixed distance from the plane, and the particle instantaneously in contact with the plane must have zero velocity in R; hence, only configurations and changes in configuration consistent with these requirements are permissible.

An equation that expresses a requirement of this sort is called a *constraint equation*. If $\mathbf{n}_x$, $\mathbf{n}_y$, $\mathbf{n}_z$ are mutually perpendicular unit vectors fixed in R, and the position vector $\mathbf{p}_i$ of a typical particle P_i of S is expressed in the form

$$\mathbf{p}_i = x_i\mathbf{n}_x + y_i\mathbf{n}_y + z_i\mathbf{n}_z \tag{2.1}$$

then a constraint equation is said to be *holonomic* when it can be expressed in the form

$$f(x_1, y_1, z_1, \cdot \cdot \cdot , x_N, y_N, z_N, t) = 0 \tag{2.2}$$

Otherwise, it is called *nonholonomic.*[1] Holonomic constraint equations are further classified as *rheonomic* and *scleronomic*, according to whether the function f does, or does not, contain t explicitly.

▪ EXAMPLE

In Figure 2.1, N designates a plane that is made to rotate with constant

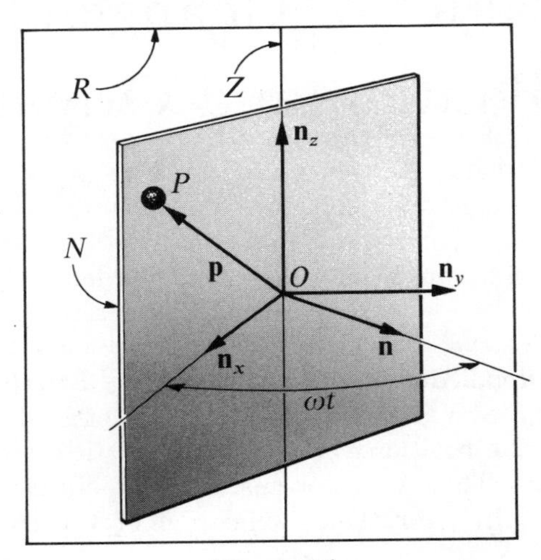

Figure 2.1

angular speed ω about a line Z fixed in N and in a reference frame R, and $\mathbf{n}$ is a unit vector normal to N. If P is a particle that remains at all times in contact with N, then the position vector $\mathbf{p}$ of P relative to O must satisfy the equation

$$\mathbf{p} \cdot \mathbf{n} = 0$$

If $\mathbf{n}_x$, $\mathbf{n}_y$, $\mathbf{n}_z$ are mutually perpendicular unit vectors fixed in R, and $\mathbf{n}_x$ coincides with $\mathbf{n}$ at time $t = 0$, then

$$\mathbf{p} = x\mathbf{n}_x + y\mathbf{n}_y + z\mathbf{n}_z$$

where x, y, and z are functions of t; and $\mathbf{n}$ can be expressed as

$$\mathbf{n} = \cos \omega t\mathbf{n}_x + \sin \omega t\mathbf{n}_y$$

Hence

$$\mathbf{p} \cdot \mathbf{n} = x \cos \omega t + y \sin \omega t$$

[1] See Section 2.22 for examples of nonholonomic constraint equations.

and the requirement that P must remain in N is expressed by the rheonomic constraint equation

$$x \cos \omega t + y \sin \omega t = f(x, y, z, t) = 0$$

2.2 Holonomic Systems, Generalized Coordinates. When a set S of N particles $P_1, \cdots, P_N$ is subject to constraints (see Section 2.1) that can be represented completely by M independent holonomic constraint equations [see Equation (2.2)], only

$$n = 3N - M \tag{2.3}$$

of the $3N$ quantities $x_1, y_1, z_1, \cdots, x_N, y_N, z_N$ are independent of each other. Consequently, it may be possible to express each of $x_1, y_1, z_1, \cdots, x_N, y_N, z_N$ as a single-valued function of n functions of t—say $q_1(t), \cdots, q_n(t)$, and t itself—in such a way that the constraint equations are satisfied for all values of $q_1, \cdots, q_n$, and t; and the position vector of each particle relative to a point fixed in R may then be regarded as a single-valued vector function of $q_1, \cdots, q_n$, and t in R. Under these circumstances, S is said to be a *holonomic system possessing n degrees of freedom in R*, and $q_1(t), \cdots, q_n(t)$ are called *generalized coordinates of S in R*.

▪ EXAMPLE

In the example in Section 2.1, $N = 1$ and $M = 1$. Hence $n = 2$; and x, y, and z can each be expressed in terms of two generalized coordinates, q_1 and q_2, and the time t in such a way that the constraint equation

$$x \cos \omega t + y \sin \omega t = 0$$

is satisfied for all values of q_1, q_2, and t. This may be accomplished by taking, for example,

$$x = -q_1 \sin \omega t \quad y = q_1 \cos \omega t \quad z = q_2$$

or

$$x = q_1 \sin \omega t \quad y = -q_1 \cos \omega t \quad z = q_2$$

Physically, the absolute value of the generalized coordinate q_1 then represents the distance from P to line Z, because this distance is equal to $(x^2 + y^2)^{1/2}$, and

$$x^2 + y^2 = q_1^2$$

The generalized coordinate q_2 is simply identical with the cartesian coordinate z.

Alternatively, generalized coordinates q_1 and q_2 may be introduced by letting

$$x = -q_2 \sin q_1 \sin \omega t$$
$$y = q_2 \sin q_1 \cos \omega t$$
$$z = q_2 \cos q_1$$

in which case q_1 and q_2 can be regarded as polar coordinates of P in the rotating plane N, as indicated in Figure 2.2.

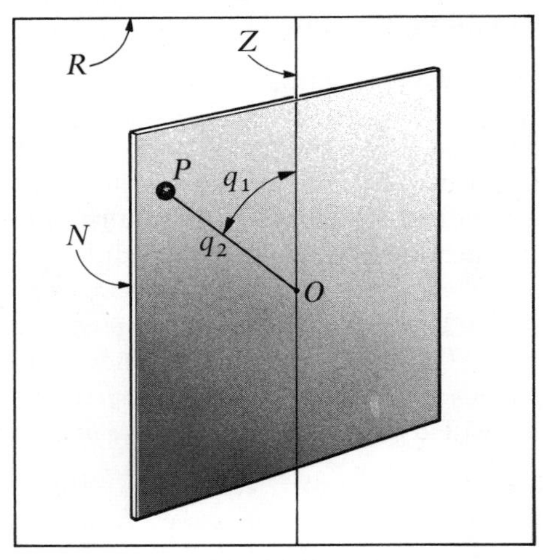

Figure 2.2

2.3 Number of Degrees of Freedom. Formally, the number n of degrees of freedom of a holonomic system S in a reference frame R (see Section 2.2) can be found only by determining the number M of constraint equations and subtracting M from $3N$. However, as n is equal to the number of generalized coordinates—that is, the number of scalar quantities required for the specification of all permissible configurations of S in R—it is frequently possible to find n by inspection.

▪ EXAMPLE

If S is a set of N particles forming a rigid body, every configuration of S in a reference frame R can be specified in terms of three cartesian coordinates of one particle of S, together with three angles that determine the orientation of the body in R. Thus it appears that S possesses six degrees of freedom in R. The same result is obtained formally by letting $\mathbf{p}_1, \cdots, \mathbf{p}_N$ be the position vectors of the particles $P_1, \cdots, P_N$ relative to a point fixed in R and noting that rigidity can be guaranteed by letting P_1, P_2, and P_3 be noncollinear particles and requiring (a) that the distances between P_1 and P_2, P_2 and P_3, and P_3 and P_1 remain constant, so that

$$\begin{aligned}(\mathbf{p}_1 - \mathbf{p}_2)^2 &= \text{const}\\ (\mathbf{p}_2 - \mathbf{p}_3)^2 &= \text{const}\\ (\mathbf{p}_3 - \mathbf{p}_1)^2 &= \text{const}\end{aligned}$$

and (b) that the distances between each of the remaining $N - 3$ particles and P_1, P_2, and P_3 remain constant; that is,

$$\begin{aligned}(\mathbf{p}_i - \mathbf{p}_1)^2 &= \text{const} \qquad i = 4, \cdots, N\\ (\mathbf{p}_i - \mathbf{p}_2)^2 &= \text{const} \qquad i = 4, \cdots, N\\ (\mathbf{p}_i - \mathbf{p}_3)^2 &= \text{const} \qquad i = 4, \cdots, N\end{aligned}$$

The number M of holonomic constraint equations is thus given by

$$M = 3 + 3(N - 3) = 3N - 6$$

and it follows that

$$n \underset{(2.3)}{=} 3N - (3N - 6) = 6$$

2.4 Partial Rates of Change of Orientation. If a rigid body B forms part of a holonomic system S that possesses n degrees of freedom in a reference frame R (see Section 2.2), and $q_1, \cdots, q_n$ are generalized coordinates of S in R, the angular velocity[2] $\boldsymbol{\omega}$ of B in R can be expressed in the form

$$\boldsymbol{\omega} = \sum_{r=1}^{n} \boldsymbol{\omega}_{\dot{q}_r}\dot{q}_r + \boldsymbol{\omega}_t \tag{2.4}$$

where $\dot{q}_r$ denotes the derivative of q_r with respect to time t, and $\boldsymbol{\omega}_{\dot{q}_r}$ and $\boldsymbol{\omega}_t$ are certain vector functions of $q_1, \cdots, q_n$, and t in R (see Section 1.1). $\boldsymbol{\omega}_{\dot{q}_r}$ is called the *partial rate of change with respect to q_r of the orientation of B in R.* (The choice of notation is explained in Section 2.5.)

$\boldsymbol{\omega}_{\dot{q}_r}$ and $\boldsymbol{\omega}_t$ may be regarded as "operators" that produce partial derivatives when crossed into any vector $\mathbf{c}$ fixed in the body B:

$$\frac{\partial \mathbf{c}}{\partial q_r} = \boldsymbol{\omega}_{\dot{q}_r} \times \mathbf{c} \qquad r = 1, \cdots, n \tag{2.5}$$

$$\frac{\partial \mathbf{c}}{\partial t} = \boldsymbol{\omega}_t \times \mathbf{c} \tag{2.6}$$

[2] The subject of angular velocity is reviewed in detail in Sections 2.7–2.10.

For later reference it is recalled that $\boldsymbol{\omega}$, itself, may be regarded as an operator that produces total time derivatives by cross multiplication:

$$\frac{d\mathbf{c}}{dt} = \boldsymbol{\omega} \times \mathbf{c} \tag{2.7}$$

Proof: Suppose that P and Q are particles of B, and let $\mathbf{p}$ and $\mathbf{q}$ be their position vectors relative to a point fixed in R. Then $\mathbf{p}$ and $\mathbf{q}$ may be regarded as vector functions of $q_1, \cdots, q_n$, and t in R (see Section 2.2). Now, every vector $\mathbf{c}$ fixed in B can be expressed as the difference between two vectors such as $\mathbf{p}$ and $\mathbf{q}$. Hence every vector $\mathbf{c}$ fixed in B may be regarded as a vector function of $q_1, \cdots, q_n$, and t in R.

Let $\mathbf{n}_1$, $\mathbf{n}_2$, $\mathbf{n}_3$ be a right-handed set of mutually perpendicular unit vectors fixed in B, and let c_1, c_2, c_3 be scalars such that

$$\mathbf{c} = c_1\mathbf{n}_1 + c_2\mathbf{n}_2 + c_3\mathbf{n}_3 \tag{a}$$

Then, since $\mathbf{c}$ is fixed in B by hypothesis, c_1, c_2, and c_3 are constants, and

$$\frac{\partial \mathbf{c}}{\partial q_r} \underset{(1.5)}{=} c_1 \frac{\partial \mathbf{n}_1}{\partial q_r} + c_2 \frac{\partial \mathbf{n}_2}{\partial q_r} + c_3 \frac{\partial \mathbf{n}_3}{\partial q_r} \tag{b}$$

The unit vectors $\mathbf{n}_1$, $\mathbf{n}_2$, and $\mathbf{n}_3$ satisfy a number of relations of interest in the sequel:

$$\mathbf{n}_1 \cdot \mathbf{n}_1 = 1 \quad \mathbf{n}_1 \cdot \mathbf{n}_3 = 0 \tag{c}$$

$$\mathbf{n}_1 \times \mathbf{n}_1 = 0 \quad \mathbf{n}_2 \times \mathbf{n}_1 = -\mathbf{n}_3 \quad \mathbf{n}_3 \times \mathbf{n}_1 = \mathbf{n}_2 \tag{d}$$

It follows that [see Equation (1.7)]

$$\frac{\partial \mathbf{n}_1}{\partial q_r} \cdot \mathbf{n}_1 \underset{(c)}{=} 0 \tag{e}$$

and

$$\frac{\partial \mathbf{n}_1}{\partial q_r} \cdot \mathbf{n}_3 + \mathbf{n}_1 \cdot \frac{\partial \mathbf{n}_3}{\partial q_r} \underset{(c)}{=} 0 \tag{f}$$

If $\boldsymbol{\omega}_{\dot{q}_r}$ is now defined by the equation

$$\boldsymbol{\omega}_{\dot{q}_r} = \frac{\partial \mathbf{n}_2}{\partial q_r} \cdot \mathbf{n}_3\mathbf{n}_1 + \frac{\partial \mathbf{n}_3}{\partial q_r} \cdot \mathbf{n}_1\mathbf{n}_2 + \frac{\partial \mathbf{n}_1}{\partial q_r} \cdot \mathbf{n}_2\mathbf{n}_3 \qquad r = 1, \cdots, n \tag{g}$$

then cross multiplication with $\mathbf{n}_1$ gives

$$\boldsymbol{\omega}_{\dot{q}_r} \times \mathbf{n}_1 \underset{(d)}{=} -\frac{\partial \mathbf{n}_3}{\partial q_r} \cdot \mathbf{n}_1\mathbf{n}_3 + \frac{\partial \mathbf{n}_1}{\partial q_r} \cdot \mathbf{n}_2\mathbf{n}_2$$

or, after elimination of $(\partial \mathbf{n}_3/\partial q_r) \cdot \mathbf{n}_1$ by use of Equation (f),

$$\boldsymbol{\omega}_{\dot{q}_r} \times \mathbf{n}_1 = \frac{\partial \mathbf{n}_1}{\partial q_r} \cdot \mathbf{n}_2\mathbf{n}_2 + \frac{\partial \mathbf{n}_1}{\partial q_r} \cdot \mathbf{n}_3\mathbf{n}_3 \tag{h}$$

Now, any vector $\mathbf{v}$ can be expressed as

$$\mathbf{v} = \mathbf{v} \cdot \mathbf{n}_1\mathbf{n}_1 + \mathbf{v} \cdot \mathbf{n}_2\mathbf{n}_2 + \mathbf{v} \cdot \mathbf{n}_3\mathbf{n}_3$$

Hence, with $\mathbf{v}$ replaced by $\partial\mathbf{n}_1/\partial q_r$,

$$\frac{\partial \mathbf{n}_1}{\partial q_r} = \frac{\partial \mathbf{n}_1}{\partial q_r} \cdot \mathbf{n}_1\mathbf{n}_1 + \frac{\partial \mathbf{n}_1}{\partial q_r} \cdot \mathbf{n}_2\mathbf{n}_2 + \frac{\partial \mathbf{n}_1}{\partial q_r} \cdot \mathbf{n}_3\mathbf{n}_3$$

In view of Equation (e), the right-hand members of this equation and of (h) are seen to be identical. Consequently,

$$\frac{\partial \mathbf{n}_1}{\partial q_r} = \boldsymbol{\omega}_{\dot{q}_r} \times \mathbf{n}_1 \tag{i}$$

Cyclic permutation of the subscripts 1, 2, and 3 in Equation (g) transforms $\boldsymbol{\omega}_{\dot{q}_r}$ into itself. Hence, cross multiplications of Equation (g) with $\mathbf{n}_2$ and $\mathbf{n}_3$, rather than with $\mathbf{n}_1$, lead to

$$\frac{\partial \mathbf{n}_2}{\partial q_r} = \boldsymbol{\omega}_{\dot{q}_r} \times \mathbf{n}_2 \tag{j}$$

and to

$$\frac{\partial \mathbf{n}_3}{\partial q_r} = \boldsymbol{\omega}_{\dot{q}_r} \times \mathbf{n}_3 \tag{k}$$

in place of Equation (i), and it follows that

$$\begin{aligned}
\frac{\partial \mathbf{c}}{\partial q_r} &\underset{(b)}{=} c_1\underset{(i)}{\boldsymbol{\omega}_{\dot{q}_r} \times \mathbf{n}_1} + c_2\underset{(j)}{\boldsymbol{\omega}_{\dot{q}_r} \times \mathbf{n}_2} + c_3\underset{(k)}{\boldsymbol{\omega}_{\dot{q}_r} \times \mathbf{n}_3} \\
&= \boldsymbol{\omega}_{\dot{q}_r} \times (c_1\mathbf{n}_1 + c_2\mathbf{n}_2 + c_3\mathbf{n}_3) \\
&\underset{(a)}{=} \boldsymbol{\omega}_{\dot{q}_r} \times \mathbf{c} \qquad r = 1, \cdots, n
\end{aligned}$$

This concludes the proof of Equation (2.5). The proof of Equation (2.6) proceeds in exactly the same way, once $\boldsymbol{\omega}_t$ has been defined as

$$\boldsymbol{\omega}_t = \frac{\partial \mathbf{n}_2}{\partial t} \cdot \mathbf{n}_3\mathbf{n}_1 + \frac{\partial \mathbf{n}_3}{\partial t} \cdot \mathbf{n}_1\mathbf{n}_2 + \frac{\partial \mathbf{n}_1}{\partial t} \cdot \mathbf{n}_2\mathbf{n}_3$$

Finally, Equation (2.4) is obtained by noting that

$$\begin{aligned}
\frac{d\mathbf{c}}{dt} &\underset{(1.11)}{=} \sum_{r=1}^{n} \frac{\partial \mathbf{c}}{\partial q_r}\dot{q}_r + \frac{\partial \mathbf{c}}{\partial t} \\
&\underset{(2.5),(2.6)}{=} \sum_{r=1}^{n} \boldsymbol{\omega}_{\dot{q}_r} \times \mathbf{c}\dot{q}_r + \boldsymbol{\omega}_t \times \mathbf{c} \\
&= \left(\sum_{r=1}^{n} \boldsymbol{\omega}_{\dot{q}_r}\dot{q}_r + \boldsymbol{\omega}_t\right) \times \mathbf{c}
\end{aligned}$$

From Equation (2.7) it then follows that

$$\boldsymbol{\omega} \times \mathbf{c} = \left(\sum_{r=1}^{n} \boldsymbol{\omega}_{\dot{q}_r} \dot{q}_r + \boldsymbol{\omega}_t \right) \times \mathbf{c}$$

and this equation is satisfied for *all* vectors **c** fixed in B if and only if Equation (2.4) is valid.

2.5 Determination by Inspection. Partial rates of change of orientation (see Section 2.4) can be found *by inspection* as soon as the angular velocity $\boldsymbol{\omega}$ has been expressed as a linear function of $\dot{q}_1, \cdots, \dot{q}_n$. $\boldsymbol{\omega}_{\dot{q}_r}$ is simply the coefficient of $\dot{q}_r$ in such an expression [see Equation (2.4)]. This is one reason for the choice of notation. A second reason is that $\boldsymbol{\omega}$ may be regarded as a vector function of the $2n + 1$ independent variables $q_1, \cdots, q_n, \dot{q}_1, \cdots, \dot{q}_n$, and t, and partial differentiation with respect to $\dot{q}_r$ then gives

$$\frac{\partial \boldsymbol{\omega}}{\partial \dot{q}_r} = \boldsymbol{\omega}_{\dot{q}_r} \qquad r = 1, \cdots, n \tag{2.8}$$

so that the subscript $\dot{q}_r$ in the symbol $\boldsymbol{\omega}_{\dot{q}_r}$ has an operational significance. Note, however, that the subscript t in the symbol $\boldsymbol{\omega}_t$ has a different character, for $\boldsymbol{\omega}_t$ is neither the coefficient of t in an expression for $\boldsymbol{\omega}$, nor is it equal to the partial derivative of $\boldsymbol{\omega}$ with respect to t.

Proof:

$$\begin{aligned}
\frac{\partial \boldsymbol{\omega}}{\partial \dot{q}_r} &\underset{(2.4)}{=} \frac{\partial}{\partial \dot{q}_r} \left(\sum_{s=1}^{n} \boldsymbol{\omega}_{\dot{q}_s} \dot{q}_s + \boldsymbol{\omega}_t \right) \\
&\underset{(1.5)}{=} \sum_{s=1}^{n} \frac{\partial}{\partial \dot{q}_r} (\boldsymbol{\omega}_{\dot{q}_s} \dot{q}_s) + \frac{\partial}{\partial \dot{q}_r} \boldsymbol{\omega}_t \\
&\underset{(1.6)}{=} \sum_{s=1}^{n} \left(\frac{\partial \boldsymbol{\omega}_{\dot{q}_s}}{\partial \dot{q}_r} \dot{q}_s + \boldsymbol{\omega}_{\dot{q}_s} \frac{\partial \dot{q}_s}{\partial \dot{q}_r} \right) + \frac{\partial \boldsymbol{\omega}_t}{\partial \dot{q}_r}
\end{aligned}$$

Now, $\boldsymbol{\omega}_{\dot{q}_s}$ and $\boldsymbol{\omega}_t$ are functions of $q_1, \cdots, q_n$, and t (see Section 2.4), but not of $\dot{q}_1, \cdots, \dot{q}_n$. Consequently,

$$\frac{\partial \boldsymbol{\omega}_{\dot{q}_s}}{\partial \dot{q}_r} = 0 \qquad r, s = 1, \cdots, n$$

and

$$\frac{\partial \boldsymbol{\omega}_t}{\partial \dot{q}_r} = 0 \qquad r = 1, \cdots, n$$

Furthermore, when $\dot{q}_1, \cdots, \dot{q}_n$ are treated as independent variables (as they must be if differentiation with respect to these quantities is to be meaningful), then

$$\frac{\partial \dot{q}_s}{\partial \dot{q}_r} = \begin{cases} 0 & s \neq r \\ 1 & s = r \end{cases}$$

Thus the only nonzero term in the right-hand member of the last expression for $\partial \boldsymbol{\omega}/\partial \dot{q}_r$ is the one corresponding to $s = r$, and Equation (2.8) follows.

2.6 Angular Velocity. Efficient determination of partial rates of change of orientation (see Sections 2.4 and 2.5) is possible only in the light of a clear understanding of the concept of angular velocity. A practically useful definition of angular velocity, and some consequences of this definition that facilitate the solutions of problems, are discussed in Sections 2.7 through 2.10.

2.7 Definition of Angular Velocity. If $\mathbf{b}_1, \mathbf{b}_2, \mathbf{b}_3$ is a right-handed set of mutually perpendicular unit vectors fixed in a rigid body B, the *angular velocity* ${}^R\boldsymbol{\omega}^B$ of B in a reference frame R may be defined as

$${}^R\boldsymbol{\omega}^B = \dot{\mathbf{b}}_2 \cdot \mathbf{b}_3\mathbf{b}_1 + \dot{\mathbf{b}}_3 \cdot \mathbf{b}_1\mathbf{b}_2 + \dot{\mathbf{b}}_1 \cdot \mathbf{b}_2\mathbf{b}_3 \tag{2.9}$$

where $\dot{\mathbf{b}}_i$ denotes the ordinary derivative of $\mathbf{b}_i$ with respect to time t in reference frame R. By means of a proof similar to that used to establish the validity of Equation (2.5), it can then be shown that, if $\mathbf{c}$ is any vector fixed in B,

$$\frac{{}^R d\mathbf{c}}{dt} = {}^R\boldsymbol{\omega}^B \times \mathbf{c} \tag{2.10}$$

2.8 Simple Angular Velocity, Angular Speed. When a rigid body B moves in a reference frame R in such a way that there exists a unit vector $\mathbf{k}$ whose orientation in both B and R is independent of t, B is said to have a *simple angular velocity* in R, and this angular velocity can be expressed as

$${}^R\boldsymbol{\omega}^B = {}^R\omega^B\,\mathbf{k} \tag{2.11}$$

where

$${}^R\omega^B = \frac{d\theta}{dt} \tag{2.12}$$

and θ is the radian measure of the angle between a line L_R whose orientation is fixed in R and a line L_B similarly fixed in B, both lines being perpendicular to $\mathbf{k}$ (see Figure 2.3) and θ being regarded as positive when the

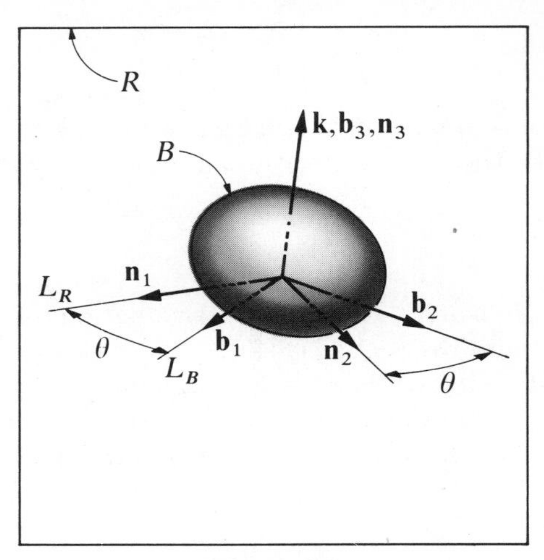

Figure 2.3

angle can be generated by a rotation of B during which a right-handed screw parallel to $\mathbf{k}$ and rigidly attached to B advances in the direction of $\mathbf{k}$. ${}^R\omega^B$ is called the *angular speed* of B in R for $\mathbf{k}$.

Proof: Let $\mathbf{b}_1$, $\mathbf{b}_2$, $\mathbf{b}_3$ be a right-handed set of mutually perpendicular unit vectors fixed in B, with $\mathbf{b}_1$ parallel to L_B and $\mathbf{b}_3 = \mathbf{k}$; and let $\mathbf{n}_1$, $\mathbf{n}_2$, $\mathbf{n}_3$ be a similar set of unit vectors fixed in R, with $\mathbf{n}_1$ parallel to L_R and $\mathbf{n}_3 = \mathbf{k}$. Then

$$\begin{aligned}\mathbf{b}_1 &= \cos\theta\mathbf{n}_1 + \sin\theta\mathbf{n}_2\\ \mathbf{b}_2 &= -\sin\theta\mathbf{n}_1 + \cos\theta\mathbf{n}_2\\ \mathbf{b}_3 &= \mathbf{n}_3\end{aligned}$$

and [see Equation (1.4)]

$$\begin{aligned}\dot{\mathbf{b}}_1 &= (-\sin\theta\mathbf{n}_1 + \cos\theta\mathbf{n}_2)\dot{\theta} = \mathbf{b}_2\dot{\theta}\\ \dot{\mathbf{b}}_2 &= -(\cos\theta\mathbf{n}_1 + \sin\theta\mathbf{n}_2)\dot{\theta} = -\mathbf{b}_1\dot{\theta}\\ \dot{\mathbf{b}}_3 &= 0\end{aligned}$$

Consequently,

$$ {}^R\omega^B \underset{(2.9)}{=} (-\mathbf{b}_1\dot{\theta})\cdot\mathbf{b}_3\mathbf{b}_1 + (\mathbf{b}_2\dot{\theta})\cdot\mathbf{b}_2\mathbf{b}_3 = \dot{\theta}\mathbf{b}_3 = \dot{\theta}\mathbf{k}$$

▪ EXAMPLE

In Figure 2.4, A designates an aircraft that carries a gyroscope consisting of an outer gimbal ring B, an inner gimbal ring C, and a rotor D.

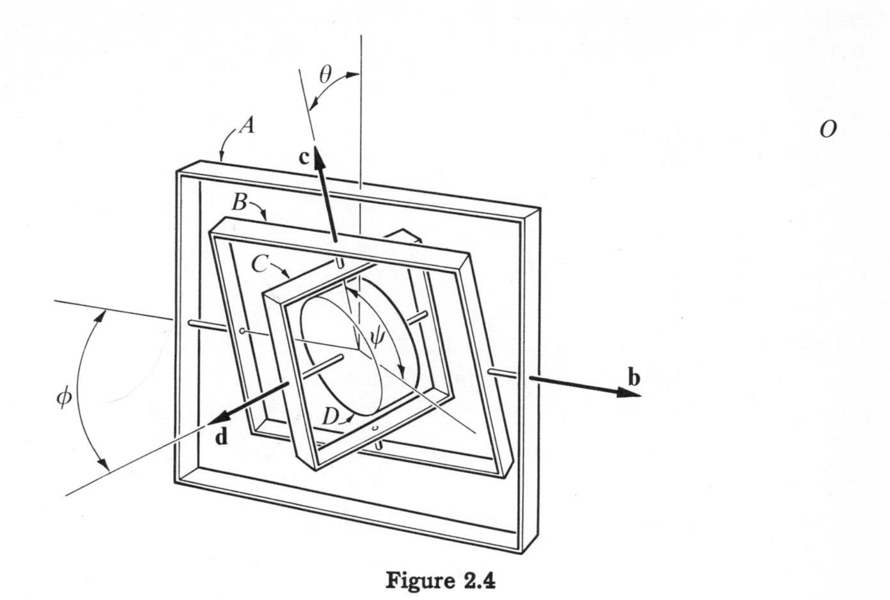

Figure 2.4

The quantities **b**, **c**, and **d** are unit vectors parallel to the outer gimbal axis, the inner gimbal axis, and the rotor axis, respectively. The angular velocities of B in A, C in B, and D in C are given by

$$\begin{aligned} {}^{A}\boldsymbol{\omega}^{B} &= \dot{\theta}\mathbf{b} \\ {}^{B}\boldsymbol{\omega}^{C} &= \dot{\phi}\mathbf{c} \\ {}^{C}\boldsymbol{\omega}^{D} &= -\dot{\psi}\mathbf{d} \end{aligned}$$

2.9 Two Reference Frames. If R and S are any two reference frames, the ordinary derivatives of a vector $\mathbf{v}$ with respect to t in R and S [see Equation (1.4)] are related to each other as follows:

$$\frac{{}^{R}d\mathbf{v}}{dt} = \frac{{}^{S}d\mathbf{v}}{dt} + {}^{R}\boldsymbol{\omega}^{S} \times \mathbf{v} \tag{2.13}$$

where ${}^{R}\boldsymbol{\omega}^{S}$ is the angular velocity of S in R (see Section 2.7).

Proof: Let $\mathbf{s}_1$, $\mathbf{s}_2$, $\mathbf{s}_3$ be a set of mutually perpendicular unit vectors fixed in S, and express $\mathbf{v}$ as (see Section 1.3)

$$\mathbf{v} = \sum_{i=1}^{3} v_i \mathbf{s}_i \tag{a}$$

Then

$$\frac{{}^R d\mathbf{v}}{dt} = \sum_{i=1}^{3} \frac{dv_i}{dt}\mathbf{s}_i + \sum_{i=1}^{3} v_i \frac{{}^R d\mathbf{s}_i}{dt}$$
$$\underset{(1.4)}{=} \frac{{}^S d\mathbf{v}}{dt} + \sum_{i=1}^{3} v_i \, \underset{(2.10)}{{}^R\boldsymbol{\omega}^S \times \mathbf{s}_i}$$
$$\underset{(a)}{=} \frac{{}^S d\mathbf{v}}{dt} + {}^R\boldsymbol{\omega}^S \times \mathbf{v}$$

2.10 Addition Theorem. The *addition theorem for angular velocities* states that, if $R_1, \cdots, R_n$ are n reference frames, then the angular velocity ${}^{R_n}\boldsymbol{\omega}^B$ of a rigid body B in reference frame R_n can be expressed as

$$ {}^{R_n}\boldsymbol{\omega}^B = {}^{R_1}\boldsymbol{\omega}^B + {}^{R_2}\boldsymbol{\omega}^{R_1} + \cdots + {}^{R_n}\boldsymbol{\omega}^{R_{n-1}} \tag{2.14}$$

Proof: Let $\mathbf{c}$ be any vector fixed in B. Then

$$\frac{{}^{R_n}d\mathbf{c}}{dt} \underset{(2.10)}{=} {}^{R_n}\boldsymbol{\omega}^B \times \mathbf{c}$$
$$\frac{{}^{R_{n-1}}d\mathbf{c}}{dt} \underset{(2.10)}{=} {}^{R_{n-1}}\boldsymbol{\omega}^B \times \mathbf{c}$$

and

$$\frac{{}^{R_n}d\mathbf{c}}{dt} \underset{(2.13)}{=} \frac{{}^{R_{n-1}}d\mathbf{c}}{dt} + {}^{R_n}\boldsymbol{\omega}^{R_{n-1}} \times \mathbf{c}$$

Consequently,

$$ {}^{R_n}\boldsymbol{\omega}^B \times \mathbf{c} = ({}^{R_{n-1}}\boldsymbol{\omega}^B + {}^{R_n}\boldsymbol{\omega}^{R_{n-1}}) \times \mathbf{c}$$

and, since this equation is satisfied for *all* $\mathbf{c}$ fixed in B,

$$ {}^{R_n}\boldsymbol{\omega}^B = {}^{R_{n-1}}\boldsymbol{\omega}^B + {}^{R_n}\boldsymbol{\omega}^{R_{n-1}}$$

Similarly,

$$ {}^{R_{n-1}}\boldsymbol{\omega}^B = {}^{R_{n-2}}\boldsymbol{\omega}^B + {}^{R_{n-1}}\boldsymbol{\omega}^{R_{n-2}}$$

so that

$$ {}^{R_n}\boldsymbol{\omega}^B = {}^{R_{n-2}}\boldsymbol{\omega}^B + {}^{R_{n-1}}\boldsymbol{\omega}^{R_{n-2}} + {}^{R_n}\boldsymbol{\omega}^{R_{n-1}}$$

and sufficient repetition leads to Equation (2.14).

▪ EXAMPLE

The angular velocity ${}^A\boldsymbol{\omega}^D$, where A and D refer to the aircraft and gyro rotor in the example in Section 2.8, is given by

$$ {}^A\boldsymbol{\omega}^D = {}^A\boldsymbol{\omega}^B + {}^B\boldsymbol{\omega}^C + {}^C\boldsymbol{\omega}^D$$
$$= \dot{\theta}\mathbf{b} + \dot{\phi}\mathbf{c} - \dot{\psi}\mathbf{d}$$

2.11 Angular Acceleration. The *angular acceleration* ${}^R\boldsymbol{\alpha}^B$ of a rigid body B in a reference frame R is defined as the time derivative in R of the angular velocity of B in R (see Section 2.7):

$$ {}^R\boldsymbol{\alpha}^B = \frac{{}^Rd\,{}^R\boldsymbol{\omega}^B}{dt} \tag{2.15} $$

If $R_1, \cdots, R_n$ are n reference frames, the angular acceleration of B in reference frame R_n is *not*, in general, equal to the sum

$$ {}^{R_1}\boldsymbol{\alpha}^B + {}^{R_2}\boldsymbol{\alpha}^{R_1} + \cdots + {}^{R_n}\boldsymbol{\alpha}^{R_{n-1}} $$

(compare with Section 2.10).

▪ EXAMPLE

The angular acceleration ${}^A\boldsymbol{\alpha}^D$ of the rotor D in the aircraft A of the example in Section 2.8 is found as follows:

$$ {}^A\boldsymbol{\alpha}^D \underset{(2.15)}{=} \frac{{}^Ad}{dt}\,{}^A\boldsymbol{\omega}^D \underset{(\text{Ex. }2.10)}{=} \frac{{}^Ad}{dt}(\dot\theta\mathbf{b} + \dot\phi\mathbf{c} - \dot\psi\mathbf{d}) $$

$$ = \ddot\theta\mathbf{b} + \dot\theta\frac{{}^Ad\mathbf{b}}{dt} + \ddot\phi\mathbf{c} + \dot\phi\frac{{}^Ad\mathbf{c}}{dt} - \ddot\psi\mathbf{d} - \dot\psi\frac{{}^Ad\mathbf{d}}{dt} $$

Since $\mathbf{b}$ is fixed in A, ${}^Ad\mathbf{b}/dt = 0$. The derivatives of $\mathbf{c}$ and $\mathbf{d}$ in A are found most conveniently by using Equation (2.10) after noting that $\mathbf{c}$ is fixed in B, and $\mathbf{d}$ in C:

$$ \frac{{}^Ad\mathbf{c}}{dt} = {}^A\boldsymbol{\omega}^B \times \mathbf{c} \underset{(\text{Ex. }2.8)}{=} \dot\theta\mathbf{b}\times\mathbf{c} $$

$$ \frac{{}^Ad\mathbf{d}}{dt} = {}^A\boldsymbol{\omega}^C \times \mathbf{d} \underset{(\text{Sec. }2.10)}{=} ({}^A\boldsymbol{\omega}^B + {}^B\boldsymbol{\omega}^C)\times\mathbf{d} $$

$$ = (\dot\theta\mathbf{b} + \dot\phi\mathbf{c})\times\mathbf{d} $$

Thus

$$ {}^A\boldsymbol{\alpha}^D = \ddot\theta\mathbf{b} + \ddot\phi\mathbf{c} - \ddot\psi\mathbf{d} + \dot\theta\dot\phi\mathbf{b}\times\mathbf{c} - \dot\psi(\dot\theta\mathbf{b} + \dot\phi\mathbf{c})\times\mathbf{d} $$

Note that

$$ {}^A\boldsymbol{\alpha}^B \underset{(2.15)}{=} \frac{{}^Ad\,{}^A\boldsymbol{\omega}^B}{dt} = \frac{{}^Ad}{dt}(\dot\theta\mathbf{b}) = \ddot\theta\mathbf{b} $$

$$ {}^B\boldsymbol{\alpha}^C \underset{(2.15)}{=} \frac{{}^Bd\,{}^B\boldsymbol{\omega}^C}{dt} = \frac{{}^Bd}{dt}(\dot\phi\mathbf{c}) = \ddot\phi\mathbf{c} $$

$$ {}^C\boldsymbol{\alpha}^D \underset{(2.15)}{=} \frac{{}^Cd\,{}^C\boldsymbol{\omega}^D}{dt} = \frac{{}^Cd}{dt}(-\dot\psi\mathbf{d}) = -\ddot\psi\mathbf{d} $$

so that

$$ {}^A\boldsymbol{\alpha}^D \neq {}^A\boldsymbol{\alpha}^B + {}^B\boldsymbol{\alpha}^C + {}^C\boldsymbol{\alpha}^D $$

which shows that, as stated in Section 2.11, the addition theorem for angular velocities (see Section 2.10) does not have an angular acceleration counterpart.

2.12 Partial Rates of Change of Position. If P is a particle of a holonomic system S possessing n degrees of freedom in a reference frame R (see Section 2.2), and $q_1, \cdots, q_n$ are generalized coordinates of S in R, the velocity $\mathbf{v}$ of P in R can be expressed in the form

$$\mathbf{v} = \sum_{r=1}^{n} \mathbf{v}_{\dot{q}_r}\dot{q}_r + \mathbf{v}_t \tag{2.16}$$

where $\dot{q}_r$ denotes the derivative of q_r with respect to time t, and $\mathbf{v}_{\dot{q}_r}$ and $\mathbf{v}_t$ are certain vector functions of $q_1, \cdots, q_n$, and t in R (see Section 1.1). The first partial derivatives of these functions satisfy the relationships

$$\frac{\partial \mathbf{v}_t}{\partial q_r} = \frac{\partial \mathbf{v}_{\dot{q}_r}}{\partial t} \qquad r = 1, \cdots, n \tag{2.17}$$

and

$$\frac{\partial \mathbf{v}_{\dot{q}_r}}{\partial q_s} = \frac{\partial \mathbf{v}_{\dot{q}_s}}{\partial q_r} \qquad r, s = 1, \cdots, n \tag{2.18}$$

$\mathbf{v}_{\dot{q}_r}$ is called the *partial rate of change with respect to* q_r *of the position of* P *in* R. (The reasons for the choice of notation are similar to those discussed in Section 2.5.)

Proof: Let $\mathbf{p}$ be the position vector of P relative to a point O fixed in R, and recall (see Section 2.2) that $\mathbf{p}$ may be regarded as a vector function of $q_1, \cdots, q_n$, and t in R. Then, by definition of velocity,

$$\mathbf{v} = \frac{d\mathbf{p}}{dt} \underset{\text{(Sec. 1.10)}}{=} \sum_{r=1}^{n} \frac{\partial \mathbf{p}}{\partial q_r}\dot{q}_r + \frac{\partial \mathbf{p}}{\partial t}$$

and Equation (2.16) follows immediately if $\mathbf{v}_t$ and $\mathbf{v}_{\dot{q}_r}$ are defined as

$$\mathbf{v}_t = \frac{\partial \mathbf{p}}{\partial t} \tag{a}$$

$$\mathbf{v}_{\dot{q}_r} = \frac{\partial \mathbf{p}}{\partial q_r} \tag{b}$$

Furthermore, $\mathbf{v}_t$ and $\mathbf{v}_{\dot{q}_r}$ are then vector functions of $q_1, \cdots, q_n$, and t (see Section 1.8), and

$$\frac{\partial}{\partial q_r}\mathbf{v}_t \underset{(a)}{=} \frac{\partial}{\partial q_r}\left(\frac{\partial \mathbf{p}}{\partial t}\right) \underset{(\text{Sec. } 1.8)}{=} \frac{\partial}{\partial t}\left(\frac{\partial \mathbf{p}}{\partial q_r}\right) \underset{(b)}{=} \frac{\partial}{\partial t}\mathbf{v}_{\dot{q}_r}$$

while

$$\frac{\partial}{\partial q_s}\mathbf{v}_{\dot{q}_r} \underset{(b)}{=} \frac{\partial}{\partial q_s}\left(\frac{\partial \mathbf{p}}{\partial q_r}\right) \underset{(\text{Sec. } 1.8)}{=} \frac{\partial}{\partial q_r}\left(\frac{\partial \mathbf{p}}{\partial q_s}\right) \underset{(b)}{=} \frac{\partial}{\partial q_r}\mathbf{v}_{\dot{q}_s}$$

2.13 Determination by Inspection. Partial rates of change of position (see Section 2.12) can be found *by inspection* as soon as the velocity $\mathbf{v}$ has been expressed as a linear function of $\dot{q}_1, \cdots, \dot{q}_n$. $\mathbf{v}_{\dot{q}_r}$ is simply the coefficient of $\dot{q}_r$ in such an expression [see Equation (2.16)]. In this respect, partial rates of change of position resemble partial rates of change of orientation (see Section 2.5), and the validity of a relationship analogous to Equation (2.8) can be established by means of a proof similar to the one used previously: If $\mathbf{v}$ is regarded as a vector function of the $2n + 1$ independent variables $q_1, \cdots, q_n, \dot{q}_1, \cdots, \dot{q}_n$, and t, then

$$\frac{\partial \mathbf{v}}{\partial \dot{q}_r} = \mathbf{v}_{\dot{q}_r} \qquad r = 1, \cdots, n \tag{2.19}$$

▪ EXAMPLE

In Figure 2.5, P designates a particle in contact with a plane N that rotates with a prescribed angular speed ω about a line fixed in N and in a reference frame R, and $\mathbf{n}_1$, $\mathbf{n}_2$, $\mathbf{n}_3$ are mutually perpendicular unit vectors fixed in N.

Expressed in terms of the two generalized coordinates q_1 and q_2 indicated in Figure 2.5, the velocity $\mathbf{v}$ of P in R is given by (see Section 2.16)

$$\mathbf{v} = \dot{q}_1\mathbf{n}_1 + \dot{q}_2\mathbf{n}_2 - \omega q_1\mathbf{n}_3$$

The coefficients of $\dot{q}_1$ and $\dot{q}_2$ are $\mathbf{n}_1$ and $\mathbf{n}_2$, respectively. Hence,

$$\mathbf{v}_{\dot{q}_1} = \mathbf{n}_1 \qquad \mathbf{v}_{\dot{q}_2} = \mathbf{n}_2$$

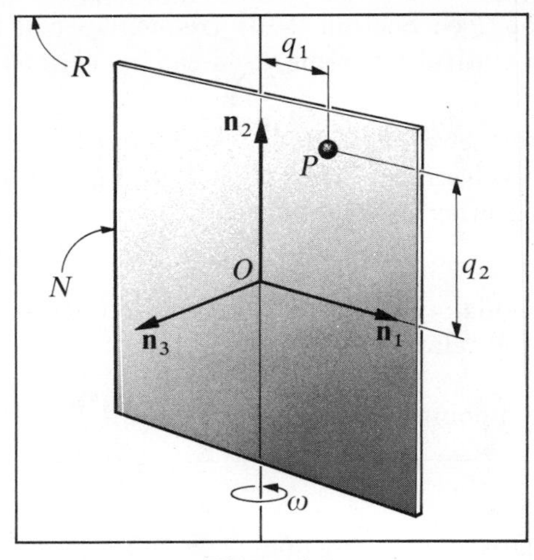

Figure 2.5

Alternatively, if q_1 and q_2 measure an angle and a distance, as shown in Figure 2.6, where $\mathbf{n}_4$ and $\mathbf{n}_5$ are unit vectors perpendicular to line OP

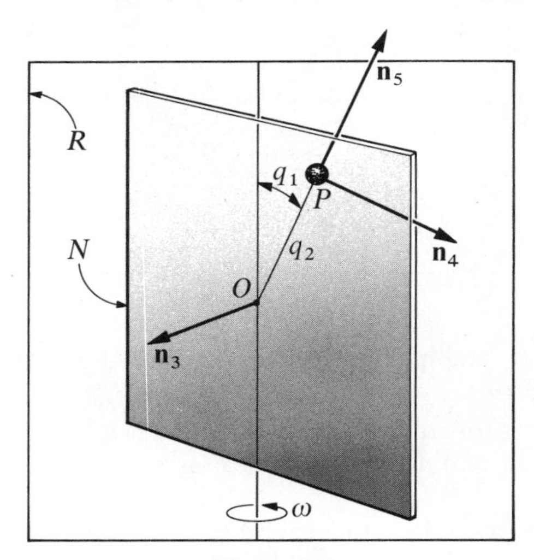

Figure 2.6

and parallel to this line, respectively, then

$$\mathbf{v} = -\omega q_2 \sin q_1 \mathbf{n}_3 + q_2 \dot{q}_1 \mathbf{n}_4 + \dot{q}_2 \mathbf{n}_5$$

and

$$\mathbf{v}_{\dot{q}_1} = q_2 \mathbf{n}_4 \qquad \mathbf{v}_{\dot{q}_2} = \mathbf{n}_5$$

2.14 Kinematical Theorems. The determination of partial rates of change of position (see Section 2.13) frequently can be facilitated by using the two kinematical theorems discussed in Sections 2.15 and 2.16.

2.15 Two Points of a Rigid Body. If P and Q are points fixed on a rigid body B, the velocities $\mathbf{v}^P$ and $\mathbf{v}^Q$ of P and Q in a reference frame R are related to each other as follows:

$$\mathbf{v}^P = \mathbf{v}^Q + \boldsymbol{\omega} \times \mathbf{r} \tag{2.20}$$

where $\boldsymbol{\omega}$ is the angular velocity of B in R (see Section 2.7) and $\mathbf{r}$ is the position vector of P relative to Q.

Proof: Let O be a point fixed in R (see Figure 2.7), $\mathbf{p}$ the position vector

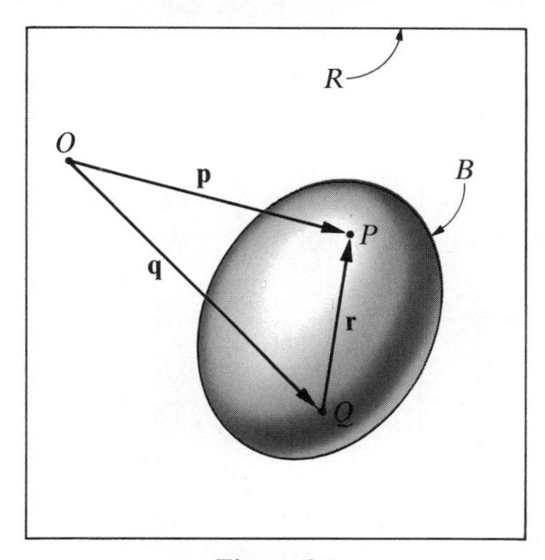

Figure 2.7

of P relative to O, and $\mathbf{q}$ the position vector of Q relative to O. Then, by definition,

$$\mathbf{v}^P = \frac{{}^R d\mathbf{p}}{dt} \tag{a}$$

and

$$\mathbf{v}^Q = \frac{{}^R d\mathbf{q}}{dt} \tag{b}$$

Now (see Figure 2.7)

$$\mathbf{p} = \mathbf{q} + \mathbf{r} \tag{c}$$

Hence

$$\mathbf{v}^P \underset{(a),(c)}{=} \frac{{}^R d\mathbf{q}}{dt} + \frac{{}^R d\mathbf{r}}{dt} \underset{(b)}{=} \mathbf{v}^Q + \frac{{}^R d\mathbf{r}}{dt} \tag{d}$$

Since $\mathbf{r}$ is a vector fixed in B,

$$\frac{^{R}d\mathbf{r}}{dt} \underset{(2.10)}{=} \boldsymbol{\omega} \times \mathbf{r} \tag{e}$$

Substitute from Equation (e) into (d).

▪ EXAMPLE

A circular disk D of radius r can rotate about an axis X fixed in a laboratory L, as shown in Figure 2.8, and a rod B is pinned to D at point Q,

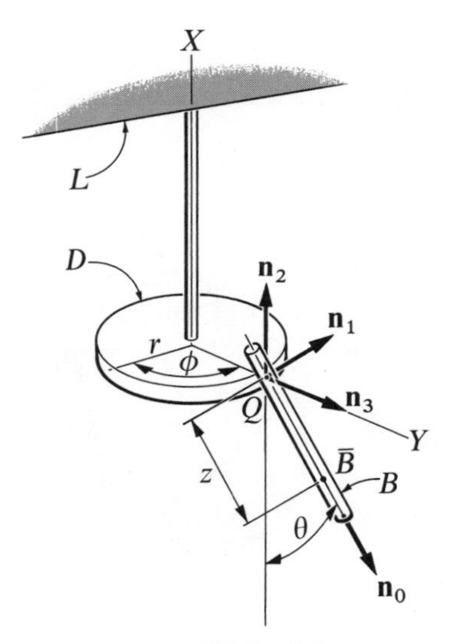

Figure 2.8

the axis Y of the pin passing through the center of D. The system comprised of D and B possesses two degrees of freedom in L. With θ and ϕ regarded as generalized coordinates, the partial rates of change with respect to θ and ϕ of the position of a point $\bar{B}$ of B are to be determined, the distance between $\bar{B}$ and point Q being equal to z.

Let $\boldsymbol{\omega}$ be the angular velocity of B in L, and $\mathbf{r}$ the position vector of $\bar{B}$ relative to Q. Then, expressed in terms of the unit vectors $\mathbf{n}_1$, $\mathbf{n}_2$, $\mathbf{n}_3$, and $\mathbf{n}_0$ shown in Figure 2.8,

$$\mathbf{v}^{Q} = r\dot{\phi}\mathbf{n}_1$$

$$\boldsymbol{\omega} \underset{(\text{Sec. } 2.8, 2.10)}{=} \dot{\phi}\mathbf{n}_2 + \dot{\theta}\mathbf{n}_3$$

and

$$\mathbf{r} = z\mathbf{n}_0$$

Hence,

$$\mathbf{v}^{\bar{B}} \underset{(2.20)}{=} r\dot{\phi}\mathbf{n}_1 + z(\dot{\phi}\mathbf{n}_2 + \dot{\theta}\mathbf{n}_3) \times \mathbf{n}_0$$

and, by inspection of the coefficients of $\dot{\theta}$ and $\dot{\phi}$ (see Section 2.13), the desired partial rates of change of position are found to be

$$\mathbf{v}_{\dot{\theta}}{}^{\bar{B}} = z\mathbf{n}_3 \times \mathbf{n}_0$$

and

$$\mathbf{v}_{\dot{\phi}}{}^{\bar{B}} = r\mathbf{n}_1 + z\mathbf{n}_2 \times \mathbf{n}_0$$

2.16 One Point Moving on a Rigid Body. If a point P moves on a rigid body B while B moves in a reference frame R, the velocity ${}^R\mathbf{v}^P$ of P in R is related to the velocity ${}^B\mathbf{v}^P$ of P in B as follows:

$${}^R\mathbf{v}^P = {}^R\mathbf{v}^{\bar{B}} + {}^B\mathbf{v}^P \tag{2.21}$$

where ${}^R\mathbf{v}^{\bar{B}}$ denotes the velocity in R of the point $\bar{B}$ of B that coincides with P at the instant under consideration.

Proof: Let B_0 be a point fixed in B, R_0 a point fixed in R, $\mathbf{b}$ the position vector of P relative to B_0, $\mathbf{r}$ the position vector of P relative to R_0, and $\mathbf{s}$ the position vector of B_0 relative to R_0, as shown in Figure 2.9. Then the

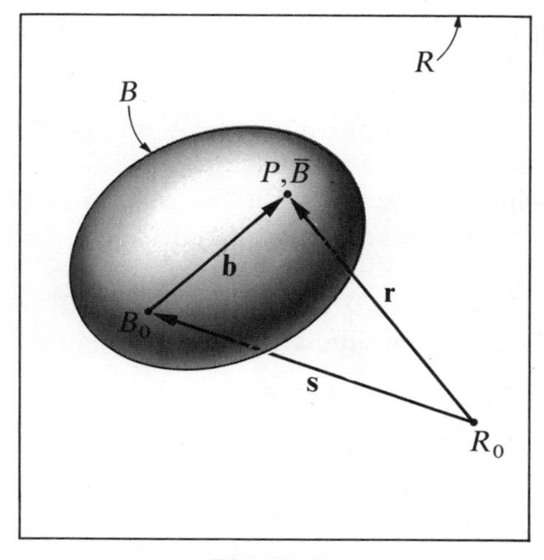

Figure 2.9

velocities ${}^R\mathbf{v}^P$, ${}^B\mathbf{v}^P$, and ${}^R\mathbf{v}^{B_0}$, each expressed as a derivative of a position vector, are given by

$$ {}^R\mathbf{v}^P = \frac{{}^R d\mathbf{r}}{dt} \tag{a} $$

$$ {}^B\mathbf{v}^P = \frac{{}^B d\mathbf{b}}{dt} \tag{b} $$

$$ {}^R\mathbf{v}^{B_0} = \frac{{}^R d\mathbf{s}}{dt} \tag{c} $$

Now (see Figure 2.9)

$$ \mathbf{r} = \mathbf{s} + \mathbf{b} \tag{d} $$

Consequently,

$$ \begin{aligned} {}^R\mathbf{v}^P &\underset{(a),(d)}{=} \frac{{}^R d\mathbf{s}}{dt} + \frac{{}^R d\mathbf{b}}{dt} \\ &= \underset{(c)}{{}^R\mathbf{v}^{B_0}} + \frac{{}^B d\mathbf{b}}{dt} \underset{(2.13)}{+} {}^R\boldsymbol{\omega}^B \times \mathbf{b} \\ &= {}^R\mathbf{v}^{B_0} + \underset{(b)}{{}^B\mathbf{v}^P} + {}^R\boldsymbol{\omega}^B \times \mathbf{b} \end{aligned} \tag{e} $$

B_0 may always be taken as $\bar{B}$—that is, as the point of B with which P coincides. Then $\mathbf{b} = 0$ and

$$ {}^R\mathbf{v}^P \underset{(e)}{=} {}^R\mathbf{v}^{\bar{B}} + {}^B\mathbf{v}^P $$

▪ EXAMPLE

A circular disk D of radius r can rotate about an axis X fixed in a laboratory L, as shown in Figure 2.10. A rod B is pinned to D at point Q, the axis Y of the pin passing through the center of D, and a particle P can slide on the rod. The system comprised of D, B, and P possesses three degrees of freedom in L. With θ, ϕ, and z regarded as the associated generalized coordinates, the partial rates of change with respect to θ, ϕ, and z of the position of P are to be determined.

Let $\bar{B}$ be the point of B that coincides with P. Then the velocity $\mathbf{v}^{\bar{B}}$ of $\bar{B}$ in L is given by (see the example in Section 2.15)

$$ \mathbf{v}^{\bar{B}} = r\dot{\phi}\mathbf{n}_1 + z(\dot{\phi}\mathbf{n}_2 + \dot{\theta}\mathbf{n}_3) \times \mathbf{n}_0 $$

The motion of P on B is rectilinear. Hence

$$ {}^B\mathbf{v}^P = \dot{z}\mathbf{n}_0 $$

and the velocity $\mathbf{v}^P$ of P in L is found to be

$$ \begin{aligned} {}^L\mathbf{v}^P &\underset{(2.21)}{=} {}^L\mathbf{v}^{\bar{B}} + {}^B\mathbf{v}^P \\ &= r\dot{\phi}\mathbf{n}_1 + z(\dot{\phi}\mathbf{n}_2 + \dot{\theta}\mathbf{n}_3) \times \mathbf{n}_0 + \dot{z}\mathbf{n}_0 \end{aligned} $$

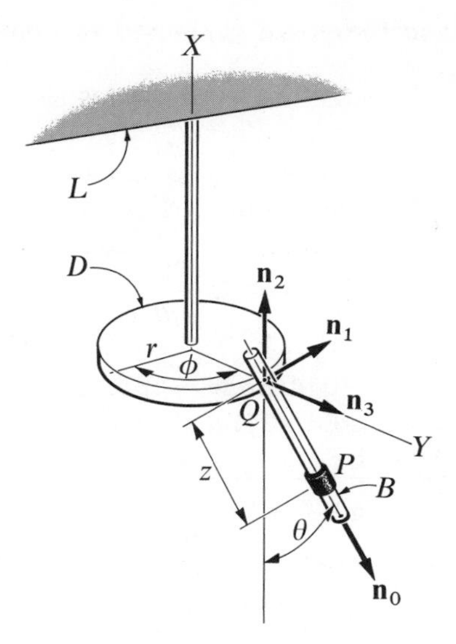

Figure 2.10

Inspection of the coefficients of $\dot{\theta}$, $\dot{\phi}$, and $\dot{z}$ then leads to the partial rates of change of position

$$\mathbf{v}_{\dot{\theta}}{}^P = z\mathbf{n}_3 \times \mathbf{n}_0$$
$$\mathbf{v}_{\dot{\phi}}{}^P = r\mathbf{n}_1 + z\mathbf{n}_2 \times \mathbf{n}_0$$
$$\mathbf{v}_{\dot{z}}{}^P = \mathbf{n}_0$$

2.17 Two Points of a Rigid Body. If P and Q are points fixed on a rigid body B that forms part of a holonomic system S possessing n degrees of freedom in a reference frame R (see Section 2.2), and $q_1, \cdots, q_n$ are generalized coordinates of S in R, partial rates of change of position of B in R (see Section 2.4) satisfy n equations analogous to Equation (2.20):

$$\mathbf{v}_{\dot{q}_r}{}^P = \mathbf{v}_{\dot{q}_r}{}^Q + \boldsymbol{\omega}_{\dot{q}_r} \times \mathbf{r} \qquad r = 1, \cdots, n \tag{2.22}$$

where $\mathbf{r}$ is the position vector of P relative to Q.

Proof: The velocities $\mathbf{v}^P$ and $\mathbf{v}^Q$ of P and Q in R and the angular velocity $\boldsymbol{\omega}$ of B in R satisfy the relationship

$$\mathbf{v}^P \underset{(2.20)}{=} \mathbf{v}^Q + \boldsymbol{\omega} \times \mathbf{r}$$
$$\underset{\text{(Secs. 2.4,2.12)}}{=} \sum_{r=1}^{n} (\mathbf{v}_{\dot{q}_r}{}^Q + \boldsymbol{\omega}_{\dot{q}_r} \times \mathbf{r})\dot{q}_r$$
$$+ \mathbf{v}_t{}^Q + \boldsymbol{\omega}_t \times \mathbf{r}$$

Examination of the coefficient of $\dot{q}_r$ shows (see Section 2.13) that Equation (2.22) is valid.

2.18 Acceleration. The solution of a certain class of problems of dynamics treated in the sequel requires the determination of accelerations. The kinematical theorems discussed in Sections 2.19 and 2.20 frequently facilitate this task.

2.19 Two Points of a Rigid Body. If P and Q are points fixed on a rigid body B, the accelerations $\mathbf{a}^P$ and $\mathbf{a}^Q$ of P and Q in a reference frame R are related to each other by

$$\mathbf{a}^P = \mathbf{a}^Q + \boldsymbol{\alpha} \times \mathbf{r} + \boldsymbol{\omega} \times (\boldsymbol{\omega} \times \mathbf{r}) \tag{2.23}$$

where $\mathbf{r}$ is the position vector of P relative to Q, $\boldsymbol{\omega}$ is the angular velocity of B in R (see Section 2.7), and $\boldsymbol{\alpha}$ is the angular acceleration of B in R (see Section 2.11).

Proof: Let $\mathbf{v}^P$ and $\mathbf{v}^Q$ be the velocities of P and Q in R. Then, by definition of acceleration,

$$\mathbf{a}^P = \frac{{}^R d\mathbf{v}^P}{dt} \tag{a}$$

and

$$\mathbf{a}^Q = \frac{{}^R d\mathbf{v}^Q}{dt} \tag{b}$$

Furthermore, $\mathbf{v}^P$ and $\mathbf{v}^Q$ are related by

$$\mathbf{v}^P \underset{(2.20)}{=} \mathbf{v}^Q + \boldsymbol{\omega} \times \mathbf{r} \tag{c}$$

and, as $\mathbf{r}$ is a vector fixed in B,

$$\frac{{}^R d\mathbf{r}}{dt} \underset{(2.10)}{=} \boldsymbol{\omega} \times \mathbf{r} \tag{d}$$

Consequently,

$$\begin{aligned} \mathbf{a}^P &\underset{(a),(c)}{=} \frac{{}^R d}{dt}(\mathbf{v}^Q + \boldsymbol{\omega} \times \mathbf{r}) \\ &= \frac{{}^R d\mathbf{v}^Q}{dt} + \frac{{}^R d\boldsymbol{\omega}}{dt} \times \mathbf{r} + \boldsymbol{\omega} \times \frac{{}^R d\mathbf{r}}{dt} \\ &= \underset{(b)}{\mathbf{a}^Q} + \underset{(\text{Sec. } 2.11)}{\boldsymbol{\alpha} \times \mathbf{r}} + \boldsymbol{\omega} \times \underset{(d)}{(\boldsymbol{\omega} \times \mathbf{r})} \end{aligned}$$

▪ EXAMPLE

Referring to the example in Section 2.15, one may find the acceleration $\mathbf{a}^{\bar{B}}$ of $\bar{B}$ in L as follows.

The acceleration $\mathbf{a}^Q$ of Q in L can be written down by inspection, because Q moves on a circle in L:

$$\mathbf{a}^Q = r\ddot{\phi}\mathbf{n}_1 - r\dot{\phi}^2\mathbf{n}_3 \tag{a}$$

(This result can also be obtained by using Equation (2.23) with P replaced by Q, $\mathbf{a}^Q = 0$, $\mathbf{r} = r\mathbf{n}_3$, $\boldsymbol{\omega} = \dot{\phi}\mathbf{n}_2$, $\boldsymbol{\alpha} = \ddot{\phi}\mathbf{n}_2$.)

The angular velocity $\boldsymbol{\omega}$ and angular acceleration $\boldsymbol{\alpha}$ of B in L are

$$\boldsymbol{\omega} \underset{\text{(Secs. 2.8,2.10)}}{=} \dot{\phi}\mathbf{n}_2 + \dot{\theta}\mathbf{n}_3 \tag{b}$$

and

$$\boldsymbol{\alpha} \underset{(2.15)}{=} \ddot{\phi}\mathbf{n}_2 + \ddot{\theta}\mathbf{n}_3 + \dot{\theta}\frac{d\mathbf{n}_3}{dt}$$

As $\mathbf{n}_3$ is fixed in D, and the angular velocity of D in L is equal to $\dot{\phi}\mathbf{n}_2$,

$$\frac{d\mathbf{n}_3}{dt} \underset{(2.10)}{=} \dot{\phi}\mathbf{n}_2 \times \mathbf{n}_3 = \dot{\phi}\mathbf{n}_1$$

Hence

$$\boldsymbol{\alpha} = \dot{\theta}\dot{\phi}\mathbf{n}_1 + \ddot{\phi}\mathbf{n}_2 + \ddot{\theta}\mathbf{n}_3 \tag{c}$$

With $\mathbf{r} = z\mathbf{n}_0$, Equation (2.23) thus gives

$$\begin{aligned}
\mathbf{a}^{\bar{B}} &= \mathbf{a}^Q + \boldsymbol{\alpha} \times \mathbf{r} + \boldsymbol{\omega} \times (\boldsymbol{\omega} \times \mathbf{r}) \\
&= \underset{(a)}{r\ddot{\phi}\mathbf{n}_1 - r\dot{\phi}^2\mathbf{n}_3} \\
&\quad + z\underset{(c)}{(\dot{\theta}\dot{\phi}\mathbf{n}_1 + \ddot{\phi}\mathbf{n}_2 + \ddot{\theta}\mathbf{n}_3)} \times \mathbf{n}_0 \\
&\quad + z\underset{(b)}{(\dot{\phi}\mathbf{n}_2 + \dot{\theta}\mathbf{n}_3)} \times [\underset{(b)}{(\dot{\phi}\mathbf{n}_2 + \dot{\theta}\mathbf{n}_3)} \times \mathbf{n}_0]
\end{aligned}$$

2.20 One Point Moving on a Rigid Body. If a point P moves on a rigid body B while B moves in a reference frame R, the acceleration ${}^R\mathbf{a}^P$ of P in R is given by

$${}^R\mathbf{a}^P = {}^B\mathbf{a}^P + {}^R\mathbf{a}^{\bar{B}} + 2\boldsymbol{\omega} \times {}^B\mathbf{v}^P \tag{2.24}$$

where ${}^B\mathbf{a}^P$ is the acceleration of P in B; ${}^R\mathbf{a}^{\bar{B}}$ denotes the acceleration in R of the point $\bar{B}$ of B that coincides with P at the instant under consideration; $\boldsymbol{\omega}$ is the angular velocity of B in R (see Section 2.7); and ${}^B\mathbf{v}^P$ is the velocity of P in B. (The term $2\boldsymbol{\omega} \times {}^B\mathbf{v}^P$ is referred to as a "Coriolis acceleration.")

Proof: Let B_0 be a point fixed in B, R_0 a point fixed in R, $\mathbf{b}$ the position vector of P relative to B_0, $\mathbf{r}$ the position vector of P relative to R_0, and $\mathbf{s}$ the position vector of B_0 relative to R_0, as shown in Figure 2.11.

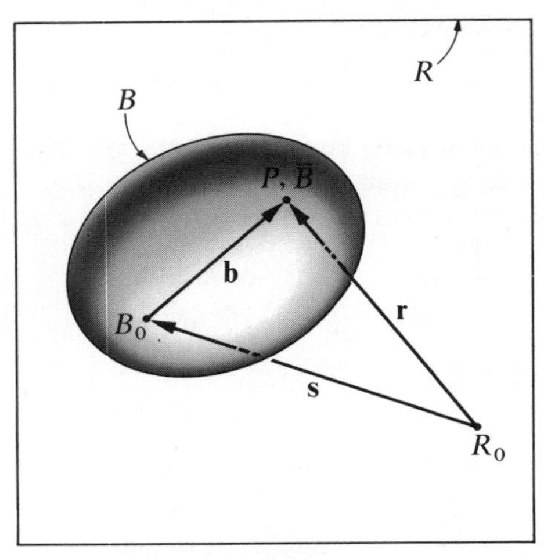

Figure 2.11

The accelerations ${}^R\mathbf{a}^P$, ${}^B\mathbf{a}^P$, and ${}^R\mathbf{a}^{B_0}$, each expressed as the derivative of a velocity vector, are given by

$$ {}^R\mathbf{a}^P = \frac{{}^Rd}{dt}\,{}^R\mathbf{v}^P \qquad \text{(a)} $$

$$ {}^B\mathbf{a}^P = \frac{{}^Bd}{dt}\,{}^B\mathbf{v}^P \qquad \text{(b)} $$

$$ {}^R\mathbf{a}^{B_0} = \frac{{}^Rd}{dt}\,{}^R\mathbf{v}^{B_0} \qquad \text{(c)} $$

Now [see Equation (e), Section 2.16]

$$ {}^R\mathbf{v}^P = {}^R\mathbf{v}^{B_0} + {}^B\mathbf{v}^P + {}^R\boldsymbol{\omega}^B \times \mathbf{b} \qquad \text{(d)} $$

Consequently,

$$ {}^R\mathbf{a}^P \underset{\text{(a),(d)}}{=} \frac{{}^Rd}{dt}\,{}^R\mathbf{v}^{B_0} + \frac{d}{dt}\,{}^B\mathbf{v}^P + \left(\frac{{}^Rd}{dt}\,{}^R\boldsymbol{\omega}^B\right) \times \mathbf{b} + {}^R\boldsymbol{\omega}^B \times \frac{{}^Rd\mathbf{b}}{dt} $$

$$ = \underset{\text{(c)}}{{}^R\mathbf{a}^{B_0}} + \frac{{}^Bd}{dt}\,{}^B\mathbf{v}^P \underset{(2.13)}{+} {}^R\boldsymbol{\omega}^B \times {}^B\mathbf{v}^P + \underset{(2.15)}{{}^R\boldsymbol{\alpha}^B} \times \mathbf{b} $$

$$ + {}^R\boldsymbol{\omega}^B \times \left(\frac{{}^Bd\mathbf{b}}{dt} \underset{(2.13)}{+} {}^R\boldsymbol{\omega}^B \times \mathbf{b}\right) \qquad \text{(e)} $$

Eqtn 2.13 $\quad \frac{{}^R d\vec{c}}{dt} = \frac{{}^S d\vec{c}}{dt} + {}^R\vec{\omega}^S \times \vec{c} = {}^R\vec{V}^B$

If B_0 is now taken as $\bar{B}$, then $\mathbf{b} = 0$ and

$$ {}^R\mathbf{a}^P \underset{\text{(e)}}{=} {}^R\mathbf{a}^{B_0} + \underset{\text{(b)}}{{}^B\mathbf{a}^P} + {}^R\boldsymbol{\omega}^B \times {}^B\mathbf{v}^P + {}^R\boldsymbol{\omega}^B \times {}^B\mathbf{v}^P $$

▪ EXAMPLE

Referring to the example in Section 2.16, one may find the acceleration $\mathbf{a}^P$ of P in L as follows:

Let $\bar{B}$ be the point of B that coincides with P. Then the acceleration $\mathbf{a}^{\bar{B}}$ of $\bar{B}$ in L is given by (see the example in Section 2.19)

$$ \begin{aligned} \mathbf{a}^{\bar{B}} = & \; r\ddot{\phi}\mathbf{n}_1 - r\dot{\phi}^2\mathbf{n}_3 \\ & + z(\dot{\theta}\dot{\phi}\mathbf{n}_1 + \ddot{\phi}\mathbf{n}_2 + \ddot{\theta}\mathbf{n}_3) \times \mathbf{n}_0 \\ & + z(\dot{\phi}\mathbf{n}_2 + \dot{\theta}\mathbf{n}_3) \times [(\dot{\phi}\mathbf{n}_2 + \dot{\theta}\mathbf{n}_3) \times \mathbf{n}_0] \end{aligned} $$

Since the motion of P on B is rectilinear,

$$ {}^B\mathbf{v}^P = \dot{z}\mathbf{n}_0 $$

and

$$ {}^B\mathbf{a}^P = \ddot{z}\mathbf{n}_0 $$

Furthermore, the angular velocity $\boldsymbol{\omega}$ of B in L is given by

$$ \boldsymbol{\omega} \underset{\text{(Secs. 2.8,2.10)}}{=} \dot{\phi}\mathbf{n}_2 + \dot{\theta}\mathbf{n}_3 $$

so that

$$ 2\boldsymbol{\omega} \times {}^B\mathbf{v}^P = 2\dot{z}(\dot{\phi}\mathbf{n}_2 + \dot{\theta}\mathbf{n}_3) \times \mathbf{n}_0 $$

Consequently,

$$ \begin{aligned} \mathbf{a}^P \underset{(2.24)}{=} & \; \ddot{z}\mathbf{n}_0 + r\ddot{\phi}\mathbf{n}_1 - r\dot{\phi}^2\mathbf{n}_3 \\ & + [z\dot{\phi}\dot{\theta}\mathbf{n}_1 + (z\ddot{\phi} + 2\dot{z}\dot{\phi})\mathbf{n}_2 + (z\ddot{\theta} + 2\dot{z}\dot{\theta})\mathbf{n}_3] \times \mathbf{n}_0 \\ & + z(\dot{\phi}\mathbf{n}_2 + \dot{\theta}\mathbf{n}_3) \times [(\dot{\phi}\mathbf{n}_2 + \dot{\theta}\mathbf{n}_3) \times \mathbf{n}_0] \end{aligned} $$

2.21 Acceleration and Partial Rates of Change of Position. If $\mathbf{v}$ is the velocity of a particle P of a holonomic system S possessing n degrees of freedom (see Section 2.2), and $q_1, \cdots, q_n$ are generalized coordinates of S, $\mathbf{v}^2$ may be regarded as a (scalar) function of the $2n + 1$ independent variables $q_1, \cdots, q_n, \dot{q}_1, \cdots, \dot{q}_n$, and t. The acceleration $\mathbf{a}$ of P is then related to the partial rates of change of position $\mathbf{v}_{\dot{q}_1}, \cdots, \mathbf{v}_{\dot{q}_n}$ (see Section 2.12) as follows[3]:

$$ \mathbf{v}_{\dot{q}_r} \cdot \mathbf{a} = \frac{1}{2}\left(\frac{d}{dt}\frac{\partial \mathbf{v}^2}{\partial \dot{q}_r} - \frac{\partial \mathbf{v}^2}{\partial q_r}\right) \qquad r = 1, \cdots, n \tag{2.25} $$

[3] This relationship is *not* valid for nonholonomic partial rates of change of position, defined in Section 2.23.

The principal reason for discussing this relationship is that it plays an essential part in the derivation of Lagrange's equations (see Sections 4.8 and 5.7). Also, it can facilitate the determination of accelerations, as will be shown by means of an example.

Proof: When $\mathbf{v}$ is expressed as in Equation (2.16), and $q_1, \cdots, q_n, \dot{q}_1, \cdots, \dot{q}_n$, and t are regarded as independent variables (so that, for example, $\partial\dot{q}_s/\partial q_r = 0$), partial differentiation of $\mathbf{v}$ with respect to q_r gives

$$\frac{\partial \mathbf{v}}{\partial q_r} \underset{(2.16)}{=} \frac{\partial}{\partial q_r}\left(\sum_{s=1}^{n} \mathbf{v}_{\dot{q}_s}\dot{q}_s + \mathbf{v}_t\right)$$

$$\underset{(\text{Secs. }1.6,1.7)}{=} \sum_{s=1}^{n}\left(\frac{\partial \mathbf{v}_{\dot{q}_s}}{\partial q_r}\right)\dot{q}_s + \frac{\partial \mathbf{v}_t}{\partial q_r}$$

$$= \sum_{s=1}^{n}\underset{(2.18)}{\left(\frac{\partial \mathbf{v}_{\dot{q}_r}}{\partial q_s}\right)}\dot{q}_s + \underset{(2.17)}{\frac{\partial \mathbf{v}_{\dot{q}_r}}{\partial t}}$$

or, with Equation (1.11) applied to $\mathbf{v}_{\dot{q}_r}$,

$$\frac{\partial \mathbf{v}}{\partial q_r} = \frac{d}{dt}\mathbf{v}_{\dot{q}_r} \tag{a}$$

Next,

$$\frac{d}{dt}(\mathbf{v}_{\dot{q}_r}\cdot\mathbf{v}) = \left(\frac{d}{dt}\mathbf{v}_{\dot{q}_r}\right)\cdot\mathbf{v} + \mathbf{v}_{\dot{q}_r}\cdot\frac{d\mathbf{v}}{dt}$$

$$\underset{(a)}{=} \frac{\partial \mathbf{v}}{\partial q_r}\cdot\mathbf{v} + \mathbf{v}_{\dot{q}_r}\cdot\frac{d\mathbf{v}}{dt}$$

which, solved for the product $\mathbf{v}_{\dot{q}_r}\cdot(d\mathbf{v}/dt)$, gives

$$\mathbf{v}_{\dot{q}_r}\cdot\frac{d\mathbf{v}}{dt} = \frac{d}{dt}(\mathbf{v}_{\dot{q}_r}\cdot\mathbf{v}) - \frac{\partial \mathbf{v}}{\partial q_r}\cdot\mathbf{v}$$

or, from Equation (2.19),

$$\mathbf{v}_{\dot{q}_r}\cdot\frac{d\mathbf{v}}{dt} = \frac{d}{dt}\left(\frac{\partial \mathbf{v}}{\partial \dot{q}_r}\cdot\mathbf{v}\right) - \frac{\partial \mathbf{v}}{\partial q_r}\cdot\mathbf{v} \tag{b}$$

From Equation (1.7),

$$\frac{\partial \mathbf{v}}{\partial \dot{q}_r}\cdot\mathbf{v} = \frac{1}{2}\frac{\partial \mathbf{v}^2}{\partial \dot{q}_r} \qquad \frac{\partial \mathbf{v}}{\partial q_r}\cdot\mathbf{v} = \frac{1}{2}\frac{\partial \mathbf{v}^2}{\partial q_r} \tag{c}$$

and, by definition of acceleration,

$$\mathbf{v}_{\dot{q}_r}\cdot\frac{d\mathbf{v}}{dt} = \mathbf{v}_{\dot{q}_r}\cdot\mathbf{a} \tag{d}$$

Substitution of Equations (c) and (d) into (b) gives Equation (2.25).

■ EXAMPLE

Figure 2.12 shows a point P, the cylindrical coordinates r, θ, and z of P,

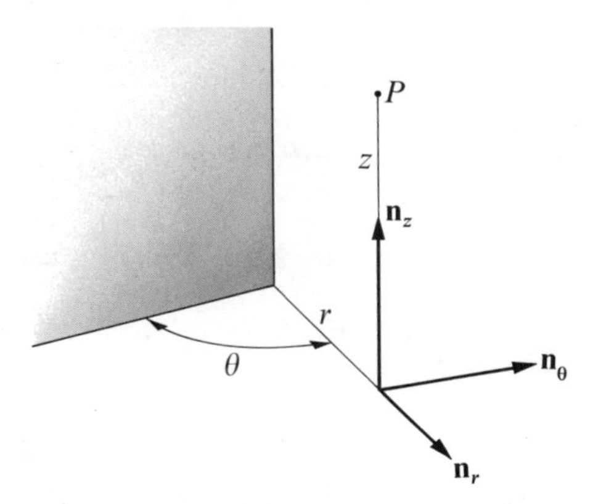

Figure 2.12

and three mutually perpendicular unit vectors $\mathbf{n}_r$, $\mathbf{n}_\theta$, and $\mathbf{n}_z$.

The velocity $\mathbf{v}$ of P is given by

$$\mathbf{v} = \dot{r}\mathbf{n}_r + r\dot{\theta}\mathbf{n}_\theta + \dot{z}\mathbf{n}_z$$

Hence, if P is regarded as a particle of a holonomic system with generalized coordinates r, θ, z (see Section 2.2), the partial rates of change of position of P (see Section 2.13) are

$$\mathbf{v}_{\dot{r}} = \mathbf{n}_r \qquad \mathbf{v}_{\dot{\theta}} = r\mathbf{n}_\theta \qquad \mathbf{v}_{\dot{z}} = \mathbf{n}_z$$

To find the acceleration $\mathbf{a}$ of P, one may express $\mathbf{a}$ as

$$\mathbf{a} = a_r\mathbf{n}_r + a_\theta\mathbf{n}_\theta + a_z\mathbf{n}_z$$

and determine the unknown scalars a_r, a_θ, and a_z by using Equation (2.25) three times:

$$\mathbf{v}_{\dot{r}} \cdot \mathbf{a} = \mathbf{n}_r \cdot \mathbf{a} = a_r = \frac{1}{2}\left(\frac{d}{dt}\frac{\partial \mathbf{v}^2}{\partial \dot{r}} - \frac{\partial \mathbf{v}^2}{\partial r}\right)$$

$$\mathbf{v}_{\dot{\theta}} \cdot \mathbf{a} = r\mathbf{n}_\theta \cdot \mathbf{a} = ra_\theta = \frac{1}{2}\left(\frac{d}{dt}\frac{\partial \mathbf{v}^2}{\partial \dot{\theta}} - \frac{\partial \mathbf{v}^2}{\partial \theta}\right)$$

$$\mathbf{v}_{\dot{z}} \cdot \mathbf{a} = \mathbf{n}_z \cdot \mathbf{a} = a_z = \frac{1}{2}\left(\frac{d}{dt}\frac{\partial \mathbf{v}^2}{\partial \dot{z}} - \frac{\partial \mathbf{v}^2}{\partial z}\right)$$

With

$$\mathbf{v}^2 = \dot{r}^2 + r^2\dot{\theta}^2 + \dot{z}^2$$

one thus obtains

$$a_r = \frac{1}{2}\left[\frac{d}{dt}(2\dot{r}) - 2r\dot{\theta}^2\right] = \ddot{r} - r\dot{\theta}^2$$

$$ra_\theta = \frac{1}{2}\left[\frac{d}{dt}(2r^2\dot{\theta}) - 0\right] = 2r\dot{r}\dot{\theta} + r^2\ddot{\theta}$$

$$a_z = \frac{1}{2}\left[\frac{d}{dt}(2\dot{z}) - 0\right] = \ddot{z}$$

Consequently,

$$\mathbf{a} = (\ddot{r} - r\dot{\theta}^2)\mathbf{n}_r + (2\dot{r}\dot{\theta} + r\ddot{\theta})\mathbf{n}_\theta + \ddot{z}\mathbf{n}_z$$

and this result has been obtained without performing explicitly any differentiations of vectors.

2.22 Nonholonomic System. Suppose that a set S of N particles $P_1, \cdots, P_N$ is subject to constraints (see Section 2.1) that can be represented only in part, rather than completely (see Section 2.2), by M holonomic constraint equations [see Equation (2.2)]. Although S is then not a holonomic system, it may nevertheless be possible to express each of the $3N$ quantities $x_1, y_1, z_1, \cdots, x_N, y_N, z_N$ (see Section 2.1) in terms of n,

$$n = 3N - M \tag{2.26}$$

functions of t, say $q_1(t), \cdots, q_n(t)$, and t itself, in such a way that these M constraint equations are satisfied for all values of $q_1, \cdots, q_n$, and t. Suppose further that the remaining constraints can then be represented by m linearly independent, nonintegrable differential equations of the form

$$\sum_{r=1}^{n} A_{sr}\dot{q}_r + B_s = 0 \qquad s = 1, \cdots, m \tag{2.27}$$

where A_{sr} and B_s are functions of $q_1, \cdots, q_n$, and t. Under these circumstances, S is said to be a *simple*[4] *nonholonomic system possessing* $n - m$ *degrees of freedom in* R, and $q_1(t), \cdots, q_n(t)$ are again called generalized coordinates of S in R. Note that every holonomic system (see Section 2.2) may be regarded as a simple nonholonomic system for which $m = 0$.

[4] The word "simple" is used to distinguish the systems here under consideration from those subject to constraints that cannot be expressed in the form of Equations (2.2) or (2.27).

▪ EXAMPLE

In Figure 2.13, C represents a uniform circular disk of radius r mounted

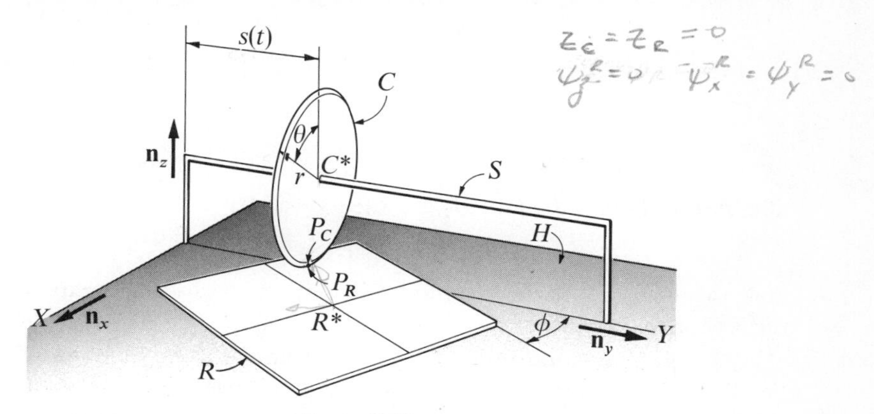

Figure 2.13

on a fixed horizontal shaft S and pressed into contact with a uniform rectangular plate R. Plate R is free to slide on a horizontal support H, and C is free to rotate on S; but C's motion along S is prescribed as a function $s(t)$ of the time t. Finally, no slip is permitted to occur at the point of contact between C and R; that is, C rolls on R.

The cartesian coordinates, x and y, of the mass center R^* of R, and the two angles θ and ϕ may be regarded as generalized coordinates of the system comprised of C and R; hence, $n = 4$. Furthermore, $m = 2$, because two equations of the form of Equations (2.27) express the requirement that no slip may occur at the point of contact between C and R; for, if P_C and P_R are the points of C and R instantaneously in contact with each other, this condition may be expressed as

$$\mathbf{v}^{P_C} = \mathbf{v}^{P_R}$$

where

$$\mathbf{v}^{P_C} \underset{(2.20)}{=} \dot{s}\mathbf{n}_y + \dot{\theta}\mathbf{n}_y \times (-r\mathbf{n}_z)$$

$$= -r\dot{\theta}\mathbf{n}_x + \dot{s}\mathbf{n}_y$$

and

$$\mathbf{v}^{P_R} \underset{(2.20)}{=} \dot{x}\mathbf{n}_x + \dot{y}\mathbf{n}_y - \dot{\phi}\mathbf{n}_z \times [-x\mathbf{n}_x - (y - s)\mathbf{n}_y]$$

$$= [\dot{x} + (s - y)\dot{\phi}]\mathbf{n}_x + (\dot{y} + x\dot{\phi})\mathbf{n}_y$$

so that

$$-r\dot{\theta} = \dot{x} + (s - y)\dot{\phi} \tag{a}$$

and

$$\dot{s} = \dot{y} + x\dot{\phi} \tag{b}$$

The system possesses $n - m = 2$ degrees of freedom. If $q_1, \cdots, q_4$ are chosen as, say,

$$q_1 = x \qquad q_2 = y \qquad q_3 = \phi \qquad q_4 = \theta$$

then the eight functions A_{sr} ($r = 1, \cdots, 4$; $s = 1, 2$) may be defined as

$$A_{11} = 1 \qquad A_{12} = 0 \qquad A_{13} = s(t) - q_2 \qquad A_{14} = r$$

$$A_{21} = 0 \qquad A_{22} = 1 \qquad A_{23} = q_1 \qquad A_{24} = 0$$

The two functions B_1 and B_2 are then given by

$$B_1 = 0 \qquad B_2 = -\dot{s}(t)$$

2.23 Nonholonomic Partial Rates of Change of Position and Orientation. If S is a simple nonholonomic system possessing $n - m$ degrees of freedom in a reference frame R (see Section 2.22), partial rates of change of position of a particle P of S, if defined as in Section 2.12, turn out to be of relatively little value, because they do not reflect the requirements imposed by Equations (2.27). The following observations now prove useful.

If Equations (2.27) are solved for m, say the last m, of $\dot{q}_1, \cdots, \dot{q}_n$ in terms of the remainder, and the results are substituted into Equation (2.16), then the velocity $\mathbf{v}$ of a typical particle P of S can be expressed uniquely as

$$\mathbf{v} = \sum_{r=1}^{n-m} \tilde{\mathbf{v}}_{\dot{q}_r} \dot{q}_r + \tilde{\mathbf{v}}_t \tag{2.28}$$

provided that this relationship be regarded as defining $\tilde{\mathbf{v}}_t$ and the $n - m$ *nonholonomic partial rates of change of the position* of P in R, $\tilde{\mathbf{v}}_{\dot{q}_1}, \cdots, \tilde{\mathbf{v}}_{\dot{q}_{n-m}}$, each of which is, in general, a function of all n generalized coordinates and the time. Similarly, if a rigid body B is a part of S, Equation (2.4) leads to the following unique expression for the angular velocity $\boldsymbol{\omega}$ of B in S:

$$\boldsymbol{\omega} = \sum_{r=1}^{n-m} \tilde{\boldsymbol{\omega}}_{\dot{q}_r} \dot{q}_r + \tilde{\boldsymbol{\omega}}_t \tag{2.29}$$

provided that this relationship be regarded as defining $\tilde{\boldsymbol{\omega}}_t$ and $n - m$ *nonholonomic partial rates of change of the orientation* of B in R, $\tilde{\boldsymbol{\omega}}_{\dot{q}_1}, \cdots, \tilde{\boldsymbol{\omega}}_{\dot{q}_{n-m}}$; and if P and Q are points fixed on B, and $\mathbf{r}$ is the position vector of P relative to Q, then [compare with Equations (2.22)]

$$\tilde{\mathbf{v}}_{\dot{q}_r}{}^P = \tilde{\mathbf{v}}_{\dot{q}_r}{}^Q + \tilde{\boldsymbol{\omega}}_{\dot{q}_r} \times \mathbf{r} \qquad r = 1, \cdots, n - m \tag{2.30}$$

■ EXAMPLE

The angular velocities $\boldsymbol{\omega}^C$ and $\boldsymbol{\omega}^R$ of the bodies C and R in the example in Section 2.22 and the velocities $\mathbf{v}^{C^*}$ and $\mathbf{v}^{R^*}$ of the mass centers of C and R are given by

$$\boldsymbol{\omega}^C = \dot{\theta}\mathbf{n}_y \quad \boldsymbol{\omega}^R = -\dot{\phi}\mathbf{n}_z$$
$$\mathbf{v}^{C^*} = \dot{s}\mathbf{n}_y \quad \mathbf{v}^{R^*} = \dot{x}\mathbf{n}_x + \dot{y}\mathbf{n}_y$$

The constraint equations (a) and (b) of the example in Section 2.22, if solved for $\dot{\theta}$ and $\dot{y}$, give

$$\dot{\theta} = \frac{1}{r}[(y - s)\dot{\phi} - \dot{x}]$$
$$\dot{y} = \dot{s} - x\dot{\phi}$$

Hence, after elimination of $\dot{\theta}$ and $\dot{y}$,

$$\boldsymbol{\omega}^C = \frac{1}{r}[(y - s)\dot{\phi} - \dot{x}]\mathbf{n}_y \qquad \boldsymbol{\omega}^R = -\dot{\phi}\mathbf{n}_z$$
$$\mathbf{v}^{C^*} = \dot{s}\mathbf{n}_y \quad \mathbf{v}^{R^*} = \dot{x}\mathbf{n}_x + (\dot{s} - x\dot{\phi})\mathbf{n}_y$$

and the associated nonholonomic partial rates of change of orientation and partial rates of change of position are

$$\tilde{\boldsymbol{\omega}}_{\dot{x}}{}^C = -\frac{1}{r}\mathbf{n}_y \qquad \tilde{\boldsymbol{\omega}}_{\dot{\phi}}{}^C = \frac{1}{r}[y - s(t)]\mathbf{n}_y$$
$$\tilde{\boldsymbol{\omega}}_{\dot{x}}{}^R = 0 \qquad \tilde{\boldsymbol{\omega}}_{\dot{\phi}}{}^R = -\mathbf{n}_z$$
$$\tilde{\mathbf{v}}_{\dot{x}}{}^{C^*} = 0 \qquad \tilde{\mathbf{v}}_{\dot{\phi}}{}^{C^*} = 0$$
$$\tilde{\mathbf{v}}_{\dot{x}}{}^{R^*} = \mathbf{n}_x \qquad \tilde{\mathbf{v}}_{\dot{\phi}}{}^{R^*} = -x\mathbf{n}_y$$

Alternatively, if the constraint equations are used to eliminate, say, $\dot{\phi}$ and $\dot{y}$, then

$$\boldsymbol{\omega}^C = \dot{\theta}\mathbf{n}_y \qquad \boldsymbol{\omega}^R = \frac{r\dot{\theta} + \dot{x}}{s - y}\mathbf{n}_z$$
$$\mathbf{v}^{C^*} = \dot{s}\mathbf{n}_y \quad \mathbf{v}^{R^*} = \dot{x}\mathbf{n}_x + \left[\dot{s} + \frac{x(r\dot{\theta} + \dot{x})}{s - y}\right]\mathbf{n}_y$$

so that

$$\tilde{\boldsymbol{\omega}}_{\dot{x}}{}^C = 0 \qquad \tilde{\boldsymbol{\omega}}_{\dot{\theta}}{}^C = \mathbf{n}_y$$
$$\tilde{\boldsymbol{\omega}}_{\dot{x}}{}^R = \frac{1}{s(t) - y}\mathbf{n}_z \qquad \tilde{\boldsymbol{\omega}}_{\dot{\theta}}{}^R = \frac{r}{s(t) - y}\mathbf{n}_z$$
$$\tilde{\mathbf{v}}_{\dot{x}}{}^{C^*} = 0 \qquad \tilde{\mathbf{v}}_{\dot{\theta}}{}^{C^*} = 0$$
$$\tilde{\mathbf{v}}_{\dot{x}}{}^{R^*} = \mathbf{n}_x + \frac{x}{s(t) - y}\mathbf{n}_y \qquad \tilde{\mathbf{v}}_{\dot{\theta}}{}^{R^*} = \frac{xr}{s(t) - y}\mathbf{n}_y$$

Note that the present expressions for $\tilde{\boldsymbol{\omega}}_{\dot{x}}{}^C$, $\tilde{\boldsymbol{\omega}}_{\dot{x}}{}^R$, and $\tilde{\mathbf{v}}_{\dot{x}}{}^{R^*}$ differ substantially from those obtained when $\dot{\theta}$ and $\dot{y}$, rather than $\dot{\phi}$ and $\dot{y}$, are eliminated.

2.24 Introduction of u's as Linear Combinations of $\dot{q}$'s. For reasons that will become apparent later, it is sometimes convenient when dealing with a simple nonholonomic system S possessing $n - m$ degrees of freedom (see Section 2.22) to introduce $n - m$ quantities $u_1, \cdots, u_{n-m}$ as linear combinations of $\dot{q}_1, \cdots, \dot{q}_{n-m}$ by means of equations of the form

$$u_r = \sum_{s=1}^{n-m} U_{rs}\dot{q}_s + U_r \qquad r = 1, \cdots, n - m \tag{2.31}$$

where U_{rs} and U_r are functions of $q_1, \cdots, q_n$, and t, and these quantities are chosen in such a way that Equations (2.31) can be solved uniquely for $\dot{q}_1, \cdots, \dot{q}_{n-m}$. From Equation (2.28) it then follows that the velocity $\mathbf{v}$ of a typical particle P of S can be expressed uniquely as

$$\mathbf{v} = \sum_{r=1}^{n-m} \tilde{\mathbf{v}}_{u_r} u_r + \tilde{\mathbf{v}}_t' \tag{2.32}$$

where $\tilde{\mathbf{v}}_{u_r}$ and $\tilde{\mathbf{v}}_t'$ are functions of $q_1, \cdots, q_n$, and t. Similarly, Equation (2.29) leads to

$$\boldsymbol{\omega} = \sum_{r=1}^{n-m} \tilde{\boldsymbol{\omega}}_{u_r} u_r + \tilde{\boldsymbol{\omega}}_t' \tag{2.33}$$

and Equations (2.30) can be replaced with

$$\tilde{\mathbf{v}}_{u_r}{}^P = \tilde{\mathbf{v}}_{u_r}{}^Q + \tilde{\boldsymbol{\omega}}_{u_r} \times \mathbf{r} \qquad r = 1, \cdots, n - m \tag{2.34}$$

If m is set equal to zero, all of the foregoing statements become applicable to a holonomic system possessing n degrees of freedom.

▪ EXAMPLE

In Figure 2.14, D designates a sharp-edged circular disk of radius r, and P represents a particle that is made to move on a line L_a fixed in D, the distance between P and the center D^* of D being a prescribed function $s(t)$ of the time t. D rolls on a plane support in which the axes X and Y of a rectangular cartesian coordinate system are embedded. Line L_1 is the tangent to the periphery of D at the point of contact C between D and the support; line L_2 passes through C and D^*; and line L_3 is normal to the middle plane of D.

The angles θ, ϕ, and ψ, together with x and y (the X and Y coordinates of D^*), constitute a set of $n = 5$ generalized coordinates of the system S comprised of D and P. As a consequence of the requirement that D must roll on the support, or, in other words, that the point of D instantaneously in contact with the support must have zero velocity, $m = 2$ constraint equations of the form of Equations (2.27) relate $\dot{x}$, $\dot{y}$, $\dot{\theta}$, $\dot{\phi}$, and $\dot{\psi}$ to each other, and these can be used to express $\dot{x}$ and $\dot{y}$ in terms of $\dot{\theta}$, $\dot{\phi}$, and $\dot{\psi}$.

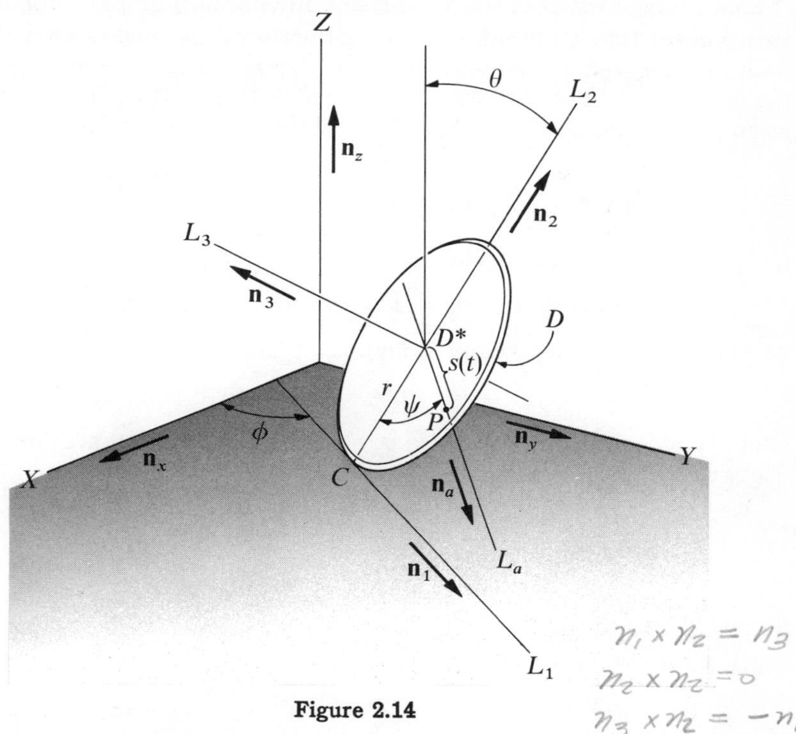

Figure 2.14

Consequently, S is a simple nonholonomic system possessing $n - m = 3$ degrees of freedom. (The constraint equations are not needed for what follows. However, they are derived for the sake of completeness.)

The angular velocity $\boldsymbol{\omega}$ of D, resolved into components parallel to the unit vectors $\mathbf{n}_1$, $\mathbf{n}_2$, $\mathbf{n}_3$, is given by (see Sections 2.8 and 2.10)

$$\boldsymbol{\omega} = -\dot{\theta}\mathbf{n}_1 + \dot{\phi}\cos\theta\,\mathbf{n}_2 + (\dot{\psi} + \dot{\phi}\sin\theta)\mathbf{n}_3 \tag{a}$$

Using this result and keeping in mind that the point of D in contact with the support has zero velocity, one obtains for the velocity of D^*

$$\mathbf{v}^{D^*} \underset{(2.20)}{=} \boldsymbol{\omega} \times (r\mathbf{n}_2) = -r[(\dot{\psi} + \dot{\phi}\sin\theta)\mathbf{n}_1 + \dot{\theta}\mathbf{n}_3] \tag{b}$$

The velocity of P is given by

$$\mathbf{v}^{P} \underset{(2.21)}{=} \mathbf{v}^{D^*} + \underset{(2.20)}{\boldsymbol{\omega}} \times (s\mathbf{n}_a) + \dot{s}\mathbf{n}_a \tag{c}$$

or

$$\mathbf{v}^{P} \underset{(c)}{=} -r[(\dot{\psi} + \dot{\phi}\sin\theta)\mathbf{n}_1 + \underset{(b)}{\dot{\theta}}\mathbf{n}_3]$$
$$+ s[-\dot{\theta}\mathbf{n}_1 + \dot{\phi}\cos\theta\,\mathbf{n}_2 \underset{(a)}{+} (\dot{\psi} + \dot{\phi}\sin\theta)\mathbf{n}_3] \times \mathbf{n}_a + \dot{s}\mathbf{n}_a$$

Nonholonomic partial rates of change of the orientation of D and of the positions of D^* and P can now be written down by inspection:

$$\tilde{\boldsymbol{\omega}}_{\dot{\theta}} = -\mathbf{n}_1 \quad \tilde{\boldsymbol{\omega}}_{\dot{\phi}} = \cos\theta\mathbf{n}_2 + \sin\theta\mathbf{n}_3 \quad \tilde{\boldsymbol{\omega}}_{\dot{\psi}} = \mathbf{n}_3$$

$$\tilde{\mathbf{v}}_{\dot{\theta}}^{D^*} = -r\mathbf{n}_3 \quad \tilde{\mathbf{v}}_{\dot{\phi}}^{D^*} = -r\sin\theta\mathbf{n}_1 \quad \tilde{\mathbf{v}}_{\dot{\psi}}^{D^*} = -r\mathbf{n}_1$$

$$\tilde{\mathbf{v}}_{\dot{\theta}}^{P} = -(r\mathbf{n}_3 + s\mathbf{n}_1 \times \mathbf{n}_a)$$

$$\tilde{\mathbf{v}}_{\dot{\phi}}^{P} = -r\sin\theta\mathbf{n}_1 + s(\cos\theta\mathbf{n}_2 + \sin\theta\mathbf{n}_3) \times \mathbf{n}_a$$

$$\tilde{\mathbf{v}}_{\dot{\psi}}^{P} = -r\mathbf{n}_1 + s\mathbf{n}_3 \times \mathbf{n}_a$$

If u_1, u_2, and u_3 are defined as

$$u_1 = -\dot{\theta} \quad u_2 = \dot{\phi}\cos\theta \quad u_3 = \dot{\psi} + \dot{\phi}\sin\theta \tag{d}$$

the expression for $\boldsymbol{\omega}$ given in Equation (a) reduces to

$$\boldsymbol{\omega} = u_1\mathbf{n}_1 + u_2\mathbf{n}_2 + u_3\mathbf{n}_3 \tag{e}$$

and the associated partial rates of change of orientation are simply [see Equation (2.33)]

$$\tilde{\boldsymbol{\omega}}_{u_1} = \mathbf{n}_1 \quad \tilde{\boldsymbol{\omega}}_{u_2} = \mathbf{n}_2 \quad \tilde{\boldsymbol{\omega}}_{u_3} = \mathbf{n}_3$$

The velocity of D^* becomes

$$\mathbf{v}^{D^*} \underset{(b),(d)}{=} -r(u_3\mathbf{n}_1 - u_1\mathbf{n}_3) \tag{f}$$

so that [see Equation (2.32)]

$$\tilde{\mathbf{v}}_{u_1}^{D^*} = r\mathbf{n}_3 \quad \tilde{\mathbf{v}}_{u_2}^{D^*} = 0 \quad \tilde{\mathbf{v}}_{u_3}^{D^*} = -r\mathbf{n}_1$$

Finally,

$$\mathbf{v}^P \underset{(c)}{=} \underbrace{-r(u_3\mathbf{n}_1 - u_1\mathbf{n}_3)}_{(f)} + s\underbrace{(u_1\mathbf{n}_1 + u_2\mathbf{n}_2 + u_3\mathbf{n}_3)}_{(e)} \times \mathbf{n}_a + \dot{s}\mathbf{n}_a$$

from which it follows that

$$\tilde{\mathbf{v}}_{u_1}^{P} = r\mathbf{n}_3 + s\mathbf{n}_1 \times \mathbf{n}_a$$

$$\tilde{\mathbf{v}}_{u_2}^{P} = s\mathbf{n}_2 \times \mathbf{n}_a$$

$$\tilde{\mathbf{v}}_{u_3}^{P} = -r\mathbf{n}_1 + s\mathbf{n}_3 \times \mathbf{n}_a$$

The constraint equations, mentioned earlier, are obtained by equating the expression

$$\dot{x}\mathbf{n}_x + \dot{y}\mathbf{n}_y + \frac{d}{dt}(r\cos\theta)\mathbf{n}_z$$

for the velocity of D^* to the right-hand member of either Equation (b) or Equation (f) and then noting that

$$\mathbf{n}_1 = \cos\phi\mathbf{n}_x + \sin\phi\mathbf{n}_y$$

$$\mathbf{n}_3 = \cos\theta\sin\phi\mathbf{n}_x - \cos\theta\cos\phi\mathbf{n}_y + \sin\theta\mathbf{n}_z$$

In the case of Equation (b), this gives

$$\begin{aligned}\dot{x} &= -r[(\dot{\psi} + \dot{\phi} \sin \theta) \cos \phi + \dot{\theta} \cos \theta \sin \phi] \\ \dot{y} &= -r[(\dot{\psi} + \dot{\phi} \sin \theta) \sin \phi - \dot{\theta} \cos \theta \cos \phi]\end{aligned}$$

whereas Equation (f) leads to

$$\begin{aligned}\dot{x} &= -r(u_3 \cos \phi - u_1 \cos \theta \sin \phi) \\ \dot{y} &= -r(u_3 \sin \phi + u_1 \cos \theta \cos \phi)\end{aligned}$$

Note that the constraint equations, as well as partial rates of change of position and orientation, are simplified through the use of u_1, u_2, and u_3 in place of $\dot{\theta}$, $\dot{\phi}$, and $\dot{\psi}$ as independent variables.

The present example provides a good opportunity to make the following observation about the quantities $u_1, \cdots, u_{n-m}$. Although u_r plays the same role relative to $\tilde{\mathbf{v}}_{u_r}$ as $\dot{q}_r$ does relative to $\tilde{\mathbf{v}}_{\dot{q}_r}$, u_r differs from $\dot{q}_r$ in one fundamental respect: *u_r is not necessarily the time derivative of any function,* whereas $\dot{q}_r$ is, by definition, the time derivative of q_r. (This statement applies both to holonomic and to nonholonomic systems.) Refer, for example, to Equations (d), and postulate the existence of three functions $f_1(\theta, \phi, \psi)$, $f_2(\theta, \phi, \psi)$, and $f_3(\theta, \phi, \psi)$, such that

$$u_r = \frac{df_r}{dt} \qquad r = 1, 2, 3$$

Equations (d) then require that

$$\begin{aligned}\frac{df_1}{dt} &= -\dot{\theta} \\ \frac{df_2}{dt} &= \dot{\phi} \cos \theta \\ \frac{df_3}{dt} &= \dot{\psi} + \dot{\phi} \sin \theta\end{aligned}$$

One can easily satisfy the first of these by taking $f_1 = -\theta$, but the second and third are nonintegrable; that is, there exist no functions f_2 and f_3 that satisfy these equations, because

$$\frac{df_2}{dt} = \frac{\partial f_2}{\partial \theta} \dot{\theta} + \frac{\partial f_2}{\partial \phi} \dot{\phi} + \frac{\partial f_2}{\partial \psi} \dot{\psi}$$

and

$$\frac{df_3}{dt} = \frac{\partial f_3}{\partial \theta} \dot{\theta} + \frac{\partial f_3}{\partial \phi} \dot{\phi} + \frac{\partial f_3}{\partial \psi} \dot{\psi}$$

Thus, since $\dot{\theta}$, $\dot{\phi}$, and $\dot{\psi}$ are independent of each other, f_2 and f_3 must be solutions of the partial differential equations

$$\frac{\partial f_2}{\partial \theta} = 0 \qquad \frac{\partial f_2}{\partial \phi} = \cos\theta \qquad \frac{\partial f_2}{\partial \psi} = 0 \tag{g}$$

$$\frac{\partial f_3}{\partial \theta} = 0 \qquad \frac{\partial f_3}{\partial \phi} = \sin\theta \qquad \frac{\partial f_3}{\partial \psi} = 1 \tag{h}$$

The first two of Equations (g) are incompatible with each other, as are the first two of Equations (h).

2.25 Virtual Velocity, Virtual Angular Velocity. If P is a particle of a simple nonholonomic system S possessing $n - m$ degrees of freedom in a reference frame R (see Section 2.22), and $\hat{q}_1, \cdots, \hat{q}_{n-m}$ are any $n - m$ quantities (positive, negative, or zero) having the dimensions of $\dot{q}_1, \cdots, \dot{q}_{n-m}$, respectively, the vector $\hat{\mathbf{v}}$ defined as

$$\hat{\mathbf{v}} = \sum_{r=1}^{n-m} \tilde{\mathbf{v}}_{\dot{q}_r}\hat{q}_r \tag{2.35}$$

(see Section 2.23 for $\tilde{\mathbf{v}}_{\dot{q}_r}$) has the dimensions of velocity and is called a *virtual velocity* of P in R. Similarly, if B is a rigid body belonging to S, the vector $\tilde{\boldsymbol{\omega}}$ defined as

$$\tilde{\boldsymbol{\omega}} = \sum_{r=1}^{n-m} \tilde{\boldsymbol{\omega}}_{\dot{q}_r}\hat{q}_r \tag{2.36}$$

has the dimensions of angular velocity and is called a *virtual angular velocity* of B in R. [To use these definitions in connection with a holonomic system possessing n degrees of freedom in R (see Section 2.2), take $m = 0$ and replace $\tilde{\mathbf{v}}_{\dot{q}_r}$ with $\mathbf{v}_{\dot{q}_r}$ (see Section 2.12) and $\tilde{\boldsymbol{\omega}}_{\dot{q}_r}$ with $\boldsymbol{\omega}_{\dot{q}_r}$ (see Section 2.4).]

Virtual velocities of several particles of S and/or virtual angular velocities of rigid bodies belonging to S are said to be *compatible* with each other when the same values of $\hat{q}_1, \cdots, \hat{q}_{n-m}$ are used to generate them.

▪ EXAMPLE

Referring to the example in Section 2.22, suppose that $r = 0.5$ feet and that at a certain instant

$$x = 2 \text{ ft} \qquad y = 5 \text{ ft} \qquad s = 3 \text{ ft} \tag{a}$$

Furthermore, let the point R^* have a virtual velocity

$$\hat{\mathbf{v}}^{R^*} = -6\mathbf{n}_x + 7\mathbf{n}_y \text{ ft/sec} \tag{b}$$

at this instant. The virtual angular velocity of C compatible with the given velocity of R^* is to be determined.

It was shown in the example in Section 2.23 that, when $\dot{x}$ and $\dot{\phi}$ are treated as independent, the two partial rates of change of the position of R^* are

$$\tilde{\mathbf{v}}_{\dot{x}}{}^{R^*} = \mathbf{n}_x \quad \tilde{\mathbf{v}}_{\dot{\phi}}{}^{R^*} = -x\mathbf{n}_y$$

Consequently, the most general virtual velocity of R^* can be expressed as

$$\begin{aligned} \hat{\mathbf{v}}^{R^*} &\underset{(2.35)}{=} \tilde{\mathbf{v}}_{\dot{x}}{}^{R^*}\hat{\dot{x}} + \tilde{\mathbf{v}}_{\dot{\phi}}{}^{R^*}\hat{\dot{\phi}} \\ &= \hat{\dot{x}}\mathbf{n}_x - x\hat{\dot{\phi}}\mathbf{n}_y \end{aligned} \tag{c}$$

and the particular value required by Equation (b) can be obtained by choosing $\hat{\dot{x}}$ and $\hat{\dot{\phi}}$ such that

$$-6\mathbf{n}_x + 7\mathbf{n}_y \underset{(b),(c)}{=} \hat{\dot{x}}\mathbf{n}_x - \underset{(a)}{2}\hat{\dot{\phi}}\mathbf{n}_y$$

that is,

$$\hat{\dot{x}} = -6 \text{ ft/sec} \qquad \hat{\dot{\phi}} = -3.5 \text{ rad/sec} \tag{d}$$

Again from the example in Section 2.23, the partial rates of change of the orientation of C are

$$\tilde{\boldsymbol{\omega}}_{\dot{x}}{}^C = -\frac{1}{r}\mathbf{n}_y \qquad \tilde{\boldsymbol{\omega}}_{\dot{\phi}}{}^C = \frac{1}{r}(y - s)\mathbf{n}_y \tag{e}$$

Hence, the most general virtual angular velocity of C is

$$\hat{\boldsymbol{\omega}}^C \underset{(2.36)}{=} \boldsymbol{\omega}_{\dot{x}}{}^C\hat{\dot{x}} + \boldsymbol{\omega}_{\dot{\phi}}{}^C\hat{\dot{\phi}} \underset{(e)}{=} \frac{-\hat{\dot{x}} + (y - s)\hat{\dot{\phi}}}{r}\mathbf{n}_y \tag{f}$$

and the virtual angular velocity of C compatible with the given virtual velocity of R^* becomes

$$\hat{\boldsymbol{\omega}}^C \underset{(f),(a),(d)}{=} \frac{6 + (5 - 3)(-3.5)}{0.5}\mathbf{n}_y = -2\mathbf{n}_y \text{ rad/sec}$$

If $\dot{x}$ and $\dot{\theta}$, rather than $\dot{x}$ and $\dot{\phi}$, are regarded as independent, $\hat{\mathbf{v}}^{R^*}$ and $\hat{\boldsymbol{\omega}}^C$ (see the example in Section 2.23 for the necessary partial rates of change of position and orientation) can be expressed as

$$\hat{\mathbf{v}}^{R^*} \underset{(2.35)}{=} \hat{\dot{x}}\mathbf{n}_x + \frac{x(\hat{\dot{x}} + r\hat{\dot{\theta}})}{s - y}\mathbf{n}_y \tag{g}$$

$$\hat{\boldsymbol{\omega}}^C \underset{(2.36)}{=} \hat{\dot{\theta}}\mathbf{n}_y \tag{h}$$

Then

$$-6\mathbf{n}_x + 7\mathbf{n}_y \underset{(b),(g)}{=} \hat{\dot{x}}\mathbf{n}_x + \frac{2(\hat{\dot{x}} + 0.5\hat{\dot{\theta}})}{3 - 5}\mathbf{n}_y$$

which requires

$$\hat{\dot{x}} = -6 \text{ ft/sec}$$

and

$$\hat{\dot{\theta}} = -2 \text{ rad/sec}$$

so that, as before,

$$\hat{\boldsymbol{\omega}}^C \underset{(h)}{=} -2\mathbf{n}_y \text{ rad/sec}$$

2.26 Two Points of a Rigid Body. If $\hat{\mathbf{v}}^P$ and $\hat{\mathbf{v}}^Q$ are virtual velocities (see Section 2.25) of two points P and Q fixed on a rigid body B, and $\hat{\boldsymbol{\omega}}$ is a virtual angular velocity of B, then, if and only if, $\hat{\mathbf{v}}^P$, $\hat{\mathbf{v}}^Q$, and $\hat{\boldsymbol{\omega}}$ are compatible with each other,

$$\hat{\mathbf{v}}^P = \hat{\mathbf{v}}^Q + \hat{\boldsymbol{\omega}} \times \mathbf{r} \tag{2.37}$$

where $\mathbf{r}$ is the position vector of P relative to Q.

Proof:

$$\begin{aligned}
\hat{\mathbf{v}}^P &\underset{(2.35)}{=} \sum_{r=1}^{n-m} \tilde{\mathbf{v}}_{\dot{q}_r}{}^P \hat{\dot{q}}_r \\
&\underset{(2.30)}{=} \sum_{r=1}^{n-m} (\tilde{\mathbf{v}}_{\dot{q}_r}{}^Q + \tilde{\boldsymbol{\omega}}_{\dot{q}_r} \times \mathbf{r})\hat{\dot{q}}_r \\
&= \sum_{r=1}^{n-m} \tilde{\mathbf{v}}_{\dot{q}_r}{}^Q \hat{\dot{q}}_r + \sum_{r=1}^{n-m} \tilde{\boldsymbol{\omega}}_{\dot{q}_r} \hat{\dot{q}}_r \times \mathbf{r} \\
&= \underset{(2.35)}{\hat{\mathbf{v}}^Q} + \underset{(2.36)}{\hat{\boldsymbol{\omega}}} \times \mathbf{r}
\end{aligned}$$

2.27 Virtual Velocities and Actual Velocities. Virtual velocities of a point P (see Section 2.25) must be carefully distinguished from actual velocities of P. In particular, it should be noted that not every velocity of P—in fact, not even every velocity that satisfies all requirements imposed by constraints—can be regarded as a virtual velocity. This can be seen by comparing the right-hand members of Equations (2.28) and (2.35): If $\tilde{\mathbf{v}}_t \neq 0$, it is in general impossible to choose $\hat{\dot{q}}_1, \cdots, \hat{\dot{q}}_{n-m}$ in such a way that $\hat{\mathbf{v}}$ becomes equal to $\mathbf{v}$. Similar remarks apply to virtual angular velocities and actual angular velocities of a rigid body.

▪ EXAMPLE

The velocity of point C^* in the example in Section 2.22 is given by

$$\mathbf{v}^{C^*} = \dot{s}\mathbf{n}_y$$

In the example in Section 2.23 it was shown that

$$\tilde{\mathbf{v}}_{\dot{x}}^{C^*} = \tilde{\mathbf{v}}_{\dot{\phi}}^{C^*} = 0$$

Hence

$$\hat{\mathbf{v}}^{C^*} \underset{(2.35)}{=} \tilde{\mathbf{v}}_{\dot{x}}^{C^*}\,\hat{\dot{x}} + \tilde{\mathbf{v}}_{\dot{\phi}}^{C^*}\,\hat{\dot{\phi}} = 0$$

for all possible choices of $\hat{\dot{x}}$ and $\hat{\dot{\phi}}$, and $\mathbf{v}^{C^*} \neq \hat{\mathbf{v}}^{C^*}$ except when $\dot{s} = 0$.

2.28 Virtual Displacement, Virtual Rotation. If P is a particle of a simple nonholonomic system S possessing $n - m$ degrees of freedom in a reference frame R (see Section 2.22), and $\delta q_1, \cdots, \delta q_{n-m}$ are any $n - m$ quantities (positive, negative, or zero) having the dimensions of $q_1, \cdots, q_{n-m}$, respectively, the vector $\delta\mathbf{p}$ defined as

$$\delta\mathbf{p} = \sum_{r=1}^{n-m} \tilde{\mathbf{v}}_{\dot{q}_r}\,\delta q_r \tag{2.38}$$

(see Section 2.23 for $\tilde{\mathbf{v}}_{\dot{q}_r}$) has the dimensions of length and is called a *virtual displacement* of P in R. Similarly, if B is a rigid body belonging to S, the vector $\delta\boldsymbol{\alpha}$ defined as

$$\delta\boldsymbol{\alpha} = \sum_{r=1}^{n-m} \tilde{\boldsymbol{\omega}}_{\dot{q}_r}\,\delta q_r \tag{2.39}$$

is called a *virtual rotation* of B in R. (For $m = 0$, these definitions become applicable to a holonomic system possessing n degrees of freedom in R.)

Virtual displacements of several particles of S and/or virtual rotations of rigid bodies belonging to S are said to be *compatible* with each other when the same values of $\delta q_1, \cdots, \delta q_{n-m}$ are used to generate them.

Note that the symbol δ, as used here, has no clearcut "operational" significance. This topic is discussed further in Section 2.32.

▪ EXAMPLE

Referring to the example in Section 2.24, suppose that, at a certain instant t^*, $s = r/2$, $\mathbf{n}_a = \mathbf{n}_1$, and a virtual rotation $\delta\boldsymbol{\alpha}$ of the disk D is given by

$$\delta\boldsymbol{\alpha}\Big|_{t^*} = -2\mathbf{n}_1 + 3\mathbf{n}_2 + 4\mathbf{n}_3$$

The virtual displacement of P compatible with this virtual rotation is to be determined.

In the example in Section 2.24, the following partial rates of change of the orientation of D were found:

$$\tilde{\boldsymbol{\omega}}_{\dot{\theta}} = -\mathbf{n}_1 \qquad \tilde{\boldsymbol{\omega}}_{\dot{\phi}} = \cos\theta\,\mathbf{n}_2 + \sin\theta\,\mathbf{n}_3 \qquad \tilde{\boldsymbol{\omega}}_{\dot{\psi}} = \mathbf{n}_3$$

The most general virtual rotation $\delta\boldsymbol{\alpha}$ of D is thus

$$\delta\boldsymbol{\alpha} \underset{(2.39)}{=} -\mathbf{n}_1\,\delta\theta + (\cos\theta\mathbf{n}_2 + \sin\theta\mathbf{n}_3)\,\delta\phi + \mathbf{n}_3\,\delta\psi$$

$$= -\delta\theta\mathbf{n}_1 + \cos\theta\,\delta\phi\mathbf{n}_2 + (\sin\theta\,\delta\phi + \delta\psi)\mathbf{n}_3$$

Consequently, at time t^*,

$$-\delta\theta = -2 \qquad \cos\theta\,\delta\phi = 3 \qquad \sin\theta\,\delta\phi + \delta\psi = 4$$

or

$$\delta\theta = 2$$

$$\delta\phi = \frac{3}{\cos\theta}$$

$$\delta\psi = 4 - 3\tan\theta$$

With $s = r/2$ and $\mathbf{n}_a = \mathbf{n}_1$, the partial rates of change of the position of P from the example in Section 2.24 reduce to

$$\tilde{\mathbf{v}}_{\dot\theta}{}^P = -r\mathbf{n}_3$$

$$\tilde{\mathbf{v}}_{\dot\phi}{}^P = -r\sin\theta\mathbf{n}_1 + \frac{r}{2}\sin\theta\mathbf{n}_2 - \frac{r}{2}\cos\theta\mathbf{n}_3$$

$$\tilde{\mathbf{v}}_{\dot\psi}{}^P = -r\mathbf{n}_1 + \frac{r}{2}\mathbf{n}_2$$

The desired virtual displacement of P is thus given by

$$\delta\mathbf{p}\Big|_{t^*} \underset{(2.38)}{=} -r\mathbf{n}_3\,\delta\theta + r(-\sin\theta\mathbf{n}_1 + \tfrac{1}{2}\sin\theta\mathbf{n}_2 - \tfrac{1}{2}\cos\theta\mathbf{n}_3)\,\delta\phi + r(-\mathbf{n}_1 + \tfrac{1}{2}\mathbf{n}_2)\,\delta\psi$$

or, using the values of $\delta\theta$, $\delta\phi$, and $\delta\psi$ found previously,

$$\delta\mathbf{p}\Big|_{t^*} = (-4\mathbf{n}_1 + 2\mathbf{n}_2 - 3.5\mathbf{n}_3)r$$

2.29 Virtual Displacements and Virtual Velocities. Virtual displacements (see Section 2.28) and virtual velocities (see Section 2.25) are related to each other as follows. If $\delta\mathbf{p}$ is a virtual displacement of a point P in a reference frame R, and δt is any quantity having the dimensions of time, then the vector $\hat{\mathbf{v}}$ given by

$$\hat{\mathbf{v}} = \frac{\delta\mathbf{p}}{\delta t} \tag{2.40}$$

is a virtual velocity of P in R. Conversely, if $\hat{\mathbf{v}}$ is a virtual velocity of P in R, then the vector $\delta\mathbf{p}$ given by

$$\delta\mathbf{p} = \hat{\mathbf{v}}\,\delta t \tag{2.41}$$

is a virtual displacement of P in R. Similarly, virtual rotations and virtual angular velocities satisfy the relationships

$$\hat{\boldsymbol{\omega}} = \frac{\delta\boldsymbol{\alpha}}{\delta t} \tag{2.42}$$

and

$$\delta\boldsymbol{\alpha} = \hat{\boldsymbol{\omega}}\,\delta t \tag{2.43}$$

All of these are immediate consequences of the definitions given in Sections 2.25 and 2.28.

2.30 Two Points of a Rigid Body. If $\delta\mathbf{p}$ and $\delta\mathbf{q}$ are virtual displacements (see Section 2.28) of two points P and Q fixed on a rigid body B, and $\delta\boldsymbol{\alpha}$ is a virtual rotation of B, then, if and only if $\delta\mathbf{p}$, $\delta\mathbf{q}$, and $\delta\boldsymbol{\alpha}$ are mutually compatible,

$$\delta\mathbf{p} = \delta\mathbf{q} + \delta\boldsymbol{\alpha} \times \mathbf{r} \tag{2.44}$$

where $\mathbf{r}$ is the position vector of P relative to Q.

Proof: Let δt be a quantity having the dimensions of time, and express $\delta\mathbf{p}$ as

$$\begin{aligned}\delta\mathbf{p} \underset{(2.41)}{=} \hat{\mathbf{v}}^P\,\delta t &\underset{(2.37)}{=} (\hat{\mathbf{v}}^Q + \hat{\boldsymbol{\omega}} \times \mathbf{r})\,\delta t \\ &= \hat{\mathbf{v}}^Q\,\delta t + (\hat{\boldsymbol{\omega}}\,\delta t) \times \mathbf{r} = \underset{(2.41)}{\delta\mathbf{q}} + \underset{(2.43)}{\delta\boldsymbol{\alpha} \times \mathbf{r}}\end{aligned}$$

2.31 Virtual and Actual Displacements. Virtual displacements (see Section 2.28) frequently have no actual counterparts; that is, actual displacements equal to the virtual displacements under consideration may violate requirements imposed by constraints.

▪ EXAMPLE

In Figure 2.15, P and Q represent points fixed on a rigid body B, and $\mathbf{r}$ designates the position vector of P relative to Q. The distance L between P and Q is thus given by

$$L = |\mathbf{r}| \tag{a}$$

and, if $\delta\boldsymbol{\alpha}$ is a virtual rotation of B, and $\delta\mathbf{p}$ and $\delta\mathbf{q}$ are virtual displacements of P and Q, respectively, these are mutually compatible whenever

$$\delta\mathbf{p} \underset{(2.44)}{=} \delta\mathbf{q} + \delta\boldsymbol{\alpha} \times \mathbf{r} \tag{b}$$

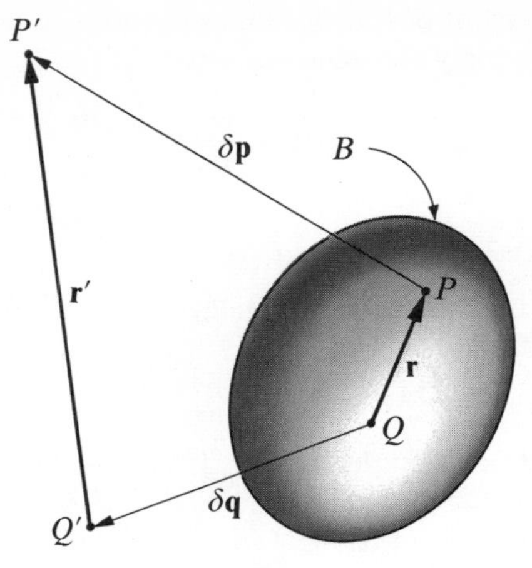

Figure 2.15

Suppose now that P and Q experience actual displacements equal to $\delta\mathbf{p}$ and $\delta\mathbf{q}$, respectively, bringing P to P' and Q to Q', as indicated in Figure 2.15. Then the distance L' between P' and Q' is given by

$$L' = |\mathbf{r}'| \tag{c}$$

where $\mathbf{r}'$ is the position vector of P' relative to Q'. Now, from Figure 2.15,

$$\begin{aligned}\mathbf{r}' &= \mathbf{r} + \delta\mathbf{p} - \delta\mathbf{q} \\ &\underset{\text{(b)}}{=} \mathbf{r} + \delta\boldsymbol{\alpha} \times \mathbf{r}\end{aligned} \tag{d}$$

Hence

$$L' \underset{\text{(c),(d)}}{=} |\mathbf{r} + \delta\boldsymbol{\alpha} \times \mathbf{r}| \tag{e}$$

and $L' \neq L$ unless

$$\delta\boldsymbol{\alpha} \times \mathbf{r} \underset{\text{(a),(e)}}{=} 0 \tag{f}$$

which is not the case for every choice of $\delta\boldsymbol{\alpha}$. But L' must be equal to L if B is a rigid body. Thus it appears that it may be impossible to regard $\delta\mathbf{p}$ and $\delta\mathbf{q}$ as actual displacements, even when they are compatible with each other.

2.32 Operational Significance of δ. The operational significance of the symbol δ appearing in the definitions of virtual displacements and rotations was mentioned at the end of Section 2.28. Examination of Equations (2.38) and (2.39) shows that δ is used in conjunction with q_r, $\mathbf{p}$, and $\boldsymbol{\alpha}$. In the case of δq_r, δ may be regarded as an operator that produces an arbitrary increment. For holonomic systems, a similar, if more complicated, interpretation is possible in connection with $\delta \mathbf{p}$, because, with $m = 0$, Equation (2.38) becomes

$$\delta \mathbf{p} = \sum_{r=1}^{n} \mathbf{v}_{\dot{q}_r} \, \delta q_r \underset{(2.19)}{=} \sum_{r=1}^{n} \frac{\partial \mathbf{p}}{\partial q_r} \, \delta q_r \tag{2.45}$$

so that δ may be regarded as an operator that produces a sum of partial derivatives, each multiplied by an arbitrary quantity having appropriate dimensions. In this respect δ is similar to the operator d of differential calculus, an important difference between the two being that, when $\mathbf{p}$ is regarded as a function of the $n + 1$ independent variables $q_1, \cdots, q_n$, and t, the total differential, $d\mathbf{p}$, of $\mathbf{p}$ is given by

$$d\mathbf{p} = \sum_{r=1}^{n} \frac{\partial \mathbf{p}}{\partial q_r} \, dq_r + \frac{\partial \mathbf{p}}{\partial t} \, dt$$

which contains one term having no counterpart in Equation (2.45). Finally, the δ in $\delta \boldsymbol{\alpha}$ has no operational significance at all, because $\boldsymbol{\alpha}$, not having been defined as a separate entity, cannot be operated upon.

Problems

▪ PROBLEM SET 1

(Sections 1.1–2.3)

1(a) Four rectangular parallelepipeds, A, B, C, and D, are arranged as shown in Figure 1(a). $\mathbf{a}_1$, $\mathbf{a}_2$, and $\mathbf{a}_3$ designate unit vectors respectively

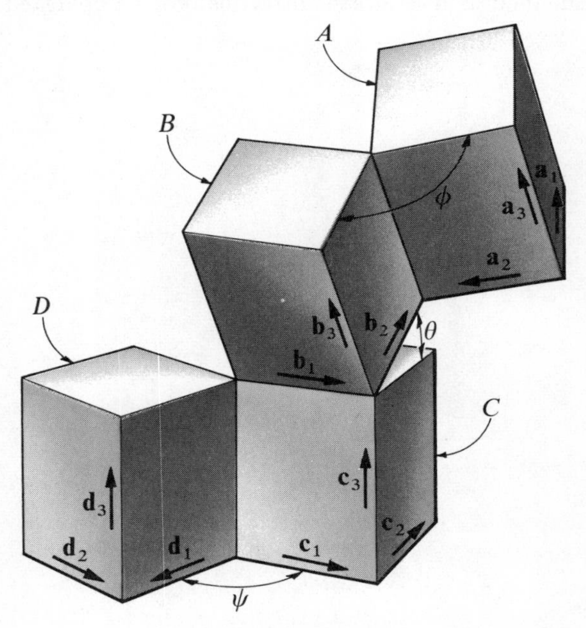

Figure 1(a)

parallel to the edges of A; $\mathbf{b}_1$, $\mathbf{b}_2$, and $\mathbf{b}_3$ are unit vectors respectively parallel to the edges of B, and so forth, and ϕ, θ, and ψ denote the radian measures of angles that determine the relative orientations of the bodies. The configuration shown is one in which ϕ, θ, and ψ are regarded as positive.

Determine the magnitude of each of the following partial derivatives: ${}^B\partial \mathbf{a}_1/\partial\phi$, ${}^B\partial \mathbf{b}_1/\partial\phi$, ${}^B\partial \mathbf{a}_3/\partial\phi$, ${}^B\partial \mathbf{b}_2/\partial\theta$, ${}^C\partial \mathbf{b}_2/\partial\theta$, ${}^D\partial \mathbf{b}_2/\partial\theta$, ${}^C\partial \mathbf{b}_2/\partial\psi$, ${}^D\partial \mathbf{b}_2/\partial\psi$, ${}^D\partial \mathbf{a}_1/\partial\psi$.

Results: 1, 0, 0, 0, 1, 1, 0, $|\cos\theta|$, $(\cos^2\phi + \sin^2\phi\cos^2\theta)^{1/2}$.

1(b) Referring to Problem 1(a), determine w_1, w_2, and w_3 such that ${}^C\partial \mathbf{a}_1/\partial\theta = w_1\mathbf{a}_1 + w_2\mathbf{a}_2 + w_3\mathbf{a}_3$.

Result: $w_1 = w_2 = 0$, $w_3 = \sin\phi$.

1(c) Referring to Problem 1(a), and assuming that ϕ, θ, and ψ are functions of the time t such that, at a certain instant t^*, $\phi = \theta = \psi = \pi/6$ rad, $\dot{\phi} = 4$ rad/sec, and $\dot{\theta} = \dot{\psi} = 6$ rad/sec, show that, at time t^*, ${}^C d\mathbf{a}_1/dt = 4\mathbf{a}_2 + 3\mathbf{a}_3$.

1(d) In Figure 1(d), N designates a plane that is made to rotate with

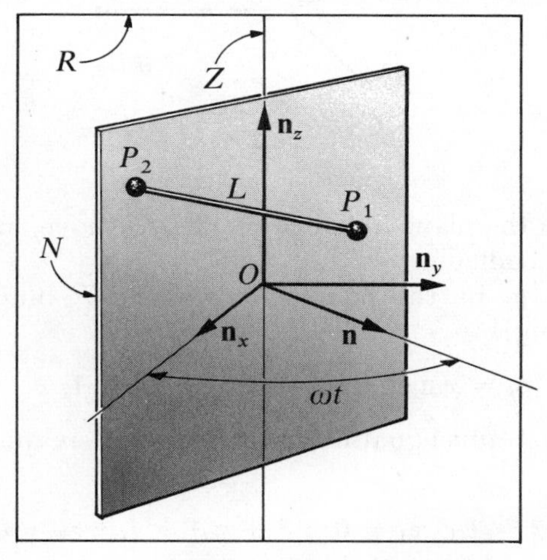

Figure 1(d)

constant angular speed ω about a line Z fixed in N and in a reference frame R. The unit vectors $\mathbf{n}_x$, $\mathbf{n}_y$, and $\mathbf{n}_z$ are mutually perpendicular and fixed in R, and $\mathbf{n}$ is a unit vector normal to N and equal to $\mathbf{n}_x$ at time $t = 0$. Finally, P_1 and P_2 represent particles connected to each other by a rigid rod of length L, these particles remaining at all times in contact with N.

Letting $\mathbf{p}_1$ and $\mathbf{p}_2$ be the position vectors of P_1 and P_2 relative to a point O fixed on line Z, and taking

$$\mathbf{p}_i = x_i\mathbf{n}_x + y_i\mathbf{n}_y + z_i\mathbf{n}_z \qquad i = 1, 2$$

determine functions $f_j(x_1, y_1, z_1, x_2, y_2, z_2, t)$, for $j = 1, 2, 3$, such that the requirements that P_1 and P_2 remain in N and be separated by the distance L can be expressed as $f_j = 0, j = 1, 2, 3$.

Result: $f_1 = x_1 \cos \omega t + y_1 \sin \omega t$, $f_2 = x_2 \cos \omega t + y_2 \sin \omega t$, $f_3 = (x_1 - x_2)^2 + (y_1 - y_2)^2 + (z_1 - z_2)^2 - L^2$.

1(e) Two particles, P_1 and P_2, are connected by a rigid rod that is free to rotate about an axis parallel to a unit vector $\mathbf{n}_z$ and passing through a point O of the rod, as shown in Figure 1(e), where $\mathbf{n}_x$ and $\mathbf{n}_y$ are unit

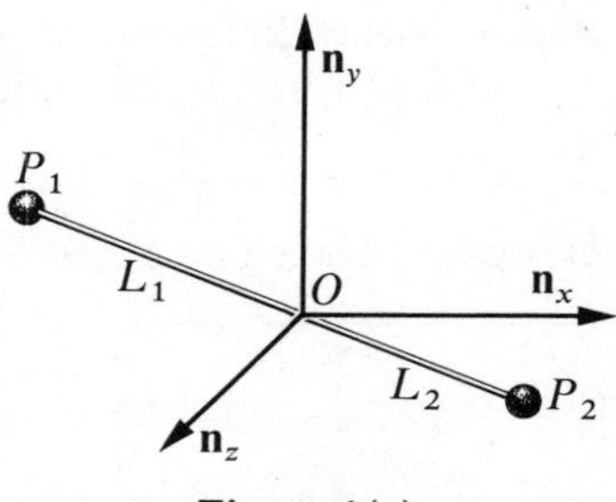

Figure 1(e)

vectors parallel to the plane in which P_1 and P_2 move, and $\mathbf{n}_x$, $\mathbf{n}_y$, and $\mathbf{n}_z$ are mutually perpendicular.

Letting $\mathbf{p}_1$ and $\mathbf{p}_2$ be the position vectors of P_1 and P_2 relative to point O, and taking

$$\mathbf{p}_i = x_i\mathbf{n}_x + y_i\mathbf{n}_y + z_i\mathbf{n}_z \qquad i = 1, 2$$

construct five constraint equations governing the six quantities x_i, y_i, z_i, for $i = 1, 2$.

Result: $x_1^2 + y_1^2 = L_1^2$, $z_1 = 0$; $x_2^2 + y_2^2 = L_2^2$, $z_2 = 0$; $x_1x_2 + y_1y_2 = -L_1L_2$.

1(f) Referring to Problem 1(d), and letting S be the set of particles P_1 and P_2, determine the number of degrees of freedom of S in R.

Result: 3.

1(g) Referring to Problem 1(e), express the six quantities x_i, y_i, z_i, with $i = 1, 2$, each as a function of a single quantity q in such a way that the five constraint equations found previously are satisfied for all values of q. (Suggestion: Let q be the radian measure of the angle between $\mathbf{n}_x$ and $\mathbf{p}_2$.)

Result: $x_1 = -L_1 \cos q$, $y_1 = L_1 \sin q$, $z_1 = 0$; $x_2 = L_2 \cos q$, $y_2 = -L_2 \sin q$, $z_2 = 0$.

1(h) Determine the number of degrees of freedom of each of the following holonomic systems: (a) Two rigid bodies attached to each other by means of a ball-and-socket connection. (b) An earth satellite carrying a rotor that is made to rotate at a prescribed rate about an axis fixed in the satellite. (c) An earth satellite carrying a rotor that is free to rotate about an axis fixed in the satellite. (d) The particles P_1 and P_2 of Problem 1(e).

Results: (a) 9, (b) 6, (c) 7, (d) 1.

▪ PROBLEM SET 2

(Sections 2.4–2.11)

2(a) In Figure 2(a), P represents a point fixed in a reference frame R,

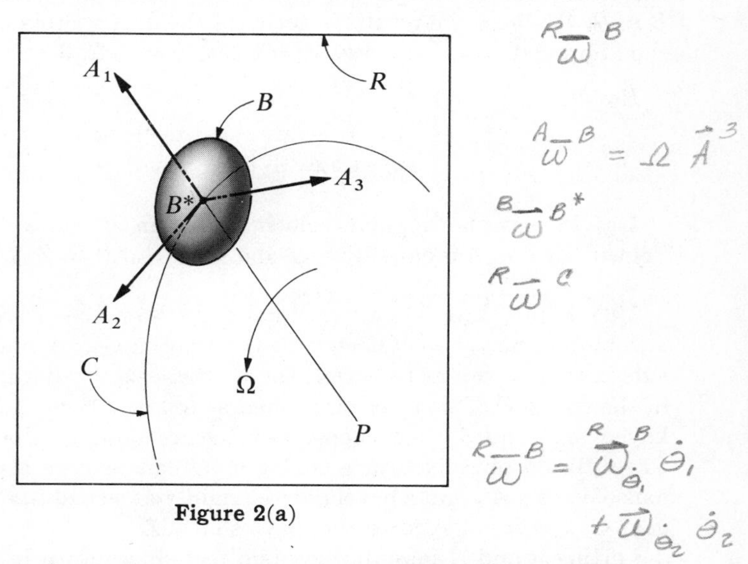

Figure 2(a)

and B^* designates the mass center of a rigid body B that moves on a circular orbit C fixed in R and centered at P. A_1, A_2, and A_3 are mutually perpendicular directed line segments, A_1 being the extension of line PB^*, A_2 pointing in the direction of motion of B^* on C, and A_3 thus being normal to the plane of the orbit of B^*.

If X_1, X_2, and X_3 are mutually perpendicular directed line segments passing through B^* and fixed in the body B, the "attitude" of B relative to A_1, A_2, A_3 can be specified in terms of three angles θ_1, θ_2, and θ_3, generated as follows: Align X_i with A_i, for $i = 1, 2, 3$, and perform successive right-handed rotations of B, of amount θ_1 about X_1, θ_2 about X_2, and θ_3 about X_3.

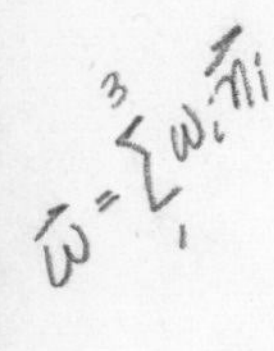

The angular velocity $\boldsymbol{\omega}$ of B in R can be expressed as $\boldsymbol{\omega} = \omega_1\mathbf{n}_1 + \omega_2\mathbf{n}_2 + \omega_3\mathbf{n}_3$, where $\mathbf{n}_i$ is a unit vector parallel to X_i, and ω_i is a function of θ_1, θ_2, θ_3, $\dot{\theta}_1$, $\dot{\theta}_2$, $\dot{\theta}_3$, and the angular speed Ω of line PB^* in R. Consequently, $\dot{\theta}_i$ can be expressed as a function f_i of θ_1, θ_2, θ_3, ω_1, ω_2, ω_3 and Ω.

Determine the functions f_1, f_2, and f_3, using the abbreviations $s_i = \sin\theta_i$, $c_i = \cos\theta_i$, for $i = 1, 2, 3$, to state the results.

Results:

$$(\omega_1 c_3 - \omega_2 s_3 + \Omega c_1 s_2)/c_2, \qquad \omega_1 s_3 + \omega_2 c_3 - \Omega s_1,$$
$$[(\omega_2 s_3 - \omega_1 c_3)s_2 + \omega_3 c_2 - \Omega c_1]/c_2$$

2(b) Referring to Problem 2(a), assume that the motion of B^* on C is prescribed, and, letting θ_1, θ_2, and θ_3 be generalized coordinates of B in R, determine the associated partial rates of change of the orientation of B in R. (Express the results in terms of the unit vectors $\mathbf{n}_1$, $\mathbf{n}_2$, $\mathbf{n}_3$, and use the abbreviations $s_i = \sin\theta_i$, $c_i = \cos\theta_i$, $i = 1, 2, 3$.)

Results:

$$\boldsymbol{\omega}_{\dot{\theta}_1} = c_2 c_3 \mathbf{n}_1 - c_2 s_3 \mathbf{n}_2 + s_2 \mathbf{n}_3$$
$$\boldsymbol{\omega}_{\dot{\theta}_2} = s_3 \mathbf{n}_1 + c_3 \mathbf{n}_2 \qquad \boldsymbol{\omega}_{\dot{\theta}_3} = \mathbf{n}_3$$

2(c) If $\boldsymbol{\omega}$ is the angular velocity of A in B, show that the angular velocity of B in A is equal to $-\boldsymbol{\omega}$, and that ${}^A d\boldsymbol{\omega}/dt = {}^B d\boldsymbol{\omega}/dt$.

2(d) Figure 2(d) shows schematically how the drive shaft D of an automobile may be connected to the two halves, A and A', of the rear axle in such a way as to permit the rear wheels to rotate at different rates in the frame F. This is accomplished as follows: Bevel gears B and B' are keyed to A and A', and engage bevel gears b and b', the latter being free to rotate on pins fixed in a casing C. C can revolve about the common axis of A and A', and a bevel gear E, rigidly attached to C, is driven by the gear G, which is keyed to the drive shaft D.

Letting $\mathbf{n}$ and $\mathbf{N}$ be unit vectors directed as shown in Figure 2(d), and expressing the angular velocities ${}^F\boldsymbol{\omega}^A$, ${}^F\boldsymbol{\omega}^{A'}$ and ${}^F\boldsymbol{\omega}^D$ of A, A' and D in F as ${}^F\boldsymbol{\omega}^A = \Omega\mathbf{N}$, ${}^F\boldsymbol{\omega}^{A'} = \Omega'\mathbf{N}$, and ${}^F\boldsymbol{\omega}^D = \omega\mathbf{n}$, show that $\omega = (a/2b)(\Omega + \Omega')$.

SUGGESTION: When each of two bevel gears has a simple angular velocity (see Section 2.8) in a certain reference frame, a relationship between angular speeds of the gears in this reference frame can be obtained

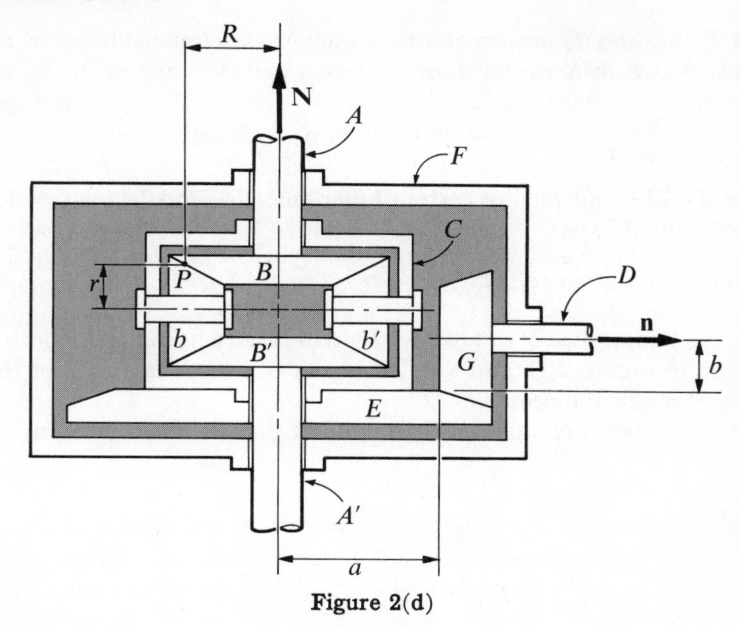

Figure 2(d)

by requiring that the velocities in this reference frame of two points of the gears that are in contact with each other be equal to each other. For example, the gears b and B have simple angular velocities in C, and, if ${}^C\omega^b$ and ${}^C\omega^B$ denote the angular speeds of b and B for $\mathbf{n}$ and $\mathbf{N}$, respectively, and r and R are the distances from point P [see Figure 2(d)] to the axes of b and B, this requirement is expressed by

$$ {}^C\omega^b\, r = -{}^C\omega^B\, R $$

Write two more equations of this sort, one for b and B', the other for E and G, and combine these with the two equations obtained by applying the addition theorem for angular velocities (see Section 2.10) to B, C, and F and to B', C, and F.

2(e) The angular velocity $\boldsymbol{\omega}$ of a rigid body B in a reference frame R is given by

$$ \boldsymbol{\omega} = \omega_1\mathbf{n}_1 + \omega_2\mathbf{n}_2 + \omega_3\mathbf{n}_3 $$

where $\mathbf{n}_1$, $\mathbf{n}_2$, $\mathbf{n}_3$ is a right-handed set of mutually perpendicular unit vectors fixed in B; and a vector $\mathbf{A}$ is defined as

$$ \mathbf{A} = I_1\omega_1\mathbf{n}_1 + I_2\omega_2\mathbf{n}_2 + I_3\omega_3\mathbf{n}_3 $$

where I_1, I_2, and I_3 are constants. Under these circumstances, the time derivative of **A** in R can be expressed as

$$\frac{^R d\mathbf{A}}{dt} = M_1\mathbf{n}_1 + M_2\mathbf{n}_2 + M_3\mathbf{n}_3$$

where M_1, M_2, and M_3 are certain functions of ω_i and I_i, for $i = 1, 2, 3$. Determine M_1.

Result: $M_1 = I_1\dot{\omega}_1 - (I_2 - I_3)\omega_2\omega_3$.

2(f) A circular disk D can rotate about an axis X fixed in a laboratory L, as shown in Figure 2(f), and a rod B is pinned to D, the axis Y of the pin passing through the center of D.

If θ and ϕ measure angles as indicated in Figure 2(f), and $\mathbf{n}_1$, $\mathbf{n}_2$, $\mathbf{n}_3$ are

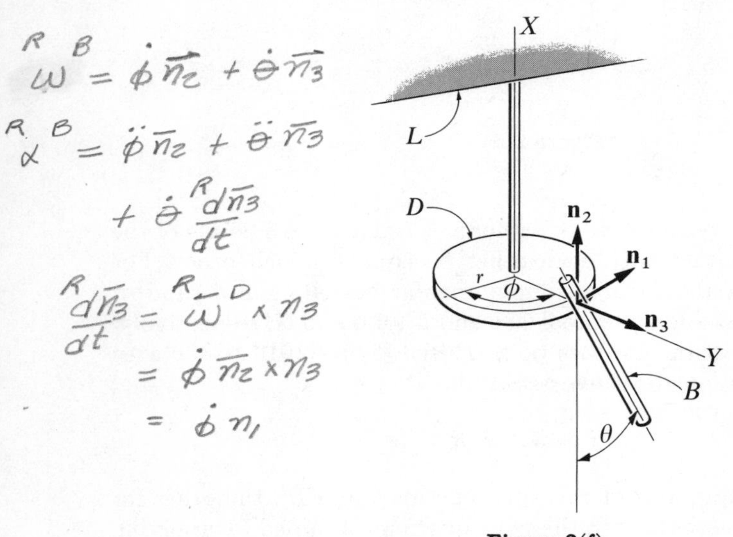

Figure 2(f)

unit vectors directed as shown, the angular acceleration of B in L can be expressed as $\alpha_1\mathbf{n}_1 + \alpha_2\mathbf{n}_2 + \alpha_3\mathbf{n}_3$, where α_1, α_2, and α_3 are functions of θ and ϕ. Determine α_1, α_2, and α_3.

Results: $\dot{\theta}\dot{\phi}$, $\ddot{\phi}$, $\ddot{\theta}$.

2(g) In Figure 2(g), D designates a sharp-edged circular disk that rolls on a plane support S fixed in a reference frame R. Line L_1 is the tangent to the periphery of D at the point of contact C between D and S; line L_2 passes through C and the center D^* of D; line L_3 is normal to the middle

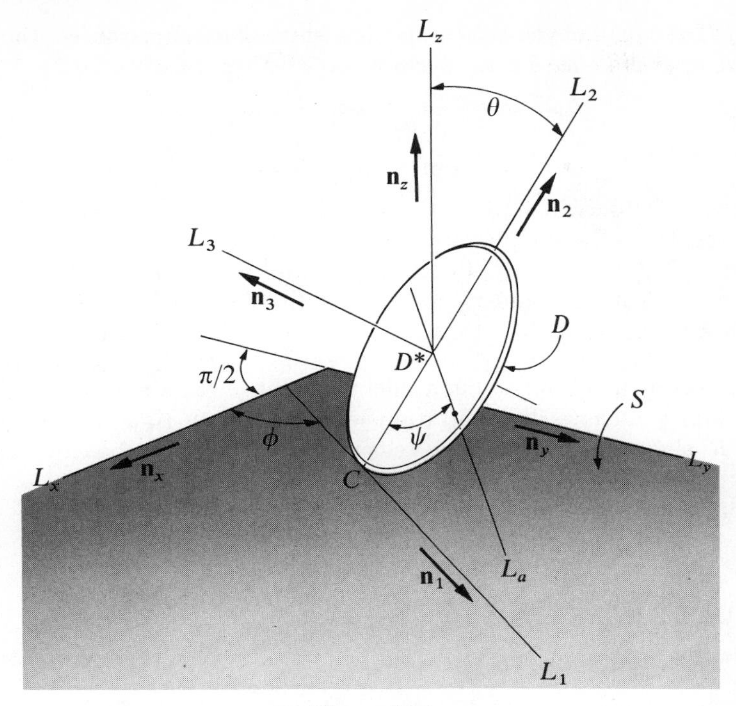

Figure 2(g)

plane of D, and line L_a lies in this plane and is fixed in D; finally, lines L_x and L_y are fixed in S, and line L_z is normal to S. Letting $\mathbf{n}_1$, $\mathbf{n}_2$, $\mathbf{n}_3$ and $\mathbf{n}_x$, $\mathbf{n}_y$, $\mathbf{n}_z$ be two sets of unit vectors as shown in Figure 2(g), and expressing the angular acceleration $\boldsymbol{\alpha}$ of D in R both as

$$\boldsymbol{\alpha} = \alpha_1\mathbf{n}_1 + \alpha_2\mathbf{n}_2 + \alpha_3\mathbf{n}_3$$

and as

$$\boldsymbol{\alpha} = \alpha_x\mathbf{n}_x + \alpha_y\mathbf{n}_y + \alpha_z\mathbf{n}_z$$

determine $\alpha_1, \cdots, \alpha_z$ as functions of the angles θ, ϕ, and ψ, and note that the expressions for α_1 and α_2 are considerably simpler than those for α_x and α_y.

Results:

$$\alpha_1 = -\ddot{\theta} + \dot{\phi}\dot{\psi}\cos\theta$$
$$\alpha_2 = \ddot{\phi}\cos\theta - \dot{\phi}\dot{\theta}\sin\theta + \dot{\theta}\dot{\psi}$$
$$\alpha_3 = \ddot{\psi} + \ddot{\phi}\sin\theta + \dot{\phi}\dot{\theta}\cos\theta$$
$$\alpha_x = -\ddot{\theta}\cos\phi + \ddot{\psi}\cos\theta\sin\phi + \dot{\theta}\dot{\phi}\sin\phi - \dot{\theta}\dot{\psi}\sin\theta\sin\phi + \dot{\psi}\dot{\phi}\cos\theta\cos\phi$$
$$\alpha_y = -\ddot{\theta}\sin\phi - \dot{\theta}\dot{\phi}\cos\phi - \ddot{\psi}\cos\theta\cos\phi + \dot{\psi}\dot{\theta}\sin\theta\cos\phi + \dot{\psi}\dot{\phi}\cos\theta\sin\phi$$
$$\alpha_z = \ddot{\phi} + \ddot{\psi}\sin\theta + \dot{\psi}\dot{\theta}\cos\theta$$

2(h) The angular velocity $\boldsymbol{\omega}$ and the angular acceleration $\boldsymbol{\alpha}$ of a rigid body B in a reference frame R can always be expressed as

$$\boldsymbol{\omega} = \omega_1\mathbf{n}_1 + \omega_2\mathbf{n}_2 + \omega_3\mathbf{n}_3$$

and

$$\boldsymbol{\alpha} = \alpha_1\mathbf{n}_1 + \alpha_2\mathbf{n}_2 + \alpha_3\mathbf{n}_3$$

where $\mathbf{n}_1$, $\mathbf{n}_2$, and $\mathbf{n}_3$ are unit vectors fixed in *any* reference frame N.

Letting $\boldsymbol{\Omega}$ be the angular velocity of N in R, show that $\alpha_i = d\omega_i/dt$, for $i = 1, 2, 3$, if and only if $\boldsymbol{\omega} \times \boldsymbol{\Omega} = 0$, and note that this condition is fulfilled when $\mathbf{n}_1$, $\mathbf{n}_2$, and $\mathbf{n}_3$ are fixed either in the body B or in the reference frame R.

2(i) Letting $\mathbf{a}$ and $\mathbf{b}$ be nonparallel vectors fixed in a rigid body B, and using dots to denote differentiation with respect to time in a reference frame R, show that the angular velocity $\boldsymbol{\omega}$ of B in R can be expressed as

$$\boldsymbol{\omega} = \frac{\dot{\mathbf{a}} \times \dot{\mathbf{b}}}{\dot{\mathbf{a}} \cdot \mathbf{b}}$$

▪ PROBLEM SET 3

(Sections 2.12–2.32)

3(a) In Figure 3(a), P_1, P_2, and P_3 designate the midpoints of three

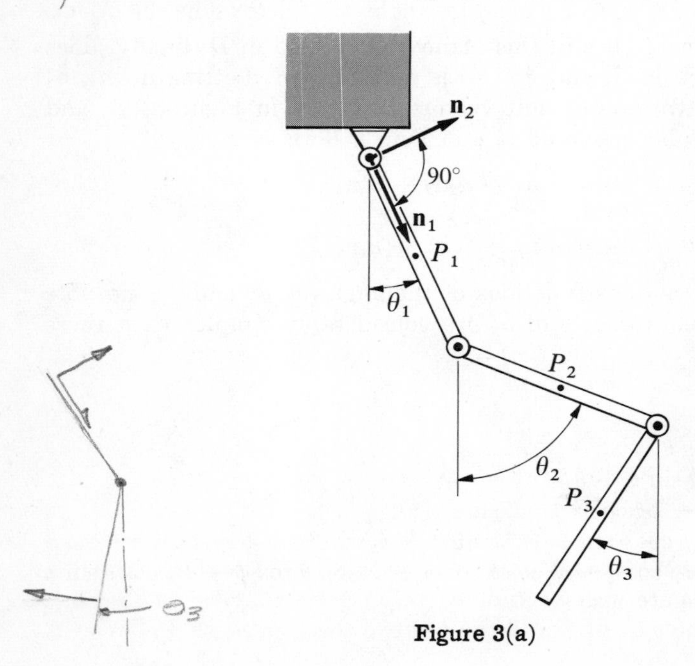

Figure 3(a)

pin-connected rods, each of length L, and θ_1, θ_2, and θ_3 measure the angles between the rods and any vertical line, all three being positive for the configuration shown. $\mathbf{n}_1$ and $\mathbf{n}_2$ are unit vectors parallel to the plane of motion of the rods.

Determine the partial rate of change of the position of each of P_1, P_2, and P_3 with respect to each of θ_1, θ_2, and θ_3 for an instant at which $\theta_1 = \theta_2 = \theta_3 = \pi/4$ radian.

Results:

	$i = 1$	$i = 2$	$i = 3$
$\mathbf{v}_{\dot{\theta}_1}{}^{P_i}$	$\frac{L}{2}\mathbf{n}_2$	$L\mathbf{n}_2$	$L\mathbf{n}_2$
$\mathbf{v}_{\dot{\theta}_2}{}^{P_i}$	0	$\frac{L}{2}\mathbf{n}_2$	$L\mathbf{n}_2$
$\mathbf{v}_{\dot{\theta}_3}{}^{P_i}$	0	0	$-\frac{L}{2}\mathbf{n}_1$

3(b) The vertex V of a right-circular cone C of height h is fixed, and a particle P is free to move on a diameter of the base of C, as indicated in Figure 3(b), where θ, ϕ, and ψ measure angles that determine the orienta-

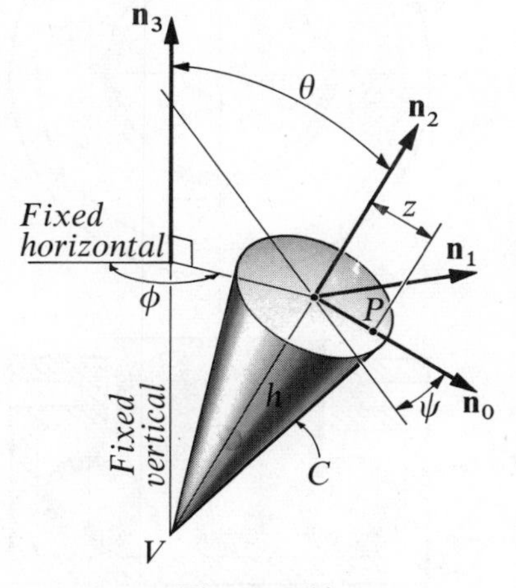

Figure 3(b)

tion of C; z represents the distance between P and the center of the base; and $\mathbf{n}_0, \cdots, \mathbf{n}_3$ are unit vectors, $\mathbf{n}_1$ being perpendicular to the plane determined by the axis of C and the vertical through point V.

Determine the partial rates of change of the position of P with respect to θ, ψ, ϕ, and z.

Results: $\mathbf{n}_1 \times \mathbf{p}$, $\mathbf{n}_2 \times \mathbf{p}$, $\mathbf{n}_3 \times \mathbf{p}$, and $\mathbf{n}_0$, where $\mathbf{p} = z\mathbf{n}_0 + h\mathbf{n}_2$.

3(c) Referring to Problem 2(g), and letting r be the radius of D, determine the magnitude of the acceleration of the particle of D that is in contact with S.

SUGGESTION: Find the velocity of D^* by making use of the fact that the velocity of the particle of D that is in contact with S is equal to zero (by definition of rolling). Next, differentiate $\mathbf{v}^{D^*}$ to obtain $\mathbf{a}^{D^*}$. Finally, use $\mathbf{a}^{D^*}$ and the results of Problem 2(g).

Result: $r|\dot{\psi}|(\dot{\psi}^2 + 2\dot{\psi}\dot{\phi}\sin\theta + \dot{\phi}^2)^{1/2}$.

3(d) A shaft, terminating in a truncated cone C of semivertical angle θ, see Figure 3(d), is supported by a thrust bearing consisting of a fixed

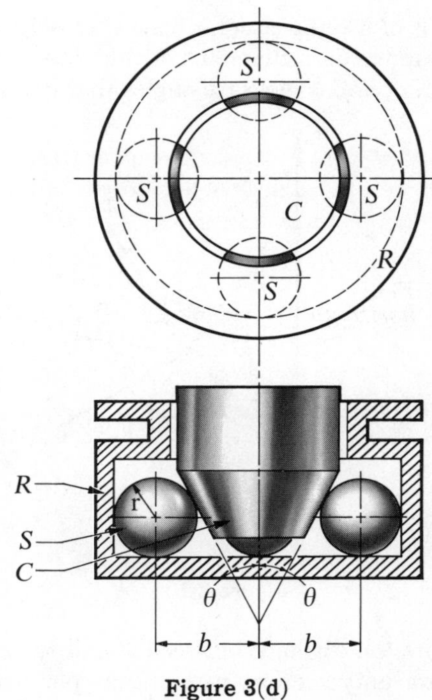

Figure 3(d)

race R and four identical spheres S of radius r. When the shaft rotates about its axis, S rolls on R at both of its points of contact with R, and C rolls on S.

Proper choice of the dimension b makes it possible to obtain "pure" rolling of C on S—that is, a motion during which not only are the velocities of the contact points of C and S equal to each other, but the angular velocity of C in S is parallel to the common tangent plane to C and S.

Determine b.

Result:

$$\frac{r(1 + \sin \theta)}{\cos \theta - \sin \theta}$$

3(e) Referring to Problem 2(a), let Q be a particle that is made to move on X_3 in such a way that the position vector $\mathbf{q}$ of Q relative to B^* is given by

$$\mathbf{q} = c(\Omega^2 t^2 - 1)\mathbf{n}_3$$

where c is a constant and Ω and $\mathbf{n}_3$ have the same meaning as before. Furthermore, let L be the (constant) distance between P and B^*, and assume that Ω is constant.

Determine the acceleration of Q in R for the instant $t = 1/\Omega$, assuming that θ_1, θ_2, and θ_3 are equal to zero at this instant.

Result: $(-L\Omega^2 + 4c\Omega\dot{\theta}_2)\mathbf{n}_1 - 4c\Omega\dot{\theta}_1\mathbf{n}_2 + 2c\Omega^2\mathbf{n}_3$.

3(f) The position of a point P in a reference frame R depends on three scalar functions $q_1(t)$, $q_2(t)$, and $q_3(t)$ of the time t in such a way that the velocity $\mathbf{v}$ of P in R can be expressed as

$$\mathbf{v} = \sum_{r=1}^{3} f_r \dot{q}_r \mathbf{n}_r$$

where f_1, f_2, and f_3 are functions of q_1, q_2, and q_3, and $\mathbf{n}_1$, $\mathbf{n}_2$, and $\mathbf{n}_3$ are mutually perpendicular unit vectors.

Determine a_r such that the acceleration $\mathbf{a}$ of P in R can be expressed as

$$\mathbf{a} = \sum_{r=1}^{3} a_r \mathbf{n}_r$$

Result:

$$\frac{1}{f_r}\left[\frac{d}{dt}(f_r^2 \dot{q}_r) - \sum_{s=1}^{3} f_s \frac{\partial f_s}{\partial q_r} \dot{q}_s^2\right]$$

3(g) In Figure 3(g), r, θ, and ϕ represent the so-called spherical

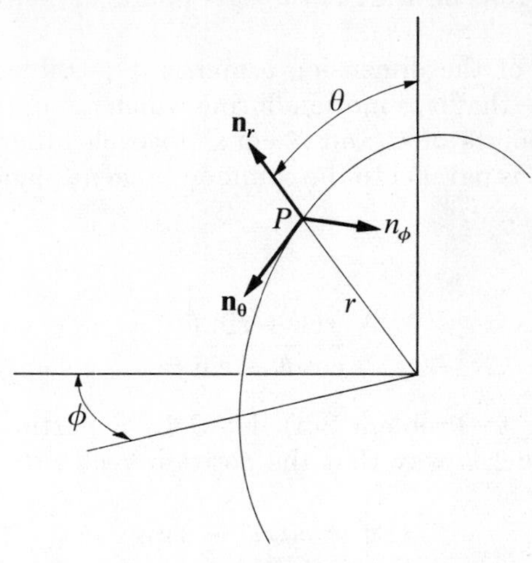

Figure 3(g)

coordinates of the point P, and $\mathbf{n}_r$, $\mathbf{n}_\theta$, and $\mathbf{n}_\phi$ are mutually perpendicular unit vectors.

Noting that the velocity $\mathbf{v}$ of P is given by

$$\mathbf{v} = \dot{r}\mathbf{n}_r + r\dot{\theta}\mathbf{n}_\theta + r \sin \theta\dot{\phi}\mathbf{n}_\phi$$

use the result obtained in Problem 3(f) to determine a_r, a_θ, and a_ϕ such that the acceleration $\mathbf{a}$ of P is given by

$$\mathbf{a} = a_r\mathbf{n}_r + a_\theta\mathbf{n}_\theta + a_\phi\mathbf{n}_\phi$$

Results:

$$\ddot{r} - r\dot{\theta}^2 - r(\dot{\phi} \sin \theta)^2 \qquad \frac{1}{r}\frac{d}{dt}(r^2\dot{\theta}) - \frac{r\dot{\phi}^2}{2} \sin 2\theta$$

$$\frac{1}{r \sin \theta}\frac{d}{dt}(\dot{\phi}r^2 \sin^2 \theta)$$

3(h) Figure 3(h) shows two sharp-edged circular disks, D and D', each of radius r, mounted at the extremities of a shaft S of length $2L$, the axis of S coinciding with those of D and D'. The disks are supported by a horizontal plane, and no slip is permitted to occur at their points of contact with this plane. Furthermore, each disk is free to rotate on S.

θ and θ' measure angles between the vertical and lines fixed in D and D', respectively. X and Y designate orthogonal axes fixed in the supporting plane, and x and y are the X and Y coordinates of the midpoint Q of S.

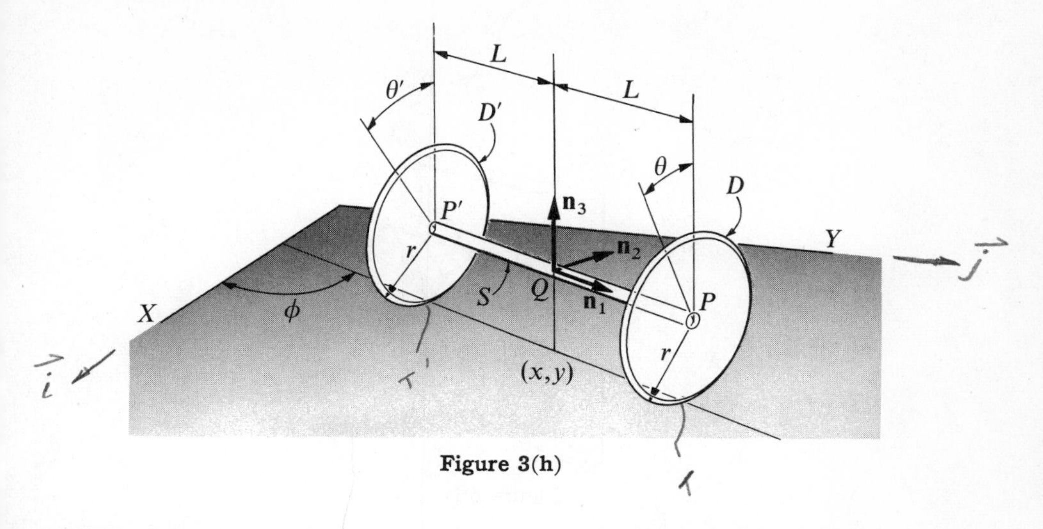

Figure 3(h)

Finally, ϕ is the angle between the X axis and the line joining the points at which D and D' touch the support.

Letting x, y, θ, θ', and ϕ be generalized coordinates of the system comprised of D, D', and S, show that the "no slip" requirement leads to three differential constraint equations, and solve these for $\dot{x}$, $\dot{y}$, and $\dot{\phi}$ as functions of $\dot{\theta}$, $\dot{\theta}'$, and the generalized coordinates. Then, treating $\dot{\theta}$ and $\dot{\theta}'$ as independent, and letting P and P' be the midpoints of D and D', respectively, determine the associated nonholonomic partial rates of change of the position of P and P' and find the nonholonomic partial rates of change of the orientation of D and D'.

Results: $\dot{x} = (r/2)(\dot{\theta} + \dot{\theta}') \sin \phi$, $\dot{y} = -(r/2)(\dot{\theta} + \dot{\theta}') \cos \phi$, $\dot{\phi} = (r/2L)(\dot{\theta}' - \dot{\theta})$,

	$\tilde{\mathbf{v}}^{P}_{(\,)}$	$\tilde{\mathbf{v}}^{P'}_{(\,)}$	$\tilde{\boldsymbol{\omega}}^{D}_{(\,)}$	$\tilde{\boldsymbol{\omega}}^{D'}_{(\,)}$
$\dot{\theta}$	$-r\mathbf{n}_2$	0	$\mathbf{n}_1 - (r/2L)\mathbf{n}_3$	$-(r/2L)\mathbf{n}_3$
$\dot{\theta}'$	0	$-r\mathbf{n}_2$	$(r/2L)\mathbf{n}_3$	$\mathbf{n}_1 + (r/2L)\mathbf{n}_3$

3(i) In Figure 3(i), X_1, X_2, and X_3 designate mutually perpendicular line segments fixed in a rigid body B that carries a gyroscope consisting of a rotor R mounted in a gimbal ring G. The gimbal axis coincides with X_1, and θ measures rotations of G relative to B. A motor (not shown) attached to G drives R with a constant angular speed in G.

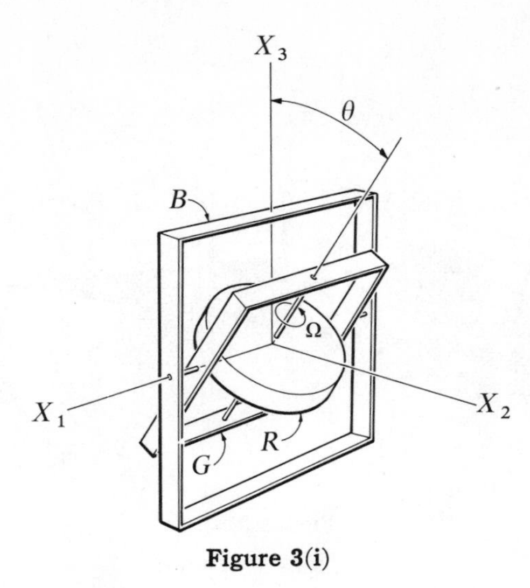

Figure 3(i)

Supposing that B is the body discussed in Problem 2(a), and taking

$$\begin{aligned} u_1 &= c_2c_3\dot{\theta}_1 + s_3\dot{\theta}_2 + \Omega(s_1s_3 - c_1s_2c_3) \\ u_2 &= -c_2s_3\dot{\theta}_1 + c_3\dot{\theta}_2 + \Omega(s_1c_3 + c_1s_2s_3) \\ u_3 &= s_2\dot{\theta}_1 + \dot{\theta}_3 + \Omega c_1c_2 \end{aligned}$$

and

$$u_4 = \dot{\theta}$$

determine $\boldsymbol{\omega}_{u_i}$ (for $i = 1, \cdots, 4$) for the rotor.

Results: $\mathbf{n}_1$, $\mathbf{n}_2$, $\mathbf{n}_3$, $-\mathbf{n}_1$.

3(**j**) Figure 3(j) shows a system comprised of two particles, P and Q, attached to pin-connected rods of lengths L, $2L$, and $3L$. Since this system possesses only one degree of freedom, a single generalized coordinate, such as θ_1, is sufficient for the specification of all admissible configurations. However, it may be convenient to introduce two "extraneous coordinates," θ_2 and θ_3, which must satisfy the "geometrical" constraint equations [see Figure 3(j)]

$$\begin{aligned} \sin\theta_1 - 2\sin\theta_2 + 3\sin\theta_3 &= 3 - \sqrt{3} \\ \cos\theta_1 - 2\cos\theta_2 + 3\cos\theta_3 &= 0 \end{aligned}$$

When these equations are differentiated with respect to time, they assume the form of Equations (2.27) of the text, and θ_1, θ_2, θ_3 can thus be regarded as generalized coordinates of a simple nonholonomic system possessing one degree of freedom.

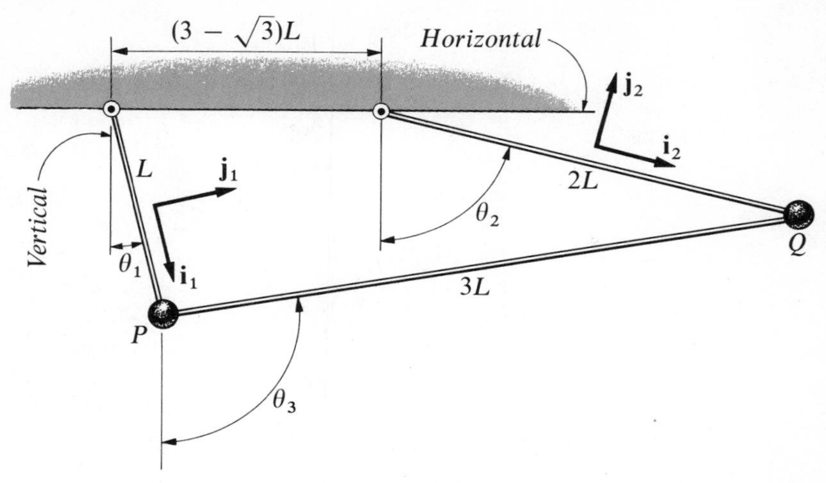

Figure 3(j)

Determine $\tilde{\mathbf{v}}_{\dot{\theta}_1}{}^P$ and $\tilde{\mathbf{v}}_{\dot{\theta}_1}{}^Q$.

Results: $L\mathbf{j}_1$, $L \dfrac{\sin(\theta_3 - \theta_1)}{\sin(\theta_3 - \theta_2)} \mathbf{j}_2$.

3(k) Referring to Problem 3(a), suppose that $\theta_1 = \theta_2 = \theta_3 = \pi/4$ at a certain instant and that virtual velocities of P_1 and P_3 for this instant are given as

$$\hat{\mathbf{v}}^{P_1} = -3v\mathbf{n}_2 \qquad \hat{\mathbf{v}}^{P_3} = 5v\mathbf{n}_1 + 4v\mathbf{n}_2$$

where v is a quantity having the dimensions of velocity.

Determine the virtual velocity of P_2 compatible with those of P_1 and P_3 at this instant.

Result: $-v\mathbf{n}_2$.

3(l) A virtual displacement $\delta\mathbf{q}$ of the particle Q of Problem 3(j) has the value $3L\mathbf{j}_2$ when $\theta_1 = 0$.

Determine the magnitude of the virtual rotation (compatible with $\delta\mathbf{q}$) of the rod connecting P and Q.

Result: $\sqrt{3}/2$ rad.

PART II

FORCE AND ENERGY

THE WORD *force,* as used in Newtonian mechanics, refers to vectors that characterize certain interactions between bodies. These interactions may involve widely separated bodies, as in the case of gravitational forces, or they may occur only when the bodies touch each other, which brings contact forces into play. The *generalized active forces* of Lagrangian mechanics are scalar quantities introduced for the purpose of analytical elimination of forces associated with constraints; and a set of scalar quantities called *generalized inertia forces* may be defined as a counterpart of the vectorial inertia forces of D'Alembert.

Chapter 3 deals with these two kinds of generalized forces. In Chapter 4 it is shown that the determination of generalized forces can sometimes be facilitated by the use of energy functions, and properties of a number of these are explored in detail.

Chapter **3**

Generalized Forces

3.1 Bound Vectors, Free Vectors, Moment, Couple, Torque, Equivalence, Replacement. In the sections that follow, such phrases as torque of a couple, equivalence of a set of bound vectors, and so forth, occur frequently. The associated concepts are reviewed briefly in the present section in order to insure full understanding of the terminology employed.

When a vector is associated with a particular point P in space, it is called a *bound* vector; otherwise, it is said to be a *free* vector. The point P is referred to as the *point of application* of the vector, and the line passing through P and parallel to the vector is called the *line of action* of the vector. Forces furnish an example of bound vectors. To specify a force, it is necessary to describe not only the *characteristics* (magnitude, orientation, and sense) of the force vector, but also its point of application.

Lines of action figure prominently in the definition of *the moment of a bound vector about a point.* If $\mathbf{v}$ is a bound vector, A a point, and B any point on the line of action of $\mathbf{v}$, the moment $\mathbf{M}$ of $\mathbf{v}$ about A is defined as

$$\mathbf{M} = \mathbf{p} \times \mathbf{v} \tag{3.1}$$

where (see Figure 3.1) $\mathbf{p}$ is the position vector of B relative to A.

The *resultant* of a set of bound vectors, like that of a set of free vectors, is defined as the vector sum of the vectors of the set, and is not, per se, a bound vector, because the definition of vector addition does not contain any rules for associating the sum of a number of vectors with a particular point. However, it is sometimes convenient to introduce a bound vector

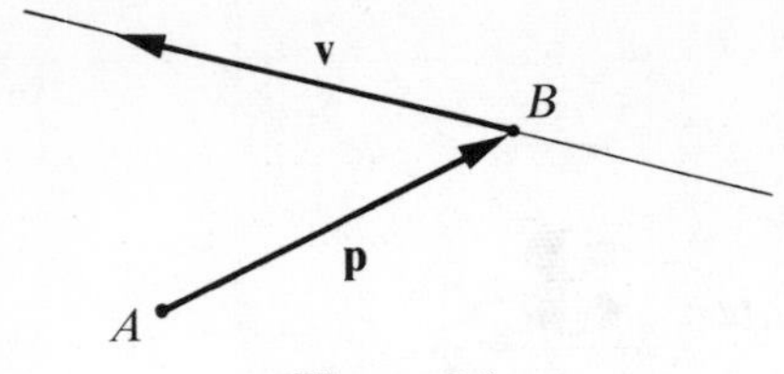

Figure 3.1

$\mathbf{v}$ that has the same characteristics as the resultant of a set of bound vectors. A phrase such as "the moment of the resultant about a point" then becomes meaningful when it is accompanied by information regarding the point of application of $\mathbf{v}$.

The moment of a set of bound vectors about a point is defined as the sum of the moments of the vectors of the set about that point. It may be verified that, if A and A' are two points, and $\mathbf{M}$ and $\mathbf{M}'$ are the moments of a set of vectors about A and A', respectively, then

$$\mathbf{M}' = \mathbf{M} + \mathbf{p} \times \mathbf{R} \tag{3.2}$$

where $\mathbf{p}$ is the position vector of A relative to A' and $\mathbf{R}$ is the resultant of the set of vectors. [The term $\mathbf{p} \times \mathbf{R}$ in Equation (3.2) may be regarded as the moment about point A' of a bound vector having the same characteristics as $\mathbf{R}$ and applied at point A.]

A set of bound vectors whose resultant is equal to zero is called a *couple*. It follows from Equation (3.2) that the moment of a couple about one point is equal to the moment of the couple about any other point. This moment is called the *torque* of the couple.

When two sets of bound vectors have equal resultants and equal moments about at least one point, the sets are said to be *equivalent* to each other, and either set may be referred to as a *replacement* of the other. It is an important consequence of these definitions that a set of bound vectors can always be replaced with a couple of torque $\mathbf{T}$ together with a bound vector $\mathbf{v}$ applied at any arbitrarily selected point P. $\mathbf{T}$ must be equal to the moment of the set about point P, and thus depends upon the position of P; $\mathbf{v}$ is independent of P, but must have the same characteristics as the resultant of the set.

▪ EXAMPLE

In Figure 3.2, P_1 and P_2 designate particles of equal mass m, and P a particle of mass M. The set of gravitational forces $\mathbf{F}_1$ and $\mathbf{F}_2$ exerted on P_1 and P_2 by P is to be replaced (a) with a couple of torque $\mathbf{T}_a$ together with a force $\mathbf{F}_a$ applied at the mass center P^* of P_1 and P_2, and (b) with a couple of torque $\mathbf{T}_b$ together with a force $\mathbf{F}_b$ applied at P.

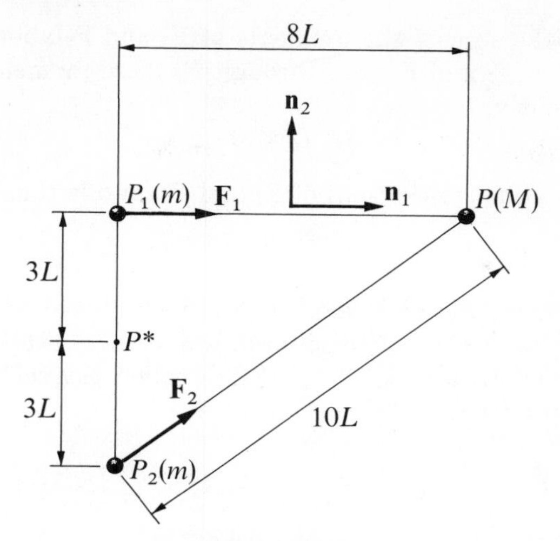

Figure 3.2

In accordance with the inverse-square law,

$$\mathbf{F}_1 = \frac{GMm}{64L^2}\,\mathbf{n}_1$$

and

$$\mathbf{F}_2 = \frac{GMm}{100L^2}\,(0.8\mathbf{n}_1 + 0.6\mathbf{n}_2)$$

where G is the universal gravitational constant and $\mathbf{n}_1$ and $\mathbf{n}_2$ are unit vectors directed as shown in Figure 3.2. As the lines of action of $\mathbf{F}_1$ and $\mathbf{F}_2$ pass through P_1 and P_2, respectively, and the position vectors of these points relative to P^* are $3L\mathbf{n}_2$ and $-3L\mathbf{n}_2$, the sum of the moments of $\mathbf{F}_1$ and $\mathbf{F}_2$ about point P^*, and hence the torque $\mathbf{T}_a$, is given by

$$\begin{aligned}\mathbf{T}_a &= 3L\mathbf{n}_2 \times \mathbf{F}_1 + (-3L\mathbf{n}_2) \times \mathbf{F}_2\\ &= \frac{3GMm}{L}\left(\frac{1}{64} - \frac{0.8}{100}\right)\mathbf{n}_2 \times \mathbf{n}_1\\ &= \frac{0.022875GMm}{L}\,\mathbf{n}_2 \times \mathbf{n}_1\end{aligned}$$

The force $\mathbf{F}_a$ is equal to sum of $\mathbf{F}_1$ and $\mathbf{F}_2$. Thus,

$$\begin{aligned}\mathbf{F}_a &= \frac{GMm}{L^2}\left[\left(\frac{1}{64} + \frac{0.8}{100}\right)\mathbf{n}_1 + \frac{0.6}{100}\,\mathbf{n}_2\right]\\ &= \frac{GMm}{L^2}\,(0.023625\mathbf{n}_1 + 0.006\mathbf{n}_2)\end{aligned}$$

$\mathbf{T}_b$ is equal to the sum of the moments of $\mathbf{F}_1$ and $\mathbf{F}_2$ about P. Since the lines of action of $\mathbf{F}_1$ and $\mathbf{F}_2$ pass through P, these moments are equal to zero. Consequently,

$$\mathbf{T}_b = 0$$

The force $\mathbf{F}_b$ is equal to the sum of $\mathbf{F}_1$ and $\mathbf{F}_2$, and is thus equal to $\mathbf{F}_a$.

3.2 Generalized Active Force. If S is a simple nonholonomic system possessing $n - m$ degrees of freedom in a reference frame R (see Section 2.22), $n - m$ quantities $F_1, \cdots, F_{n-m}$, called *generalized active forces* for S in R, are defined as

$$F_r = \sum_{i=1}^{N} \tilde{\mathbf{v}}_{\dot{q}_r}{}^{P_i} \cdot \mathbf{F}_i \qquad r = 1, \cdots, n - m \tag{3.3}$$

where N is the number of particles comprising S, P_i is a typical particle, $\tilde{\mathbf{v}}_{\dot{q}_r}{}^{P_i}$ is a nonholonomic partial rate of change of the position of P_i in R (see Section 2.23), and $\mathbf{F}_i$ is the resultant of all contact and body forces (for example, gravitational forces, magnetic forces) acting on P_i. [Sometimes it is convenient to replace $\tilde{\mathbf{v}}_{\dot{q}_r}{}^{P_i}$ with $\tilde{\mathbf{v}}_{u_r}{}^{P_i}$ (see Section 2.24); and, for a holonomic system possessing n degrees of freedom in R (see Section 2.2), take $m = 0$ and use $\mathbf{v}_{\dot{q}_r}{}^{P_i}$ in place of $\tilde{\mathbf{v}}_{\dot{q}_r}{}^{P_i}$, or $\mathbf{v}_{u_r}{}^{P_i}$ in place of $\tilde{\mathbf{v}}_{u_r}{}^{P_i}$.]

▪ EXAMPLE

In Figure 3.3, P_1 and P_2 designate particles of mass m_1 and m_2 that can

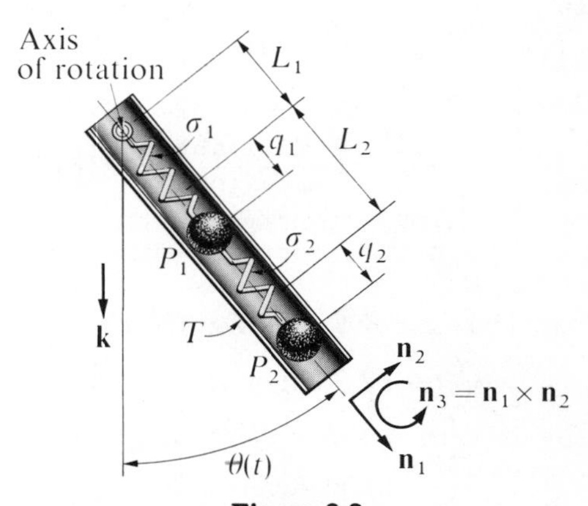

Figure 3.3

slide in a smooth tube T and are attached to linear springs σ_1 and σ_2 having spring constants k_1 and k_2 and "natural" lengths L_1 and L_2. T is made to rotate about a fixed horizontal axis passing through one end of T, in such a way that the angle $\theta(t)$ between the vertical and the axis of T is a prescribed function of time t. The generalized active forces F_1 and F_2 are to be determined, q_1 and q_2 designating the displacements of P_1 and P_2 from the positions occupied by the particles when the springs are "relaxed."

The velocities $\mathbf{v}^{P_1}$ and $\mathbf{v}^{P_2}$ of P_1 and P_2 are (see Figure 3.3 for the unit vectors $\mathbf{n}_1$ and $\mathbf{n}_2$)

$$\mathbf{v}^{P_1} \underset{(2.21)}{=} \dot{q}_1\mathbf{n}_1 + (L_1 + q_1)\dot{\theta}\mathbf{n}_2$$

$$\mathbf{v}^{P_2} \underset{(2.21)}{=} \dot{q}_2\mathbf{n}_1 + (L_1 + L_2 + q_2)\dot{\theta}\mathbf{n}_2$$

Consequently (see Section 2.13),

$$\mathbf{v}_{\dot{q}_1}{}^{P_1} = \mathbf{n}_1 \qquad \mathbf{v}_{\dot{q}_1}{}^{P_2} = 0$$
$$\mathbf{v}_{\dot{q}_2}{}^{P_1} = 0 \qquad \mathbf{v}_{\dot{q}_2}{}^{P_2} = \mathbf{n}_1$$

Contact forces are exerted on P_1 by σ_1, σ_2, and T. The forces exerted by σ_1 and σ_2 are

$$-k_1q_1\mathbf{n}_1 \quad \text{and} \quad k_2(q_2 - q_1)\mathbf{n}_1$$

respectively, and the force exerted by T can be expressed as

$$R_{12}\mathbf{n}_2 + R_{13}\mathbf{n}_3$$

where R_{12} and R_{13} are unknown scalars. No component parallel to $\mathbf{n}_1$ is included because T is presumed to be smooth.

The only body force acting on P_1 is the gravitational force

$$m_1g\mathbf{k}$$

where g is the "acceleration of gravity" and $\mathbf{k}$ is a unit vector directed vertically downward. The resultant $\mathbf{F}_1$ of all contact and body forces acting on P_1 is thus given by

$$\mathbf{F}_1 = [-k_1q_1 + k_2(q_2 - q_1)]\mathbf{n}_1 + R_{12}\mathbf{n}_2 + R_{13}\mathbf{n}_3 + m_1g\mathbf{k}$$

Similarly, the resultant $\mathbf{F}_2$ of all contact and gravitational forces acting on P_2 can be expressed as

$$\mathbf{F}_2 = -k_2(q_2 - q_1)\mathbf{n}_1 + R_{22}\mathbf{n}_2 + R_{23}\mathbf{n}_3 + m_2g\mathbf{k}$$

The generalized active force F_1 is now given by

$$F_1 \underset{(3.3)}{=} \mathbf{v}_{\dot{q}_1}{}^{P_1} \cdot \mathbf{F}_1 + \mathbf{v}_{\dot{q}_1}{}^{P_2} \cdot \mathbf{F}_2 = \mathbf{n}_1 \cdot \mathbf{F}_1$$
$$= -k_1q_1 + k_2(q_2 - q_1) + m_1g\cos\theta$$

and the generalized active force F_2 by

$$F_2 \underset{(3.3)}{=} \mathbf{v}_{\dot{q}_2}{}^{P_1} \cdot \mathbf{F}_1 + \mathbf{v}_{\dot{q}_2}{}^{P_2} \cdot \mathbf{F}_2 = \mathbf{n}_1 \cdot \mathbf{F}_2$$

$$= -k_2(q_2 - q_1) + m_2 g \cos \theta$$

Note that the (unknown) forces exerted on P_1 and P_2 by T contribute nothing to the generalized active forces F_1 and F_2.

3.3 Force Exerted by a Smooth Body. See Section 3.2: If $\mathbf{F}$ is a contact force exerted on a particle P of S by a smooth rigid body B whose motion in R is prescribed as a function of time (for example, B may be at rest in R), then $\mathbf{F}$ contributes nothing to F_r.

Proof: The velocity ${}^R\mathbf{v}^P$ of P in R can be expressed as

$${}^R\mathbf{v}^P \underset{(2.21)}{=} {}^R\mathbf{v}^{\bar{B}} + {}^B\mathbf{v}^P$$

If the motion of B in R is prescribed as a function of time, generalized coordinates $q_1, \cdots, q_n$ of S in R can always be chosen in such a way that ${}^B\mathbf{v}^P$ depends on $\dot{q}_1, \cdots, \dot{q}_{n-m}$, whereas ${}^R\mathbf{v}^{\bar{B}}$ is independent of these quantities. ${}^R\tilde{\mathbf{v}}_{\dot{q}_r}{}^P$ then depends only on ${}^B\mathbf{v}^P$, not on ${}^R\mathbf{v}^{\bar{B}}$, so that

$${}^R\tilde{\mathbf{v}}_{\dot{q}_r}{}^P = {}^B\tilde{\mathbf{v}}_{\dot{q}_r}{}^P \tag{a}$$

Next, consider an instant such that all of $\dot{q}_1, \cdots, \dot{q}_{n-m}$ vanish, except one, say $\dot{q}_r$. Under these circumstances,

$${}^B\mathbf{v}^P \underset{(2.28)}{=} {}^B\tilde{\mathbf{v}}_{\dot{q}_r}{}^P \dot{q}_r$$

and if P remains in contact with B, and $\mathbf{n}$ is a unit vector normal to the surface of B at P, then ${}^B\mathbf{v}^P$, and hence ${}^B\tilde{\mathbf{v}}_{\dot{q}_r}{}^P$, must be perpendicular to $\mathbf{n}$; therefore,

$${}^B\tilde{\mathbf{v}}_{\dot{q}_r}{}^P \cdot \mathbf{n} = 0 \tag{b}$$

Furthermore, if B is smooth, any force $\mathbf{F}$ exerted on P by B is normal to the surface of B at P. In other words,

$$\mathbf{F} = \mathbf{n}F \tag{c}$$

where F is some scalar. The contribution of $\mathbf{F}$ to F_r is thus given by

$${}^R\tilde{\mathbf{v}}_{\dot{q}_r}{}^P \cdot \mathbf{F} \underset{(a),(c)}{=} {}^B\tilde{\mathbf{v}}_{\dot{q}_r}{}^P \cdot \mathbf{n}F \underset{(b)}{=} 0$$

▪ EXAMPLE

In the example in Section 3.2, the forces exerted by T on P_1 and P_2 contribute nothing to F_1 and F_2.

3.4 Forces Acting on a Rigid Body. See Section 3.2: If B is a rigid body belonging to S, the particles comprising B may exert contact and body forces on each other; the total contribution of these forces to F_r is equal to zero; and if $\mathbf{F}_1, \cdots, \mathbf{F}_{N'}$ are forces acting respectively on particles $P_1, \cdots, P_{N'}$ of B, and this set of forces is equivalent to (see Section 3.1) a couple of torque $\mathbf{T}$ together with a force $\mathbf{F}$ applied at a point Q of B, so that

$$\left.\begin{aligned} \mathbf{F} &= \sum_{i=1}^{N'} \mathbf{F}_i \\ \mathbf{T} &= \sum_{i=1}^{N'} \mathbf{r}_i \times \mathbf{F}_i \end{aligned}\right\} \tag{3.4}$$

where $\mathbf{r}_i$ is the position vector of P_i relative to Q (see Figure 3.4), then $(F_r)_B$, the contribution to F_r of the forces $\mathbf{F}_1, \cdots, \mathbf{F}_{N'}$, is given by

$$\boxed{(F_r)_B = \tilde{\mathbf{v}}_{\dot{q}_r} \cdot \mathbf{F} + \tilde{\boldsymbol{\omega}}_{\dot{q}_r} \cdot \mathbf{T}} \tag{3.5}$$

where $\tilde{\mathbf{v}}_{\dot{q}_r}$ and $\tilde{\boldsymbol{\omega}}_{\dot{q}_r}$ are nonholonomic partial rates of change of the position of Q and of the orientation of B in R (see Section 2.23), respectively.

Proof: In Figure 3.4, P_i and P_j designate particles of B. $\mathbf{F}_{ij}$ is the

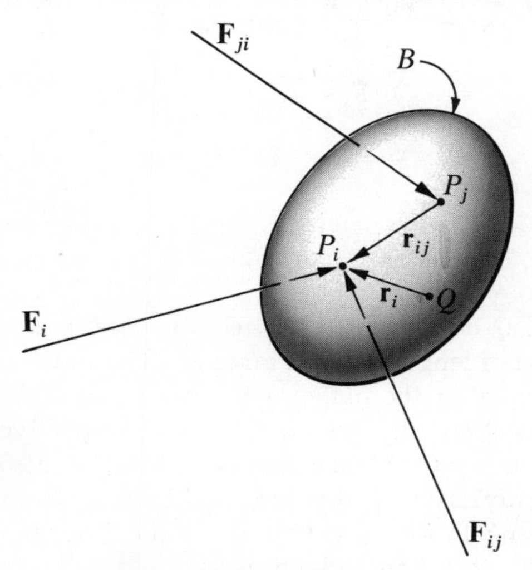

Figure 3.4

resultant of all contact and gravitational forces exerted on P_i by P_j, and $\mathbf{F}_{ji}$ is the resultant of all contact and gravitational forces exerted on P_j by P_i. Finally, $\mathbf{F}_i$ represents a force acting on P_i, but not necessarily exerted on P_i by P_j.

It will be shown first that the total contribution of $\mathbf{F}_{ij}$ and $\mathbf{F}_{ji}$ to F_r is equal to zero.

The Law of Action and Reaction asserts that (1) $\mathbf{F}_{ij}$ and $\mathbf{F}_{ji}$ have equal magnitudes and opposite directions, and (2) the lines of action of $\mathbf{F}_{ij}$ and $\mathbf{F}_{ji}$ coincide. Furthermore, the line of action of $\mathbf{F}_{ij}$ must pass through P_i and that of $\mathbf{F}_{ji}$ through P_j. It follows that

$$\mathbf{F}_{ji} = -\mathbf{F}_{ij} \tag{a}$$

and that $\mathbf{r}_{ij}$ is parallel to $\mathbf{F}_{ij}$; thus,

$$\mathbf{r}_{ij} \times \mathbf{F}_{ij} = 0 \tag{b}$$

The total contributions of $\mathbf{F}_{ij}$ and $\mathbf{F}_{ji}$ to F_r is given by [see Equation (3.3)]

$$\begin{aligned}\tilde{\mathbf{v}}_{\dot{q}_r}{}^{P_i} \cdot \mathbf{F}_{ij} + \tilde{\mathbf{v}}_{\dot{q}_r}{}^{P_j} \cdot \mathbf{F}_{ji} \underset{(a)}{=} (\tilde{\mathbf{v}}_{\dot{q}_r}{}^{P_i} - \tilde{\mathbf{v}}_{\dot{q}_r}{}^{P_j}) \cdot \mathbf{F}_{ij} &\underset{(2.30)}{=} (\tilde{\mathbf{v}}_{\dot{q}_r}{}^{P_j} \\ &\quad + \tilde{\boldsymbol{\omega}}_{\dot{q}_r} \times \mathbf{r}_{ij} - \tilde{\mathbf{v}}_{\dot{q}_r}{}^{P_j}) \cdot \mathbf{F}_{ij} \\ &= \tilde{\boldsymbol{\omega}}_{\dot{q}_r} \cdot \mathbf{r}_{ij} \times \mathbf{F}_{ij} \underset{(b)}{=} 0\end{aligned}$$

Next, to establish the validity of Equation (3.5), note that, by definition of $(F_r)_B$,

$$\begin{aligned}(F_r)_B &\underset{(3.3)}{=} \sum_{i=1}^{N'} \tilde{\mathbf{v}}_{\dot{q}_r}{}^{P_i} \cdot \mathbf{F}_i \underset{(2.30)}{=} \sum_{i=1}^{N'} (\tilde{\mathbf{v}}_{\dot{q}_r} + \tilde{\boldsymbol{\omega}}_{\dot{q}_r} \times \mathbf{r}_i) \cdot \mathbf{F}_i \\ &= \tilde{\mathbf{v}}_{\dot{q}_r} \cdot \sum_{i=1}^{N'} \mathbf{F}_i + \tilde{\boldsymbol{\omega}}_{\dot{q}_r} \cdot \sum_{i=1}^{N'} \mathbf{r}_i \times \mathbf{F}_i \\ &\underset{(3.4)}{=} \tilde{\mathbf{v}}_{\dot{q}_r} \cdot \mathbf{F} + \tilde{\boldsymbol{\omega}}_{\dot{q}_r} \cdot \mathbf{T}\end{aligned}$$

▪ EXAMPLE

In Figure 3.5, Q designates a particle of mass m, and B represents a uniform, thin rod of length $2L$ and mass M. The rod B is presumed to be constrained to move in the plane of the paper, so that R, θ, and ψ may serve as generalized coordinates q_1, q_2, and q_3, respectively.

It can be shown [see Problem 4(1)] that, when the ratio L/R approaches zero, the set of gravitational forces exerted by Q on B approaches equivalence (see Section 3.1) with a couple of torque $\mathbf{T}$ together with a force $\mathbf{F}$ applied at B^*, $\mathbf{T}$ and $\mathbf{F}$ being given by

$$\mathbf{T} = \frac{GmML^2}{2R^3} \sin 2\psi \, \mathbf{n}_\theta \times \mathbf{n}_r$$

$$\mathbf{F} = -\frac{GmM}{R^2} \mathbf{n}_r$$

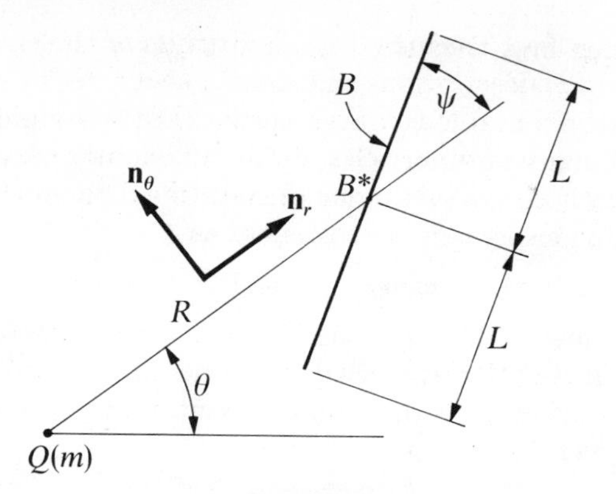

Figure 3.5

where G is the universal gravitational constant and $\mathbf{n}_r$ and $\mathbf{n}_\theta$ are unit vectors directed as shown in the sketch. The contributions $(F_1)_B$, $(F_2)_B$, and $(F_3)_B$ to F_1, F_2, and F_3 of gravitational forces exerted on B by Q are to be determined.

The velocity $\mathbf{v}$ of B^* can be expressed as

$$\mathbf{v} = \dot{R}\mathbf{n}_r + R\dot{\theta}\mathbf{n}_\theta$$

and the angular velocity $\boldsymbol{\omega}$ of B is given by

$$\boldsymbol{\omega} = (\dot{\theta} + \dot{\psi})\mathbf{n}_r \times \mathbf{n}_\theta$$

Hence (see Sections 2.13 and 2.5)

$$\mathbf{v}_{\dot{R}} = \mathbf{n}_r \qquad \mathbf{v}_{\dot{\theta}} = R\mathbf{n}_\theta \qquad \mathbf{v}_{\dot{\psi}} = 0$$

and

$$\boldsymbol{\omega}_{\dot{R}} = 0 \qquad \boldsymbol{\omega}_{\dot{\theta}} = \mathbf{n}_r \times \mathbf{n}_\theta \qquad \boldsymbol{\omega}_{\dot{\psi}} = \mathbf{n}_r \times \mathbf{n}_\theta$$

Consequently,

$$(F_1)_B \underset{(3.5)}{=} \mathbf{v}_{\dot{R}} \cdot \mathbf{F} + \boldsymbol{\omega}_{\dot{R}} \cdot \mathbf{T} = -\frac{GmM}{R^2}$$

$$(F_2)_B \underset{(3.5)}{=} \mathbf{v}_{\dot{\theta}} \cdot \mathbf{F} + \boldsymbol{\omega}_{\dot{\theta}} \cdot \mathbf{T} = -\frac{GmML^2}{2R^3}\sin 2\psi$$

$$(F_3)_B \underset{(3.5)}{=} \mathbf{v}_{\dot{\psi}} \cdot \mathbf{F} + \boldsymbol{\omega}_{\dot{\psi}} \cdot \mathbf{T} = -\frac{GmML^2}{2R^3}\sin 2\psi$$

3.5 Gravitational Forces. The gravitational forces exerted by the earth on the particles $P_1, \cdots, P_N$ of a system S comprise a set of forces that cannot be described with complete accuracy by reference only to the inverse-square law, because the constitution of the earth is not known precisely. However, descriptions sufficiently accurate for specific purposes can sometimes be obtained rather easily. For example, if the largest distance between any two particles of S is sufficiently small in comparison with the diameter of the earth, the gravitational force $\mathbf{G}_i$ exerted on the particle P_i by the earth can be expressed as

$$\mathbf{G}_i = m_i g \mathbf{k} \qquad i = 1, \cdots, N \tag{3.6}$$

where m_i is the mass of P_i, g is the (local) acceleration of gravity, and $\mathbf{k}$ is a unit vector directed vertically downward. To this order of approximation, the contribution $(F_r)_G$ to F_r of all gravitational forces exerted by the earth on S is given by

$$(F_r)_G = mg\mathbf{k} \cdot \tilde{\mathbf{v}}_{\dot{q}_r}{}^{P^*} \tag{3.7}$$

where m is the total mass of S and P^* designates the mass center of S.

Proof: It follows from the definition of the mass center of a set of particles that

$$\sum_{i=1}^{N} m_i \mathbf{v}^{P_i} = m\mathbf{v}^{P^*}$$

and, consequently, that

$$\sum_{i=1}^{N} m_i \tilde{\mathbf{v}}_{\dot{q}_r}{}^{P_i} \underset{(2.19)}{=} m\tilde{\mathbf{v}}_{\dot{q}_r}{}^{P^*} \tag{a}$$

By definition,

$$\begin{aligned}(F_r)_G &\underset{(3.3)}{=} \sum_{i=1}^{N} \tilde{\mathbf{v}}_{\dot{q}_r}{}^{P_i} \cdot \mathbf{G}_i \\ &\underset{(3.6)}{=} g\mathbf{k} \cdot \sum_{i=1}^{N} m_i \tilde{\mathbf{v}}_{\dot{q}_r}{}^{P_i} \end{aligned} \tag{b}$$

Hence,

$$(F_r)_G \underset{\text{(a),(b)}}{=} mg\mathbf{k} \cdot \tilde{\mathbf{v}}_{\dot{q}_r}{}^{P^*}$$

▪ EXAMPLE

Figure 3.6 shows nine uniform, pin-connected links, each of length L and mass M. Two links are pinned to a fixed support at A, and another is pinned to a slider that must remain in a horizontal slot. The system thus has but one degree of freedom, and the contribution $(F)_G$ of gravitational forces to the generalized active force can be found as follows.

It appears at a glance that point B is the mass center of the nine links. The velocity $\mathbf{v}$ of B is given by

$$\mathbf{v} = \frac{d}{dt}\left(\tfrac{3}{2}L \sin\theta\right)\mathbf{k} + \cdots$$

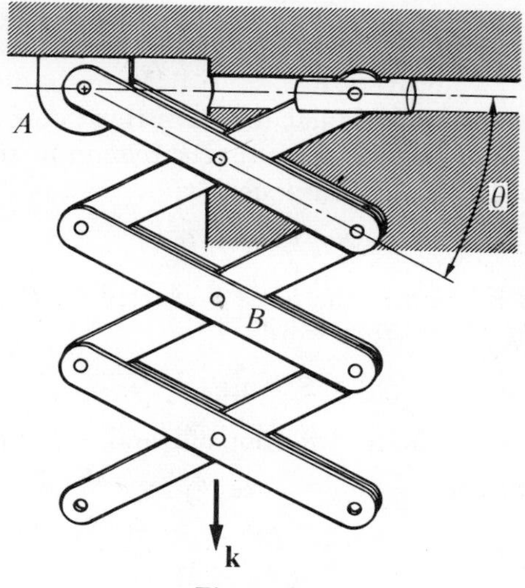

Figure 3.6

where the dots represent a vector perpendicular to **k**. [It can be seen from Equation (3.7) that it is unnecessary to consider this vector further.] Consequently,

$$\mathbf{v}_{\dot{\theta}} = \tfrac{3}{2}L \cos\theta\mathbf{k} + \cdots$$

and

$$(F)_G \underset{(3.7)}{=} 9Mg\mathbf{k} \cdot (\tfrac{3}{2}L \cos\theta\mathbf{k} + \cdots) = 13.5MgL \cos\theta$$

3.6 Forces of Interaction between a Smooth Body and a Particle. Referring to Section 3.2, suppose that B is a rigid body belonging to S, P is a particle of S that remains in contact with a smooth surface of B (P may be a particle of another rigid body that belongs to S), and $\mathbf{F}$ is the contact force exerted by B on P (and hence $-\mathbf{F}$ the contact force exerted by P on B). Then the total contribution of $\mathbf{F}$ and $-\mathbf{F}$ to F_r is equal to zero. (In Section 3.3, where a similar situation was considered, B did not belong to S.)

Proof: The velocity of P in R can be expressed as

$$^R\mathbf{v}^P \underset{\text{(Sec. 2.16)}}{=} {}^R\mathbf{v}^{\bar{B}} + {}^B\mathbf{v}^P$$

where $^R\mathbf{v}^{\bar{B}}$ denotes the velocity in R of the point $\bar{B}$ of B that is in contact with P. Hence,

$$^R\mathbf{v}^P - {}^R\mathbf{v}^{\bar{B}} = {}^B\mathbf{v}^P$$

and

$$^R\tilde{\mathbf{v}}_{\dot{q}_r}{}^P - {}^R\tilde{\mathbf{v}}_{\dot{q}_r}{}^{\bar{B}} = {}^B\tilde{\mathbf{v}}_{\dot{q}_r}{}^P \tag{a}$$

Next, if $\mathbf{n}$ is a unit vector normal to the surface of B at P, then $^B\tilde{\mathbf{v}}_{\dot{q}_r}{}^P$ must be perpendicular to $\mathbf{n}$ in order that P remain in contact with B for all values of $\dot{q}_1, \cdots, \dot{q}_{n-m}$. Consequently,

$$^B\tilde{\mathbf{v}}_{\dot{q}_r}{}^P \cdot \mathbf{n} = 0 \tag{b}$$

Furthermore, if B is smooth, the force $\mathbf{F}$ exerted by B on P is normal to the surface of B at P. In other words,

$$\mathbf{F} = \mathbf{n}F \tag{c}$$

where F is some scalar. The contribution of $\mathbf{F}$ and $-\mathbf{F}$ to F_r is thus given by

$$^R\tilde{\mathbf{v}}_{\dot{q}_r}{}^P \cdot \mathbf{F} + {}^R\tilde{\mathbf{v}}_{\dot{q}_r}{}^{\bar{B}} \cdot (-\mathbf{F}) = ({}^R\tilde{\mathbf{v}}_{\dot{q}_r}{}^P - {}^R\mathbf{v}_{\dot{q}_r}{}^{\bar{B}}) \cdot \mathbf{F}$$
$$\underset{\text{(a)}}{=} {}^B\tilde{\mathbf{v}}_{\dot{q}_r}{}^P \cdot \mathbf{F} \underset{\text{(c)}}{=} {}^B\tilde{\mathbf{v}}_{\dot{q}_r}{}^P \cdot \mathbf{n}F \underset{\text{(b)}}{=} 0$$

▪ EXAMPLE

In Figure 3.7, XX and YY are lines fixed in a reference frame R. A, B,

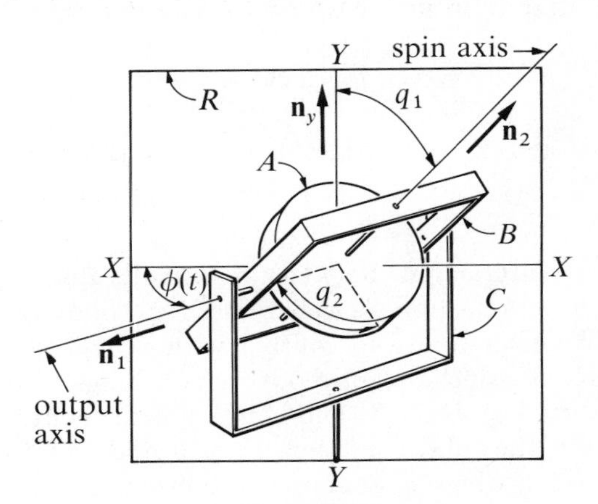

Figure 3.7

and C designate, respectively, the rotor, the inner gimbal ring, and the outer gimbal ring of a gyroscope. C is driven (by a motor, not shown) in such a way that the angle ϕ between the output axis (see sketch) and XX is a prescribed function of time, but B is free to rotate relative to C, and A can rotate relative to B. Under these circumstances, the angle q_1 between YY and the spin axis (see sketch), and the angle q_2 between the output axis and a line fixed in A are convenient generalized coordinates of the system comprised of A and B; and, when evaluating the associated generalized active forces, one must take into account all forces exerted by A and B on each other across the bearing surfaces at which these two bodies come into contact. If these surfaces are smooth, the total contribution of all such forces to F_1 and F_2 is equal to zero. Note also that the contributions to F_1 and F_2 of forces exerted on B by C are equal to zero) (see Section 3.3).

3.7 Forces of Interaction That Contribute to Generalized Active Forces. Sections 3.4 and 3.6 deal with forces exerted by one part of a system on another; and in both cases the forces in question contribute nothing to generalized active forces. However, in some situations forces of interaction *do* contribute to generalized active forces. For example, when two particles of a system are not rigidly connected, the gravitational forces exerted by the particles on each other generally make such contributions, and these forces may be neglected in the solutions of particular problems only when they are small in comparison with other forces. Bodies connected by springs furnish further examples (see the example in Section 3.2), as do bodies connected by certain energy-dissipating devices.

▪ EXAMPLE

Figure 3.8 shows a double pendulum consisting of two rigid rods A

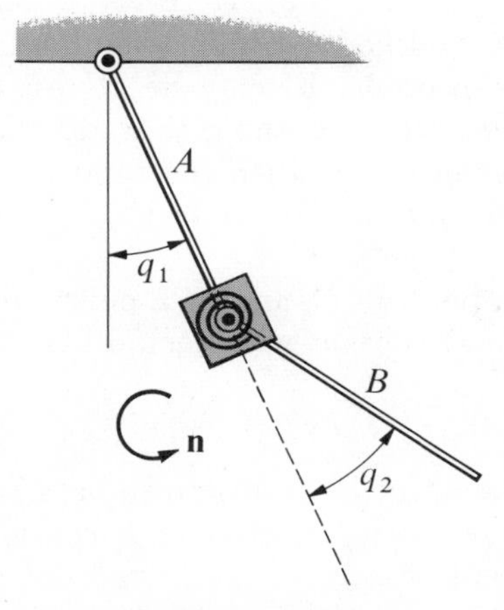

Figure 3.8

and B. Rod A is hinged to a fixed support, and A and B are pin-connected. Relative motion of the rods is resisted by a torsion spring of modulus σ and by a viscous fluid damper with damping constant δ, the spring being undeformed when A and B are aligned. In other words, the system of forces exerted on A by B through the spring and damper is equivalent to a couple whose torque $\mathbf{T}^A$ is given by

$$\mathbf{T}^A = (\sigma q_2 + \delta \dot{q}_2)\mathbf{n}$$

where q_2 is the angle shown in Figure 3.8 and $\mathbf{n}$ is a unit vector directed as indicated; and the system of forces exerted by A on B through the spring and damper is equivalent to a torque $\mathbf{T}^B$ given by

$$\mathbf{T}^B = -\mathbf{T}^A$$

The contributions f_1 and f_2 of the spring and damper forces to the generalized active forces F_1 and F_2 for this system, can be found as follows:

$$f_1 \underset{(3.5)}{=} \boldsymbol{\omega}_{\dot{q}_1}{}^A \cdot \mathbf{T}^A + \boldsymbol{\omega}_{\dot{q}_1}{}^B \cdot \mathbf{T}^B$$

$$f_2 \underset{(3.5)}{=} \boldsymbol{\omega}_{\dot{q}_2}{}^A \cdot \mathbf{T}^A + \boldsymbol{\omega}_{\dot{q}_2}{}^B \cdot \mathbf{T}^B$$

Now (see Section 2.5)

$$\boldsymbol{\omega}_{\dot{q}_1}{}^A = \mathbf{n} \qquad \boldsymbol{\omega}_{\dot{q}_1}{}^B = \mathbf{n}$$

$$\boldsymbol{\omega}_{\dot{q}_2}{}^A = 0 \qquad \boldsymbol{\omega}_{\dot{q}_2}{}^B = \mathbf{n}$$

Hence,

$$f_1 = \sigma q_2 + \delta \dot{q}_2 - (\sigma q_2 + \delta \dot{q}_2) = 0$$

and

$$f_2 = -(\sigma q_2 + \delta \dot{q}_2)$$

3.8 Rolling. See Section 3.2: If B is a rigid body belonging to S, and B rolls on a rigid body B' whose motion in R is prescribed as a function of time (for example, B' may be at rest in R), contact forces exerted on B by B' contribute nothing to F_r. If the motion of B' is not prescribed, but B' belongs to S, then the total contribution to F_r of all contact forces exerted by B and B' on each other is equal to zero.

Proof: Let P be a point of B, and P' a point of B', and suppose that P and P' are in (rolling) contact with each other. Then, by definition of rolling,

$$^R\mathbf{v}^P = {}^R\mathbf{v}^{P'} \tag{a}$$

Now, if the motion of B' in R is prescribed as a function of time, generalized coordinates $q_1, \cdots, q_n$ of S in R can always be chosen such that $^R\mathbf{v}^{P'}$ is independent of $q_1, \cdots, q_n$. $^R\tilde{\mathbf{v}}_{\dot{q}_r}{}^P$, for $r = 1, \cdots, n - m$, are then all equal to zero at the instant of contact, and it follows from

Equation (3.3) that a contact force exerted by B' on B at P contributes nothing to F_r. If, on the other hand, the motion of B' is not prescribed, but B' belongs to S, then

$$^R\tilde{\mathbf{v}}_{\dot{q}_r}{}^P \underset{(a)}{=} {}^R\mathbf{v}_{\dot{q}_r}{}^{P'} \qquad r = 1, \cdots, n - m \tag{b}$$

so that, if $\mathbf{F}$ is the contact force exerted on B by B', and $-\mathbf{F}$ the contact force exerted on B' by B, then the total contribution to F_r of all contact forces exerted by B and B' on each other becomes

$$^R\tilde{\mathbf{v}}_{r\dot{q}}{}^P \cdot \mathbf{F} + {}^R\tilde{\mathbf{v}}_{\dot{q}_r}{}^{P'} \cdot (-\mathbf{F}) \underset{(b)}{=} 0$$

▪ EXAMPLE

Referring to the example in Section 2.22, suppose that x, y, θ, and ϕ are regarded as generalized coordinates $q_1, \cdots, q_4$ of the system comprised of C and R. Then forces exerted by C and R on each other contribute nothing to the generalized active forces $F_1, \cdots, F_4$.

3.9 Generalized Inertia Force. If S is a simple nonholonomic system possessing $n - m$ degrees of freedom in a reference frame R (see Section 2.22), $n - m$ quantities $F_1^*, \cdots, F_{n-m}^*$, called *generalized inertia forces* for S in R, are defined as

$$F_r^* = \sum_{i=1}^{N} \tilde{\mathbf{v}}_{\dot{q}_r}{}^{P_i} \cdot \mathbf{F}_i^* \qquad r = 1, \cdots, n - m \tag{3.8}$$

where N is the number of particles comprising S, P_i is a typical particle, $\tilde{\mathbf{v}}_{\dot{q}_r}{}^{P_i}$ is a nonholonomic partial rate of change of the position of P_i in R (see Section 2.23), and $\mathbf{F}_i^*$ is the *inertia force* for P_i in R; that is,

$$\mathbf{F}_i^* = -m_i\mathbf{a}_i \tag{3.9}$$

where m_i is the mass of P_i and $\mathbf{a}_i$ is the acceleration of P_i in R. [Sometimes it is convenient to replece $\tilde{\mathbf{v}}_{\dot{q}_r}{}^{P_i}$ with $\tilde{\mathbf{v}}_{u_r}{}^{P_i}$ (see Section 2.24); and, for a holonomic system possessing n degrees of freedom in R (see Section 2.2), take $m = 0$ and use $\mathbf{v}_{\dot{q}_r}{}^{P_i}$ in place of $\tilde{\mathbf{v}}_{\dot{q}_r}{}^{P_i}$ in place of $\tilde{\mathbf{v}}_{u_r}{}^{P_i}$.]

▪ EXAMPLE

Figure 3.9 shows the system previously described in the example in Section 3.2. As before, q_1 and q_2 measure the displacements of the particles P_1 and P_2 from the positions occupied by these particles when the springs σ_1 and σ_2 have their natural lengths L_1 and L_2, and it is presumed that the

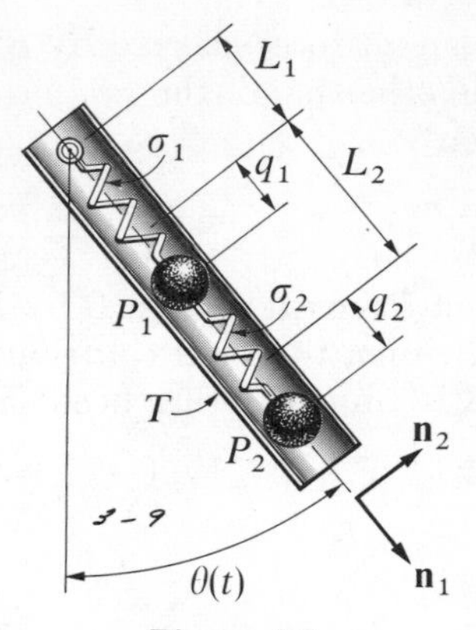

Figure 3.9

angle $\theta(t)$ between the vertical and the axis of the tube T is a prescribed function of time. The generalized inertia forces $F_1{}^*$ and $F_2{}^*$ are to be determined.

The accelerations $\mathbf{a}_1$ and $\mathbf{a}_2$ of P_1 and P_2 are (see Figure 3.9 for the unit vectors $\mathbf{n}_1$ and $\mathbf{n}_2$)

$$\mathbf{a}_1 = [\ddot{q}_1 - (L_1 + q_1)\dot{\theta}^2]\mathbf{n}_1 + [2\dot{q}_1\dot{\theta} + (L_1 + q_1)\ddot{\theta}]\mathbf{n}_2$$

and

$$\mathbf{a}_2 = [\ddot{q}_2 - (L_1 + L_2 + q_2)\dot{\theta}^2]\mathbf{n}_1 + [2\dot{q}_2\dot{\theta} + (L_1 + L_2 + q_2)\ddot{\theta}]\mathbf{n}_2$$

The partial rates of change of position of P_1 and P_2 with respect to q_1 and q_2 were found in the example in Section 3.2 to be

$$\mathbf{v}_{\dot{q}_1}{}^{P_1} = \mathbf{n}_1 \qquad \mathbf{v}_{\dot{q}_1}{}^{P_2} = 0$$
$$\mathbf{v}_{\dot{q}_2}{}^{P_1} = 0 \qquad \mathbf{v}_{\dot{q}_2}{}^{P_2} = \mathbf{n}_1$$

Consequently,

$$F_1{}^* \underset{(3.8),(3.9)}{=} -m_1\mathbf{v}_{\dot{q}_1}{}^{P_1} \cdot \mathbf{a}_1 - m_2\mathbf{v}_{\dot{q}_1}{}^{P_2} \cdot \mathbf{a}_2$$
$$= -m_1[\ddot{q}_1 - (L_1 + q_1)\dot{\theta}^2]$$

and

$$F_2{}^* \underset{(3.8),(3.9)}{=} -m_1\mathbf{v}_{\dot{q}_2}{}^{P_1} \cdot \mathbf{a}_1 - m_2\mathbf{v}_{\dot{q}_2}{}^{P_2} \cdot \mathbf{a}_2$$
$$= -m_2[\ddot{q}_2 - (L_1 + L_2 + q_2)\dot{\theta}^2]$$

3.10 Rigid-Body Contribution. Referring to Section 3.9, $(F_r{}^*)_B$, the total contribution to $F_r{}^*$ of all inertia forces for the particles of a rigid body B belonging to S, can be expressed as

$$(F_r{}^*)_B = \tilde{\mathbf{v}}_{\dot{q}_r} \cdot \mathbf{F}^* + \tilde{\boldsymbol{\omega}}_{\dot{q}_r} \cdot \mathbf{T}^* \qquad r = 1, \cdots, n - m \tag{3.10}$$

where $\tilde{\mathbf{v}}_{\dot{q}_r}$ is the nonholonomic partial rate of change with respect to q_r of the position of the mass center of B in R; $\tilde{\boldsymbol{\omega}}_{\dot{q}_r}$ is the nonholonomic partial rate of change with respect to q_r of the orientation of B in R (see Section 2.23); $\mathbf{F}^*$, called the *inertia force* for B in R, is defined in terms of the mass m of B and the acceleration $\mathbf{a}^*$ of the mass center of B in R as

$$\mathbf{F}^* = -m\mathbf{a}^* \tag{3.11}$$

and $\mathbf{T}^*$, called the *inertia torque* for B in R, is defined as

$$\mathbf{T}^* = -\sum_{i=1}^{\bar{N}} m_i\mathbf{r}_i \times \mathbf{a}_i \tag{3.12}$$

where m_i is the mass of a typical particle P_i of B, $\bar{N}$ is the number of particles comprising B, $\mathbf{r}_i$ is the position vector of P_i relative to the mass center of B, and $\mathbf{a}_i$ is the acceleration of P_i in R.

Proof: With m_i and $\mathbf{r}_i$ as defined,

$$\sum_{i=1}^{\bar{N}} m_i\mathbf{r}_i = 0 \tag{a}$$

$\mathbf{a}_i$ can be expressed in terms of the acceleration $\mathbf{a}^*$ of the mass center of B, the angular acceleration $\boldsymbol{\alpha}$ of B in R, and the angular velocity $\boldsymbol{\omega}$ of B in R, as follows:

$$\mathbf{a}_i \underset{(2.23)}{=} \mathbf{a}^* + \boldsymbol{\alpha} \times \mathbf{r}_i + \boldsymbol{\omega} \times (\boldsymbol{\omega} \times \mathbf{r}_i) \tag{b}$$

It follows that

$$\begin{aligned}\sum_{i=1}^{\bar{N}} m_i\mathbf{a}_i &\underset{(b)}{=} \left(\sum_{i=1}^{\bar{N}} m_i\right)\mathbf{a}^* + \boldsymbol{\alpha} \times \sum_{i=1}^{\bar{N}} m_i\mathbf{r}_i \\ &\qquad + \boldsymbol{\omega} \times \left(\boldsymbol{\omega} \times \sum_{i=1}^{\bar{N}} m_i\mathbf{r}_i\right) \\ &\underset{(a)}{=} m\mathbf{a}^* \underset{(3.11)}{=} -\mathbf{F}^*\end{aligned} \tag{c}$$

Consequently,

$$\begin{aligned}(F_r^*)_B &\underset{(3.8),(3.9)}{=} -\sum_{i=1}^{\bar{N}} m_i\tilde{\mathbf{v}}_{\dot{q}_r}{}^{P_i} \cdot \mathbf{a}_i \\ &\underset{(2.30)}{=} -\sum_{i=1}^{\bar{N}} m_i(\tilde{\mathbf{v}}_{\dot{q}_r} + \tilde{\boldsymbol{\omega}}_{\dot{q}_r} \times \mathbf{r}_i) \cdot \mathbf{a}_i \\ &= -\tilde{\mathbf{v}}_{\dot{q}_r} \cdot \sum_{i=1}^{\bar{N}} m_i\mathbf{a}_i - \tilde{\boldsymbol{\omega}}_{\dot{q}_r} \cdot \sum_{i=1}^{\bar{N}} m_i\mathbf{r}_i \times \mathbf{a}_i \\ &= \tilde{\mathbf{v}}_{\dot{q}_r} \underset{(c)}{\cdot} \mathbf{F}^* + \tilde{\boldsymbol{\omega}}_{\dot{q}_r} \underset{(3.12)}{\cdot} \mathbf{T}^*\end{aligned}$$

3.11 Inertia Properties. The inertia torque $\mathbf{T}^*$ [see Equation (3.12)] depends on the acceleration, the mass, and the position within B of every particle of B. Speaking more broadly, one may say that $\mathbf{T}^*$ depends jointly on the motion and on the constitution of B, and this dependence suggests that certain calculations may be facilitated by a clear-cut separation of kinematical considerations from those concerned with the inertia properties of the body.

In order to render the present work self-contained, the subject of inertia properties is treated briefly in the sections that follow, with primary emphasis on definitions and theorems specifically needed in the sequel, rather than on details of computation with which the reader is probably already familiar.

3.12 Second Moment, Product of Inertia, Moment of Inertia, Radius of Gyration. In Figure 3.10, P_i represents a typical particle of a set S of

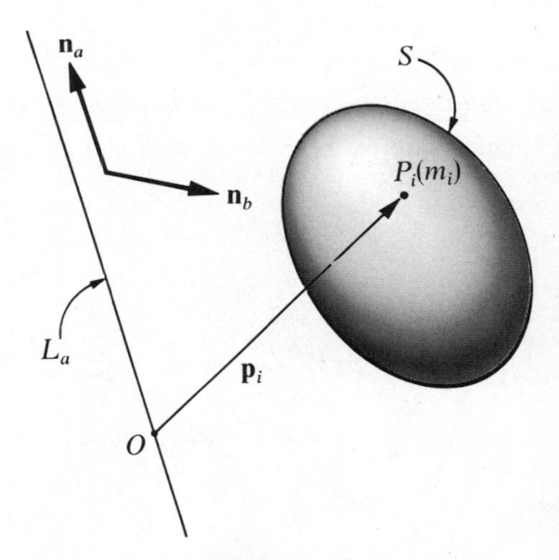

Figure 3.10

N particles, m_i designates the mass of P_i, $\mathbf{n}_a$ and $\mathbf{n}_b$ are unit vectors, O is a point, and L_a is the straight line passing through O and parallel to $\mathbf{n}_a$; finally, $\mathbf{p}_i$ is the position vector of P_i relative to O.

A vector $\mathbf{I}_a$, called the *second moment* of S relative to O for $\mathbf{n}_a$, is defined as

$$\mathbf{I}_a = \sum_{i=1}^{N} m_i \mathbf{p}_i \times (\mathbf{n}_a \times \mathbf{p}_i) \tag{3.13}$$

and this vector is used to form a scalar denoted by I_{ab}, called the *product of inertia* of S relative to O for $\mathbf{n}_a$ and $\mathbf{n}_b$, and defined as

$$I_{ab} = \mathbf{I}_a \cdot \mathbf{n}_b \tag{3.14}$$

This definition does not require that $\mathbf{n}_a$ and $\mathbf{n}_b$ differ from each other. When $\mathbf{n}_b$ is equal to $\mathbf{n}_a$, the corresponding product of inertia is denoted either by I_{aa} or by I_a, and is called the *moment of inertia* of S about line L_a; that is,

$$I_a = I_{aa} = \mathbf{I}_a \cdot \mathbf{n}_a \tag{3.15}$$

The reason for mentioning line L_a in this case is that

$$I_a \underset{(3.13),(3.15)}{=} \sum_{i=1}^{N} m_i(\mathbf{n}_a \times \mathbf{p}_i)^2 = \sum_{i=1}^{N} m_i l_i^2 \tag{3.16}$$

where l_i is the distance from P_i to L_a.

I_a can always be expressed as

$$I_a = \left(\sum_{i=1}^{N} m_i\right) k_a^2 \tag{3.17}$$

where k_a is a real, non-negative quantity called the *radius of gyration* of S with respect to line L_a.

Second moments, products of inertia, moments of inertia, and radii of gyration of a continuous body B can be defined analogously. In Figure 3.11, P represents a generic point of B; ρ designates the mass density of B

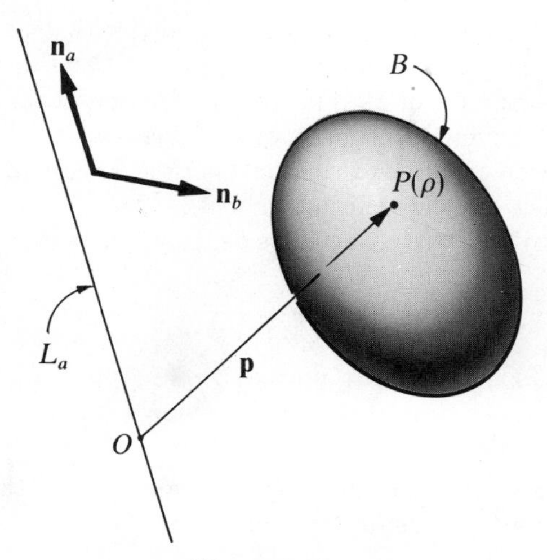

Figure 3.11

at P; $\mathbf{n}_a$, $\mathbf{n}_b$, O, and L_a have the same meaning as before; and $\mathbf{p}$ is the position vector of P relative to O. The definition given in Equation (3.13) is now replaced with

$$\mathbf{I}_a = \int \rho \mathbf{p} \times (\mathbf{n}_a \times \mathbf{p})\, d\tau \tag{3.18}$$

where $d\tau$ is the length, area, or volume of a differential element of the figure (curve, surface, or solid) occupied by B; Equations (3.14), (3.15), and (3.16) then apply to B as well as to S.

▪ EXAMPLE

In Figure 3.12, B represents a uniform, thin rod of length $2L$ and

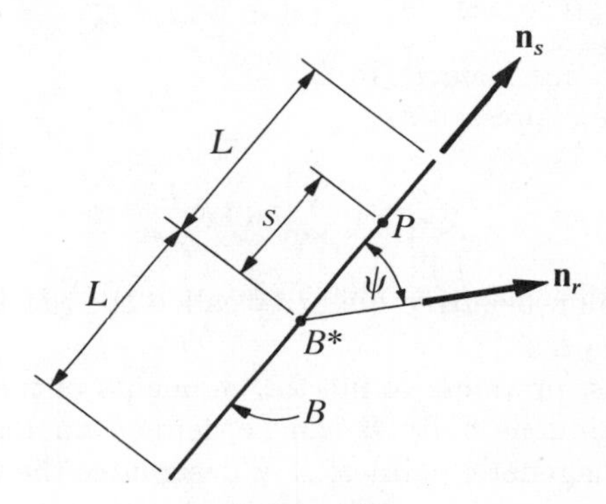

Figure 3.12

mass M; $\mathbf{n}_r$ is a unit vector making an angle ψ with B; $\mathbf{n}_s$ is a unit vector parallel to B; P is a generic point of B; and s is the distance between P and the mass center B^* of B. The second moment $\mathbf{I}_r$ of B relative to B^* for $\mathbf{n}_r$ will be determined, and it will be shown, incidentally, that the torque $\mathbf{T}$ discussed in the example in Section 3.4 assumes a particularly simple form when expressed in terms of $\mathbf{I}_r$.

Since B is a uniform body, the mass density ρ of B at P is given by

$$\rho = \frac{M}{2L}$$

The position vector $\mathbf{p}$ of P relative to B^* can be expressed as

$$\mathbf{p} = s\mathbf{n}_s$$

and, if B is regarded as occupying a straight line, the length $d\tau$ of a differential element of this figure is given by

$$d\tau = ds$$

Consequently,

$$\rho \mathbf{p} \times (\mathbf{n}_r \times \mathbf{p})\, ds = \frac{M}{2L} \mathbf{n}_s \times (\mathbf{n}_r \times \mathbf{n}_s) s^2\, ds$$

and

$$\mathbf{I}_r \underset{(3.18)}{=} \frac{M}{2L} \mathbf{n}_s \times (\mathbf{n}_r \times \mathbf{n}_s) \int_{-L}^{L} s^2\, ds = \frac{ML^2}{3} \mathbf{n}_s \times (\mathbf{n}_r \times \mathbf{n}_s)$$

Consider the cross product of $\mathbf{n}_r$ and $\mathbf{I}_r$:

$$\mathbf{n}_r \times \mathbf{I}_r = \frac{ML^2}{3} \mathbf{n}_r \times [\mathbf{n}_s \times (\mathbf{n}_r \times \mathbf{n}_s)]$$

$$= \frac{ML^2}{3} \mathbf{n}_s \cdot \mathbf{n}_r \mathbf{n}_s \times \mathbf{n}_r$$

Now

$$\mathbf{n}_s \cdot \mathbf{n}_r = \cos \psi$$

and

$$\mathbf{n}_s \times \mathbf{n}_r = \sin \psi \mathbf{n}_\theta \times \mathbf{n}_r$$

where $\mathbf{n}_\theta$ is the unit vector previously shown in Figure 3.5. Hence,

$$\mathbf{n}_r \times \mathbf{I}_r = \frac{ML^2}{3} \cos \psi \sin \psi \mathbf{n}_\theta \times \mathbf{n}_r = \frac{ML^2}{3} \frac{\sin 2\psi}{2} \mathbf{n}_\theta \times \mathbf{n}_r$$

and the torque $\mathbf{T}$ described in Section 3.4 can be expressed as

$$\mathbf{T} = \frac{3Gm}{R^3} \mathbf{n}_r \times \mathbf{I}_r$$

This expression is of interest because it can be shown to be valid not only for thin rods, but for bodies of any shape whatsoever.

3.13 Mutually Perpendicular Unit Vectors. If $\mathbf{n}_j$, $j = 1, 2, 3$, are mutually perpendicular unit vectors, and $\mathbf{n}_a$ and $\mathbf{n}_b$ are unit vectors given by

$$\mathbf{n}_a = \sum_{j=1}^{3} a_j \mathbf{n}_j \tag{3.19}$$

and

$$\mathbf{n}_b = \sum_{j=1}^{3} b_j \mathbf{n}_j \tag{3.20}$$

respectively, the second moments $\mathbf{I}_j$ and $\mathbf{I}_a$ of a set S of particles relative to a point O (see Section 3.12) can be expressed as

$$\mathbf{I}_j = \sum_{k=1}^{3} I_{jk}\mathbf{n}_k \qquad j = 1, 2, 3 \tag{3.21}$$

$$\mathbf{I}_a = \sum_{j=1}^{3} a_j\mathbf{I}_j \tag{3.22}$$

and

$$\mathbf{I}_a = \sum_{j=1}^{3}\sum_{k=1}^{3} a_j I_{jk}\mathbf{n}_k \tag{3.23}$$

where I_{jk} is the product of inertia of S relative to O for $\mathbf{n}_j$ and $\mathbf{n}_k$ (see Section 3.12). The product of inertia I_{ab} is given by

$$I_{ab} = \sum_{j=1}^{3}\sum_{k=1}^{3} a_j I_{jk} b_k \tag{3.24}$$

The last of these relationships is frequently used in the solution of practical problems, whereas the first five serve primarily to facilitate later proofs.

Proofs: The following is an identity:

$$\mathbf{I}_j = \sum_{k=1}^{3} \mathbf{I}_j \cdot \mathbf{n}_k\mathbf{n}_k$$

But

$$\mathbf{I}_j \cdot \mathbf{n}_k \underset{(3.14)}{=} I_{jk}$$

Hence Equation (3.21) follows immediately.

Next,

$$\begin{aligned}
\mathbf{I}_a &\underset{(3.13)}{=} \sum_{i=1}^{N} m_i\mathbf{p}_i \times (\mathbf{n}_a \times \mathbf{p}_i) \\
&\underset{(3.19)}{=} \sum_{i=1}^{N} m_i\mathbf{p}_i \times \left(\sum_{j=1}^{3} a_j\mathbf{n}_j \times \mathbf{p}_i\right) \\
&= \sum_{j=1}^{3} a_j \sum_{i=1}^{N} m_i\mathbf{p}_i \times (\mathbf{n}_j \times \mathbf{p}_i) \\
&\underset{(3.13)}{=} \sum_{j=1}^{3} a_j\mathbf{I}_j
\end{aligned}$$

which concludes the proof of Equation (3.22). Finally, Equation (3.23) is obtained by substitution from (3.21) into (3.22); and Equation (3.24) follows from (3.14), (3.23), and (3.20).

3.14 Symmetry. Products of inertia (see Section 3.12) possess the following "symmetry" property: The order of the subscripts in a symbol denoting a product of inertia is immaterial; that is,

$$I_{ab} = I_{ba} \tag{3.25}$$

Proof:

$$I_{ab} \underset{(3.14),(3.13)}{=} \sum_{i=1}^{N} m_i \mathbf{p}_i \times (\mathbf{n}_a \times \mathbf{p}_i) \cdot \mathbf{n}_b$$

$$= \sum_{i=1}^{N} m_i (\mathbf{n}_a \times \mathbf{p}_i) \cdot (\mathbf{n}_b \times \mathbf{p}_i)$$

and this expression remains unaltered when $\mathbf{n}_a$ and $\mathbf{n}_b$ are interchanged.

3.15 Parallel-Axes Theorem. In Figure 3.13, P_i represents a typical

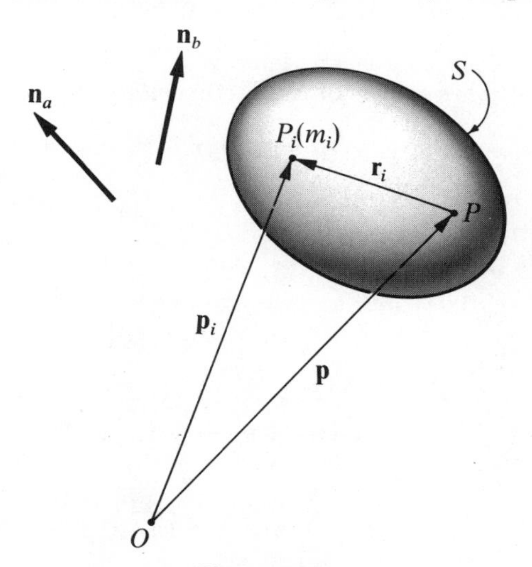

Figure 3.13

particle of a set S of N particles, m_i designates the mass of P_i, $\mathbf{n}_a$ and $\mathbf{n}_b$ are unit vectors, O is a point, P is the mass center of S, and $\mathbf{p}_i$ and $\mathbf{p}$ are the position vectors of P_i and P, respectively, relative to O.

Suppose that $\mathbf{I}_a^{S/O}$ and $\mathbf{I}_a^{S/P}$ designate second moments (see Section 3.12) of S relative to O and P, respectively, and that $\mathbf{I}_a^{P/O}$ denotes the second moment relative to O for $\mathbf{n}_a$ of a (fictitious) particle that is situated at P and has a mass equal to the total mass of S; that is,

$$\mathbf{I}_a^{P/O} = \left(\sum_{i=1}^{N} m_i\right) \mathbf{p} \times (\mathbf{n}_a \times \mathbf{p}) \tag{3.26}$$

Then

$$\mathbf{I}_a^{S/O} = \mathbf{I}_a^{S/P} + \mathbf{I}_a^{P/O} \tag{3.27}$$

Furthermore,

$$I_{ab}^{S/O} = I_{ab}^{S/P} + I_{ab}^{P/O} \tag{3.28}$$

and

$$I_a^{S/O} = I_a^{S/P} + I_a^{P/O} \tag{3.29}$$

The last of these equations is known as the *Parallel-Axes Theorem* because it establishes the relationship between the moments of inertia of S about two lines that are parallel to each other and pass through O and P, respectively.

Proof: Let $\mathbf{r}_i$ be the position vector of P_i relative to P, and note that

$$\sum_{i=1}^{N} m_i \mathbf{r}_i = 0 \tag{a}$$

because P is the mass center of S. Furthermore

$$\mathbf{p}_i = \mathbf{p} + \mathbf{r}_i \tag{b}$$

Then

$$\begin{aligned}
\mathbf{I}_a^{S/O} &\underset{(3.13)}{=} \sum_{i=1}^{N} m_i \mathbf{p}_i \times (\mathbf{n}_a \times \mathbf{p}_i) \\
&\underset{(b)}{=} \sum_{i=1}^{N} m_i (\mathbf{p} + \mathbf{r}_i) \times [\mathbf{n}_a \times (\mathbf{p} + \mathbf{r}_i)] \\
&= \left(\sum_{i=1}^{N} m_i\right) \mathbf{p} \times (\mathbf{n}_a \times \mathbf{p}) + \mathbf{p} \times \left(\mathbf{n}_a \times \sum_{i=1}^{N} m_i \mathbf{r}_i\right) \\
&\qquad + \left(\sum_{i=1}^{N} m_i \mathbf{r}_i\right) \times (\mathbf{n}_a \times \mathbf{p}) + \sum_{i=1}^{N} m_i \mathbf{r}_i \times (\mathbf{n}_a \times \mathbf{r}_i) \\
&= \underset{(3.26)}{\mathbf{I}_a^{P/O}} + \underset{(a)}{0} + \underset{(a)}{0} + \underset{(3.13)}{\mathbf{I}_a^{S/P}}
\end{aligned}$$

Next,

$$\begin{aligned}
I_{ab}^{S/O} &\underset{(3.14)}{=} \mathbf{I}_a^{S/O} \cdot \mathbf{n}_b \underset{(3.27)}{=} \mathbf{I}_a^{S/P} \cdot \mathbf{n}_b + \mathbf{I}_a^{P/O} \cdot \mathbf{n}_b \\
&\underset{(3.14)}{=} I_{ab}^{S/P} + I_{ab}^{P/O}
\end{aligned}$$

Finally, Equation (3.29) is a special case of (3.28).

▪ EXAMPLE

In Figure 3.14, $\mathbf{n}_1$, $\mathbf{n}_2$, $\mathbf{n}_3$ are mutually perpendicular unit vectors; Q designates the mass center of a body B having a mass of 12 slugs; products of inertia of B with respect to point O for $\mathbf{n}_1$, $\mathbf{n}_2$, $\mathbf{n}_3$ are tabulated in units of slug-ft²; and $\mathbf{n}_a$ and $\mathbf{n}_b$ are unit vectors. The product of inertia $I_{ab}{}^{B/N}$ is to be determined.

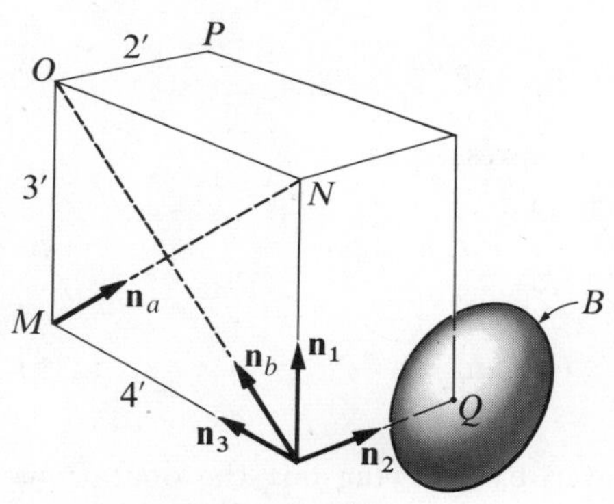

I_{jk}	1	2	3
1	260	72	−144
2	72	325	96
3	−144	96	169

Figure 3.14

As Q is the mass center of B, $I_a{}^{B/N}$ is given by

$$I_{ab}{}^{B/N} \underset{(3.28)}{=} I_{ab}{}^{B/Q} + I_{ab}{}^{Q/N} \tag{a}$$

and $I_{ab}{}^{B/Q}$ can be expressed as

$$I_{ab}{}^{B/Q} \underset{(3.28)}{=} I_{ab}{}^{B/O} - I_{ab}{}^{Q/O} \tag{b}$$

where $I_{ab}{}^{Q/N}$ and $I_{ab}{}^{Q/O}$ can be found from

$$I_{ab}{}^{Q/N} \underset{(3.14)}{=} \mathbf{I}_a{}^{Q/N} \cdot \mathbf{n}_b \tag{c}$$

and

$$I_{ab}{}^{Q/O} \underset{(3.14)}{=} \mathbf{I}_a{}^{Q/O} \cdot \mathbf{n}_b \tag{d}$$

Furthermore, if $\mathbf{p}^{Q/N}$ and $\mathbf{p}^{Q/O}$ are the position vectors of Q relative to N and O, respectively, and if m denotes the mass of B, then

$$\mathbf{I}_a{}^{Q/N} \underset{(3.26)}{=} m\mathbf{p}^{Q/N} \times (\mathbf{n}_a \times \mathbf{p}^{Q/N}) \tag{e}$$

and

$$\mathbf{I}_a{}^{Q/C} \underset{(3.26)}{=} m\mathbf{p}^{Q/O} \times (\mathbf{n}_a \times \mathbf{p}^{Q/O}) \tag{f}$$

Consequently,

$$I_{ab}{}^{Q/N} \underset{(c)}{=} m\mathbf{p}^{Q/N} \times \underset{(e)}{(\mathbf{n}_a} \times \mathbf{p}^{Q/N}) \cdot \mathbf{n}_b \tag{g}$$

or, equivalently,

$$I_{ab}{}^{Q/N} \underset{(g)}{=} m(\mathbf{p}^{Q/N} \times \mathbf{n}_a) \cdot (\mathbf{p}^{Q/N} \times \mathbf{n}_b) \tag{h}$$

Similarly,

$$I_{ab}{}^{Q/O} \underset{(d),(f)}{=} m(\mathbf{p}^{Q/O} \times \mathbf{n}_a) \cdot (\mathbf{p}^{Q/O} \times \mathbf{n}_b) \tag{i}$$

The product of inertia $I_{ab}{}^{B/O}$ can be expressed as

$$I_{ab}{}^{B/O} \underset{(3.24)}{=} \sum_{j=1}^{3} \sum_{k=1}^{3} a_j I_{jk} b_k \tag{j}$$

where

$$a_j = \mathbf{n}_a \cdot \mathbf{n}_j \qquad b_k = \mathbf{n}_b \cdot \mathbf{n}_k \tag{k}$$

and I_{jk} is given in Figure 3.14.

The solution may now be completed by carrying out the operations indicated in Equations (h) through (j) and then substituting into Equations (b) and (a). For example,

$$\mathbf{n}_a = \tfrac{3}{5}\mathbf{n}_1 - \tfrac{4}{5}\mathbf{n}_3 \tag{l}$$

$$\mathbf{n}_b = \tfrac{3}{5}\mathbf{n}_1 + \tfrac{4}{5}\mathbf{n}_3 \tag{m}$$

$$\mathbf{p}^{Q/N} = -3\mathbf{n}_1 + 2\mathbf{n}_2 \tag{n}$$

$$\mathbf{p}^{Q/N} \times \mathbf{n}_a \underset{(l),(n)}{=} -\tfrac{1}{5}(8\mathbf{n}_1 + 12\mathbf{n}_2 + 6\mathbf{n}_3) \tag{o}$$

$$\mathbf{p}^{Q/N} \times \mathbf{n}_b \underset{(m),(n)}{=} \tfrac{1}{5}(8\mathbf{n}_1 + 12\mathbf{n}_2 - 6\mathbf{n}_3) \tag{p}$$

and

$$(\mathbf{p}^{Q/N} \times \mathbf{n}_a) \cdot (\mathbf{p}^{Q/N} \times \mathbf{n}_b) \underset{(o),(p)}{=} -\tfrac{1}{25}(64 + 144 - 36) = -\tfrac{172}{25} \tag{q}$$

so that, with $m = 12$,

$$I_{ab}{}^{Q/N} \underset{(h),(q)}{=} -\frac{(12)(172)}{25} = -\frac{2064}{25} \tag{r}$$

Similarly,

$$I_{ab}{}^{Q/O} \underset{(i)}{=} -\frac{336}{25} \tag{s}$$

From Equations (k) through (m),

$$a_1 = \tfrac{3}{5} \qquad a_2 = 0 \qquad a_3 = -\tfrac{4}{5}$$
$$b_1 = \tfrac{3}{5} \qquad b_2 = 0 \qquad b_3 = \tfrac{4}{5}$$

so that

$$I_{ab}{}^{B/O} \underset{(j)}{=} -\frac{364}{25} \tag{t}$$

Thus,

$$I_{ab}{}^{B/Q} \underset{(b)}{=} -\underset{(t)}{\frac{364}{25}} + \underset{(s)}{\frac{336}{25}} = -\frac{28}{25} \tag{u}$$

and

$$I_{ab}{}^{B/N} \underset{(a)}{=} -\underset{(u)}{\frac{28}{25}} - \underset{(r)}{\frac{2064}{25}} = -\frac{2092}{25} \text{ slug-ft}^2$$

3.16 Principal Moments of Inertia. In general, the second moment $\mathbf{I}_a$ (see Section 3.12) is not parallel to $\mathbf{n}_a$. When $\mathbf{I}_a$ is parallel to $\mathbf{n}_a$, the line L_a passing through O and parallel to $\mathbf{n}_a$ is said to be a *principal axis of inertia* of S for O; the moment of inertia I_a [see Equation (3.15)] is called a *principal moment of inertia* of S for O; and the radius of gyration of S with respect to L_a is called a *principal radius of gyration* of S for O.

When $\mathbf{n}_a$ is parallel to a principal axis of inertia of S for O,

$$\mathbf{I}_a = I_a \mathbf{n}_a \tag{3.30}$$

and if $\mathbf{n}_b$ is any unit vector perpendicular to $\mathbf{n}_a$,

$$I_{ab} = 0 \tag{3.31}$$

If $\mathbf{n}_1$, $\mathbf{n}_2$, $\mathbf{n}_3$ are mutually perpendicular unit vectors, each parallel to a principal axis of S for O, and $\mathbf{n}_z$ is a unit vector given by

$$\mathbf{n}_z = \sum_{j=1}^{3} z_j \mathbf{n}_j \tag{3.32}$$

then

$$\mathbf{I}_z = \sum_{j=1}^{3} z_j I_j \mathbf{n}_j \tag{3.33}$$

and

$$I_z = \sum_{j=1}^{3} z_j{}^2 I_j \tag{3.34}$$

Principal axes can frequently be found by symmetry considerations. For example, when all particles of a set lie in a plane, every normal to the plane is a principal axis of the set for the point of intersection of the normal with the plane.

Proofs: When $\mathbf{I}_a$ is parallel to $\mathbf{n}_a$, there exists a quantity λ such that

$$\lambda \mathbf{n}_a = \mathbf{I}_a$$

Dot multiplication of both sides of this equation with $\mathbf{n}_a$ gives

$$\lambda = \mathbf{I}_a \cdot \mathbf{n}_a \underset{(3.15)}{=} I_a$$

Consequently,

$$I_a \mathbf{n}_a = \mathbf{I}_a$$

Next, dot-multiply both sides of Equation (3.30) by $\mathbf{n}_b$:

$$\mathbf{I}_a \cdot \mathbf{n}_b = I_a \mathbf{n}_a \cdot \mathbf{n}_b$$

But

$$\mathbf{I}_a \cdot \mathbf{n}_b \underset{(3.14)}{=} I_{ab}$$

and, if $\mathbf{n}_b$ is perpendicular to $\mathbf{n}_a$,

$$\mathbf{n}_a \cdot \mathbf{n}_b = 0$$

Consequently,

$$I_{ab} = 0$$

If $\mathbf{n}_1$, $\mathbf{n}_2$, and $\mathbf{n}_3$ are each parallel to a principal axis of S for O, it now appears that

$$I_{jk} \underset{(3.31)}{=} 0 \qquad k \neq j$$

Hence

$$\mathbf{I}_z \underset{(3.23)}{=} \sum_{j=1}^{3} z_j I_{jj} \mathbf{n}_j \underset{(3.15)}{=} \sum_{j=1}^{3} z_j I_j \mathbf{n}_j$$

and

$$I_z \underset{(3.15),(3.33)}{=} \sum_{j=1}^{3} z_j^2 I_j$$

3.17 Location of Principal Axes and Determination of Principal Moments of Inertia. For every set of particles there exists at least one set of three mutually perpendicular principal axes of inertia (see Section 3.15) for every point in space.

Let S be a set of particles; O any point in space (see Figure 3.15); $\mathbf{n}_1$, $\mathbf{n}_2$, $\mathbf{n}_3$ a set of mutually perpendicular unit vectors; I_{ij} the product of

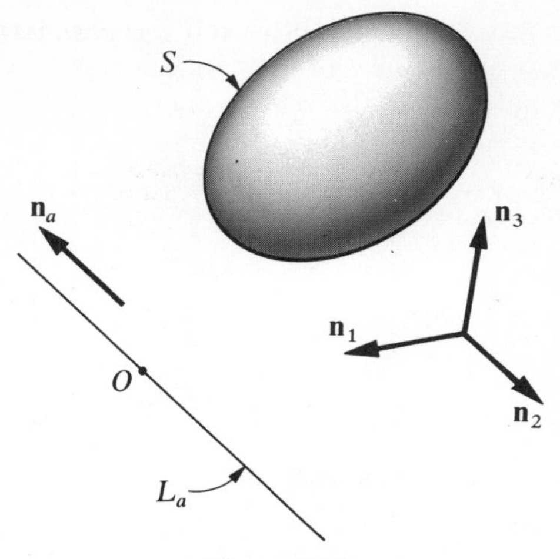

Figure 3.15

inertia of S relative to O for $\mathbf{n}_i$ and $\mathbf{n}_j$ (see Section 3.12); L_a a line passing through O; I_a the moment of inertia of S about L_a; and $\mathbf{n}_a$ a unit vector parallel to L_a. Then I_a is a principal moment of inertia of S for O if it is a solution of the equation

$$\begin{vmatrix} (I_1 - I_a) & I_{21} & I_{31} \\ I_{12} & (I_2 - I_a) & I_{32} \\ I_{13} & I_{23} & (I_3 - I_a) \end{vmatrix} = 0 \tag{3.35}$$

and $\mathbf{n}_a$ is parallel to the corresponding principal axis if

$$\mathbf{n}_a = \sum_{k=1}^{3} a_k \mathbf{n}_k \tag{3.36}$$

where a_1, a_2, a_3 satisfy the four equations

$$\sum_{j=1}^{3} a_j I_{jk} = I_a a_k \qquad k = 1, 2, 3 \tag{3.37}$$

and

$$a_1^2 + a_2^2 + a_3^2 = 1 \tag{3.38}$$

When Equation (3.35) has two equal roots, all lines passing through O and lying in a certain plane are principal axes of inertia of S for O; and when (3.35) has three equal roots, every line passing through O is a principal axis of inertia of S for O.

Proof: L_a is a principal axis of S for O if $\mathbf{I}_a$ is parallel to $\mathbf{n}_a$ (see Section 3.15), and $\mathbf{I}_a$ is then given by

$$\mathbf{I}_a \underset{(3.30)}{=} I_a \mathbf{n}_a \tag{a}$$

To find a unit vector $\mathbf{n}_a$ that satisfies this equation, set

$$\mathbf{n}_a = \sum_{k=1}^{3} a_k \mathbf{n}_k \tag{b}$$

and express $\mathbf{I}_a$ as

$$\mathbf{I}_a \underset{(3.23)}{=} \sum_{j=1}^{3} \sum_{k=1}^{3} a_j I_{jk} \mathbf{n}_k \tag{c}$$

Note that a_1, a_2, a_3, and I_a must then satisfy the vector equation

$$\sum_{j=1}^{3} \sum_{k=1}^{3} \underset{(c)}{a_j I_{jk} \mathbf{n}_k} \underset{(a)}{=} I_a \sum_{k=1}^{3} \underset{(b)}{a_k \mathbf{n}_k} \tag{d}$$

or, equivalently, the three scalar equations

$$\sum_{j=1}^{3} a_j I_{jk} - I_a a_k = 0 \qquad k = 1, 2, 3 \tag{e}$$

Equations (e) are linear and homogeneous in a_1, a_2, and a_3. As a_1, a_2, and a_3 cannot all be equal to zero [see Equation (b)], it follows that a_1, a_2, and a_3 can exist only if the determinant of their coefficients in Equations (e) vanishes: that is, if

$$\begin{vmatrix} (I_1 - I_a) & I_{21} & I_{31} \\ I_{12} & (I_2 - I_a) & I_{32} \\ I_{13} & I_{23} & (I_3 - I_a) \end{vmatrix} = 0 \tag{f}$$

Equation (f), called a "characteristic" equation, is cubic in I_a. It possesses three roots; moreover, once a real root of Equation (f) has been found, an associated principal axis is determined by any three numbers a_1, a_2, and a_3 that satisfy both Equations (e) and the equation

$$a_1^2 + a_2^2 + a_3^2 \underset{(b)}{=} 1 \tag{g}$$

It will now be shown that all roots of (f) are real.

Let A, B, α_j, and β_j be real quantities such that

$$I_a = A + iB \tag{h}$$

and

$$a_j = \alpha_j + i\beta_j \tag{i}$$

where

$$i = \sqrt{-1}$$

Then

$$\sum_{j=1}^{3} \underset{(\text{i})}{(\alpha_j + i\beta_j)} I_{jk} \underset{(\text{e})}{=} \underset{(\text{h})}{(A + iB)} \underset{(\text{i})}{(\alpha_k + i\beta_k)}$$

or, after separating the real and imaginary parts of this equation,

$$\sum_{j=1}^{3} \alpha_j I_{jk} = A\alpha_k - B\beta_k \tag{j}$$

$$\sum_{j=1}^{3} \beta_j I_{jk} = B\alpha_k + A\beta_k \tag{k}$$

Multiply Equation (k) by α_k and Equation (j) by β_k, and subtract:

$$\sum_{j=1}^{3} \alpha_k \beta_j I_{jk} - \sum_{j=1}^{3} \alpha_j \beta_k I_{jk} = B(\alpha_k^2 + \beta_k^2) \tag{l}$$

Note that Equation (l) represents three equations ($k = 1, 2, 3$), and add these:

$$\sum_{k=1}^{3} \sum_{j=1}^{3} \alpha_k \beta_j I_{jk} - \sum_{k=1}^{3} \sum_{j=1}^{3} \alpha_j \beta_k I_{jk} = B \sum_{k=1}^{3} (\alpha_k^2 + \beta_k^2) \tag{m}$$

Now

$$\sum_{k=1}^{3} \sum_{j=1}^{3} \alpha_k \beta_j I_{jk} = \sum_{k=1}^{3} \sum_{j=1}^{3} \alpha_j \beta_k I_{kj} \underset{(3.25)}{=} \sum_{k=1}^{3} \sum_{j=1}^{3} \alpha_j \beta_k I_{jk} \tag{n}$$

Consequently, the left-hand member of Equation (m) is equal to zero, and

$$B \sum_{k=1}^{3} (\alpha_k^2 + \beta_k^2) \underset{(\text{m}),(\text{n})}{=} 0$$

The quantities α_j and β_j, for $j = 1, 2, 3$, cannot all vanish, because [see Equations (i) and (b)] this would imply that $\mathbf{n}_a = 0$, which is impossible since $\mathbf{n}_a$ is a unit vector. It follows that $B = 0$ and that [see Equation (h)] I_a and a_1, a_2, a_3 [see Equations (e)] are real.

So far it has been shown that S possesses three principal moments of inertia for O, each of these corresponding to one of the roots of Equation (f). Suppose that two of the roots, say I_x and I_y, are distinct from each other. Then, with self-explanatory notation,

$$\mathbf{I}_x \underset{(3.30)}{=} I_x \mathbf{n}_x \tag{o}$$

$$\mathbf{I}_y \underset{(3.30)}{=} I_y \mathbf{n}_y \tag{p}$$

and

$$I_{xy} \underset{(3.25)}{=} I_{yx} \tag{q}$$

Dot-multiply Equation (o) by $\mathbf{n}_y$, and Equation (p) by $\mathbf{n}_x$, and subtract as follows:

$$\mathbf{I}_x \cdot \mathbf{n}_y - \mathbf{I}_y \cdot \mathbf{n}_x = (I_x - I_y)\mathbf{n}_x \cdot \mathbf{n}_y \tag{r}$$

But

$$\mathbf{I}_x \cdot \mathbf{n}_y \underset{(3.14)}{=} I_{xy} \underset{(q)}{=} I_{yx} \underset{(3.14)}{=} \mathbf{I}_y \cdot \mathbf{n}_x$$

so that

$$\mathbf{I}_x \cdot \mathbf{n}_y - \mathbf{I}_y \cdot \mathbf{n}_x = 0 \tag{s}$$

Hence,

$$(I_x - I_y)\mathbf{n}_x \cdot \mathbf{n}_y \underset{(r),(s)}{=} 0$$

and, since I_x and I_y differ from each other by hypothesis,

$$\mathbf{n}_x \cdot \mathbf{n}_y = 0$$

This shows that, whenever two principal moments of inertia are unequal, the associated principal axes are perpendicular to each other. Consequently, when Equation (f) has three distinct roots, a unique principal axis corresponds to each root and these three principal axes are mutually perpendicular.

Suppose that I_z is one of the roots of Equation (f), that $\mathbf{n}_z$ is a unit vector parallel to the associated principal axis, and that the remaining two roots of Equation (f) are equal to each other (and possibly to I_z). Let $\mathbf{n}_x$ and $\mathbf{n}_y$ be unit vectors perpendicular to each other and to $\mathbf{n}_z$, and note that [see Equations (3.31) and (3.25)]

$$I_{yz} = I_{zy} = 0 \qquad I_{zx} = I_{xz} = 0$$

In Equation (f), the subscripts 1, 2, and 3 can then be replaced with x, y, and z, respectively, and Equation (f) reduces to

$$(I_z - I_a)[I_a^2 - (I_x + I_y)I_a + I_xI_y - I_{xy}^2] = 0$$

which has the three roots

$$I_a = \frac{I_x + I_y}{2} \pm \left[\left(\frac{I_x - I_y}{2}\right)^2 + I_{xy}^2\right]^{1/2}, I_z \tag{t}$$

The first two of these can be equal to each other only if

$$\left(\frac{I_x - I_y}{2}\right)^2 + I_{xy}^2 = 0$$

that is, if

$$I_y = I_x$$

and

$$I_{xy} = 0$$

However, I_x is a root of Equation (f); that is,

$$I_a \underset{(t)}{=} I_x, I_x, I_z$$

which means that I_x is a principal moment of inertia of S for O. Now, $\mathbf{n}_x$ was restricted only to the extent of being required to be perpendicular to $\mathbf{n}_z$. Thus it appears that, when Equation (f) has two equal roots, every line passing through O and normal to $\mathbf{n}_z$ is a principal axis of S for O; and from this it follows that, when Equation (f) has three equal roots, every line passing through O is a principal axis of S for O.

▪ EXAMPLE

A uniform, rectangular plate R has a mass m and sides of length b and c. The principal moments of inertia I_x, I_y, I_z and the corresponding principal axes X, Y, Z for one corner O of the plate are to be found.

Let $\mathbf{n}_1$, $\mathbf{n}_2$, and $\mathbf{n}_3$ be unit vectors oriented relative to R as shown in Figure 3.16, and let P designate the mass center of R. Then, from sym-

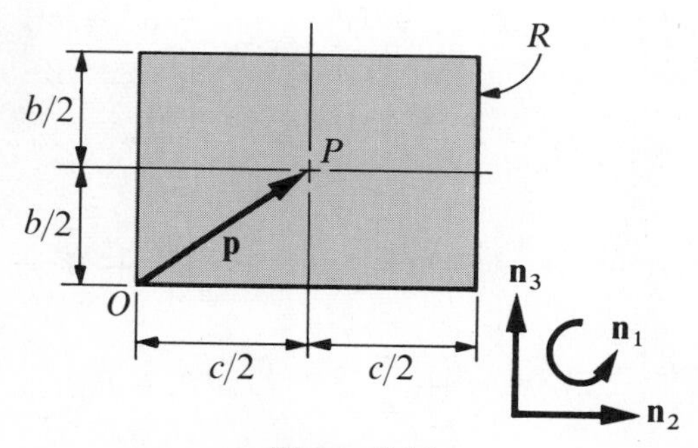

Figure 3.16

metry considerations, the three lines passing through P and parallel to $\mathbf{n}_1$, $\mathbf{n}_2$, and $\mathbf{n}_3$ are seen to be principal axes of R for P. The associated principal moments of inertia have the values

$$I_1^{R/P} = \frac{m}{12}(b^2 + c^2) \qquad I_2^{R/P} = \frac{m}{12}b^2 \qquad I_3^{R/P} = \frac{m}{12}c^2 \tag{a}$$

The moments and products of inertia for point O of a particle that is situated at P and has a mass m are found as follows. With $N = 1$ and

$$\mathbf{p} = \frac{c}{2}\mathbf{n}_2 + \frac{b}{2}\mathbf{n}_3$$

substitution into Equation (3.13) gives

$$\mathbf{I}_1^{P/O} = m\mathbf{p} \times (\mathbf{n}_1 \times \mathbf{p}) = \frac{m}{4}(b^2 + c^2)\mathbf{n}_1$$

$$\mathbf{I}_2^{P/O} = m\mathbf{p} \times (\mathbf{n}_2 \times \mathbf{p}) = \frac{m}{4}(b^2\mathbf{n}_2 - bc\mathbf{n}_3)$$

$$\mathbf{I}_3^{P/O} = m\mathbf{p} \times (\mathbf{n}_3 \times \mathbf{p}) = \frac{m}{4}(-bc\mathbf{n}_2 + c^2\mathbf{n}_3)$$

and Equation (3.14) then leads to

$$I_1^{P/O} = \frac{m}{4}(b^2 + c^2) \qquad I_2^{P/O} = \frac{m}{4}b^2 \qquad I_3^{P/O} = \frac{m}{4}c^2$$
$$I_{12}^{P/O} = 0 \qquad I_{23}^{P/O} = -\frac{m}{4}bc \qquad I_{31}^{P/O} = 0 \tag{b}$$

The moments and products of inertia of R relative to O are now seen to be

$$I_1^{R/O} \underset{(3.29)}{=} I_1^{R/P} + I_1^{P/O} \underset{(a),(b)}{=} \frac{m}{3}(b^2 + c^2)$$
$$I_2^{R/O} = \frac{m}{3}b^2 \qquad I_3^{R/O} = \frac{m}{3}c^2 \tag{c}$$

and

$$I_{12}^{R/O} \underset{(3.28)}{=} I_{12}^{R/P} + I_{12}^{P/O} = \underset{(3.31)}{0} + \underset{(b)}{0} = 0$$
$$I_{23}^{R/O} = -\frac{m}{4}bc \qquad I_{31}^{R/O} = 0 \tag{d}$$

(Alternatively, these could be found by direct integration.) Next,

$$\begin{vmatrix} \left[\frac{m}{3}(b^2 + c^2) - I_a\right] & 0 & 0 \\ 0 & \left(\frac{m}{3}b^2 - I_a\right) & -\frac{m}{4}bc \\ 0 & -\frac{m}{4}bc & \left(\frac{m}{3}c^2 - I_a\right) \end{vmatrix} \underset{(3.35)}{=} 0$$

or

$$\left[\frac{m}{3}(b^2 + c^2) - I_a\right]\left[I_a^2 - \frac{m}{3}(b^2 + c^2)I_a + \frac{7}{144}m^2b^2c^2\right] = 0$$

and, if I_x, I_y, I_z denote the three roots of this equation—that is, the principal moments of inertia of R for O—then

$$I_x = \frac{m}{3}(b^2 + c^2) \tag{e}$$

$$I_y = \frac{m}{6}(b^2 + c^2) - \frac{m}{12}(4b^4 + b^2c^2 + 4c^4)^{1/2} \tag{f}$$

$$I_z = \frac{m}{6}(b^2 + c^2) + \frac{m}{12}(4b^4 + b^2c^2 + 4c^4)^{1/2} \tag{g}$$

Let $\mathbf{n}_x$ be a unit vector parallel to X, the principal axis corresponding to I_x, and express $\mathbf{n}_x$ as

$$\mathbf{n}_x = x_1\mathbf{n}_1 + x_2\mathbf{n}_2 + x_3\mathbf{n}_3 \tag{h}$$

Then Equations (3.37) require that x_1, x_2, x_3 be solutions of

$$\left[\frac{m}{3}(b^2 + c^2) - I_x\right] x_1 = 0 \tag{i}$$

$$\left(\frac{m}{3}b^2 - I_x\right) x_2 - \frac{m}{4} bcx_3 = 0 \tag{j}$$

$$-\frac{m}{4} bcx_2 + \left(\frac{m}{3}c^2 - I_x\right) x_3 = 0 \tag{k}$$

Furthermore,

$$x_1^2 + x_2^2 + x_3^2 \underset{(3.38)}{=} 1 \tag{l}$$

Equation (i) is satisfied for all values of x_1, as a consequence of Equation (e). Equations (j) and (k) can be satisfied simultaneously only if

$$x_2 = x_3 = 0$$

and Equation (l) thus shows that

$$x_1 = \pm 1$$

so that

$$\mathbf{n}_x \underset{(h)}{=} \pm\mathbf{n}_1$$

In other words, the principal axis X is perpendicular to R. (This can also be deduced from the fact that $I_x = I_1^{R/O}$.)

To find Y, the principal axis corresponding to I_y, let $\mathbf{n}_y$ be the unit vector

$$\mathbf{n}_y = y_1\mathbf{n}_1 + y_2\mathbf{n}_2 + y_3\mathbf{n}_3$$

and determine y_1, y_2, and y_3 such that

$$\left[\frac{m}{3}(b^2 + c^2) - I_y\right] y_1 = 0 \tag{m}$$

$$\left(\frac{m}{3}b^2 - I_y\right) y_2 - \frac{m}{4} bcy_3 = 0 \tag{n}$$

$$-\frac{m}{4} bcy_2 + \left(\frac{m}{3}c^2 - I_y\right) y_3 = 0 \tag{o}$$

From Equations (m) and (f),

$$y_1 = 0$$

and from Equations (n) and (f) [or (o) and (f)],

$$\frac{y_3}{y_2} = \frac{2(b^2 - c^2) + (4b^4 + b^2c^2 + 4c^4)^{1/2}}{3bc} \tag{p}$$

Y thus lies in the plane of R, and the tangent of the angle θ between Y and the edge of R parallel to $\mathbf{n}_2$ (see Figure 3.17) is given by Equation (p).

Finally, Z, the principal axis of R for O corresponding to I_z, is perpendicular to both X and Y, as shown in Figure 3.17.

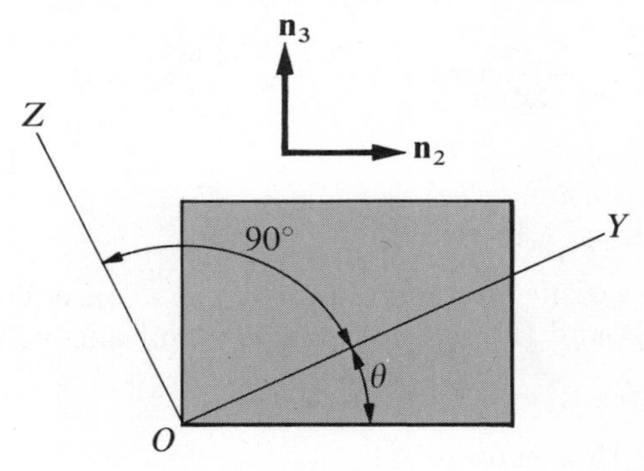

Figure 3.17

3.18 Inertia Ellipsoid. The locus E of points P whose distance R from a point O is inversely proportional to the square root of the moment of inertia of a set S of particles about line OP is an ellipsoid. It is called an *inertia ellipsoid* of S for O.

Proof: In Figure 3.18, X_1, X_2, X_3 designate mutually perpendicular coordinate axes passing through O, and each of these is a principal axis of inertia of S for O (see Section 3.16); x_1, x_2, x_3 are the coordinates of

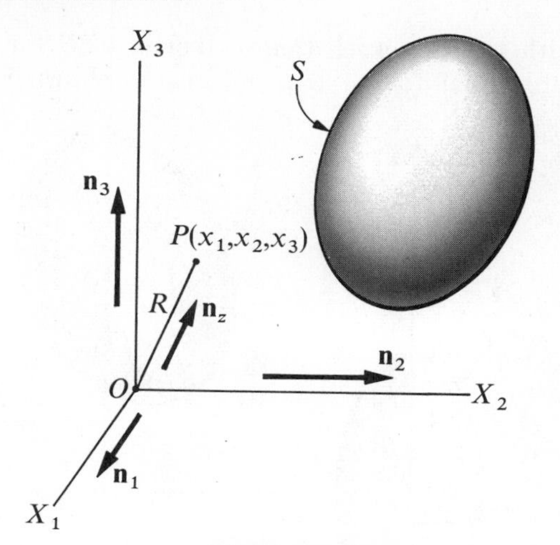

Figure 3.18

P; and $\mathbf{n}_z$, a unit vector parallel to line OP, is given by

$$\mathbf{n}_z = z_1\mathbf{n}_1 + z_2\mathbf{n}_2 + z_3\mathbf{n}_3 \tag{a}$$

where $\mathbf{n}_1$, $\mathbf{n}_2$, $\mathbf{n}_3$ are unit vectors directed as indicated.

It is to be shown that x_1, x_2, x_3 satisfy the equation of an ellipsoid whenever

$$R = CI_z^{-1/2} \tag{b}$$

where C is a constant.

Note that the position vector of P relative to O can be expressed both as $R\mathbf{n}_z$ and as $x_1\mathbf{n}_1 + x_2\mathbf{n}_2 + x_3\mathbf{n}_3$. Consequently [see Equation (a)],

$$z_1 = \frac{x_1}{R} \qquad z_2 = \frac{x_2}{R} \qquad z_3 = \frac{x_3}{R} \tag{c}$$

Next,

$$I_z \underset{(3.34)}{=} z_1^2 I_1 + z_2^2 I_2 + z_3^2 I_3 \tag{d}$$

Hence,

$$I_z \underset{(c),(d)}{=} \frac{1}{R^2}(x_1^2 I_1 + x_2^2 I_2 + x_3^2 I_3)$$

and, after elimination of R by means of Equation (b),

$$\frac{x_1^2}{A_1^2} + \frac{x_2^2}{A_2^2} + \frac{x_3^2}{A_3^2} = 1 \tag{e}$$

where

$$A_j = C(I_j)^{-1/2} \qquad j = 1, 2, 3 \tag{f}$$

Equation (e) is the equation of an ellipsoid whose center is at O, whose axes coincide with the principal axes of inertia of S for O, and whose semidiameters are equal to A_1, A_2, and A_3, as shown in Figure 3.19.

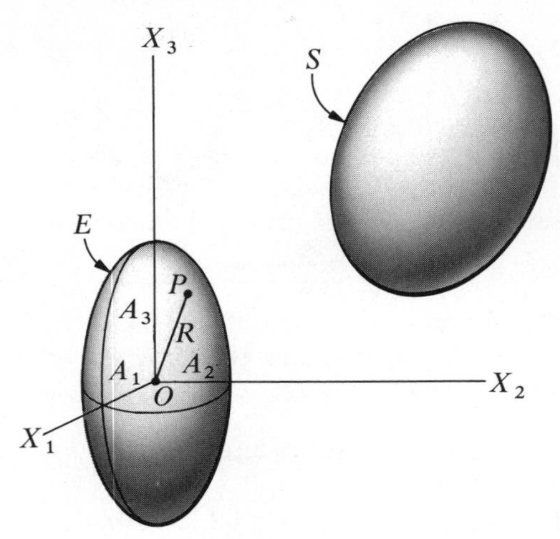

Figure 3.19

(Note that not every ellipsoid can be an inertia ellipsoid. The fact that the sum of two principal moments of inertia is never smaller than the third imposes certain restrictions on the relative magnitudes of A_1, A_2, and A_3.)

3.19 Maximum and Minimum Moments of Inertia. Of all lines passing through a point O, those with respect to which a set S of particles has a larger, or smaller, moment of inertia than it has with respect to all other lines passing through O are principal axes of inertia of S for O (see Section 3.16).

Proof: Referring to Section 3.18; the numbering of X_1, X_2, X_3 can always be arranged in such a way that

$$I_1 \geq I_2 \geq I_3$$

or, in view of Equation (f),

$$A_1 \leq A_2 \leq A_3$$

The distance R from the center O to any point P of the ellipsoid (see Figure 3.19) is never smaller than the smallest semidiameter, and never larger than the largest; that is,

$$A_1 \leq R \leq A_3$$

Using Equations (f) and (b) to eliminate A_1, A_3, and R gives

$$C(I_1)^{-1/2} \leq CI_z^{-1/2} \leq C(I_3)^{-1/2}$$

or, equivalently,

$$I_1 \geq I_z \geq I_3$$

which shows that the moment of inertia of S about a line that passes through O and is not a principal axis of S for O cannot be smaller than the smallest or larger than the largest principal moment of inertia of S for O.

3.20 Absolute Minimum Moment of Inertia. No moment of inertia of a set of particles is smaller than the smallest principal moment of inertia for the mass center of the set.

Proof: The moment of inertia of a set of particles about a line that does not pass through the mass center is always larger than the moment of inertia about a parallel line that does pass through the mass center [see Equation (3.29)]. Hence the smallest moment of inertia must be one about a line passing through the mass center; and, in accordance with Section 3.19, it must be a principal moment of inertia for the mass center.

3.21 Inertia Dyadic. Inertia properties of a set of particles have so far been discussed in terms of certain scalars (moments and products of inertia) and vectors (second moments). Dyadics may be brought into the discussion as follows.

In Section 3.13 it was shown that, if $\mathbf{n}_j$, with $j = 1, 2, 3$, is a set of mutually perpendicular unit vectors, and $\mathbf{n}_a$ is a unit vector given by

$$\mathbf{n}_a = \sum_{j=1}^{3} a_j \mathbf{n}_j \tag{a}$$

then the second moment $\mathbf{I}_a$ of a set S of particles relative to a point O for $\mathbf{n}_a$ (see Section 3.12) can be expressed as

$$\mathbf{I}_a \underset{(3.23)}{=} \sum_{j=1}^{3} \sum_{k=1}^{3} a_j I_{jk} \mathbf{n}_k \tag{b}$$

where I_{jk} is the product of inertia of S relative to O for $\mathbf{n}_j$ and $\mathbf{n}_k$. Now, a_j may be expressed as

$$a_j \underset{(a)}{=} \mathbf{n}_a \cdot \mathbf{n}_j \tag{c}$$

Consequently,

$$\mathbf{I}_a \underset{(b),(c)}{=} \sum_{j=1}^{3} \sum_{k=1}^{3} \mathbf{n}_a \cdot \mathbf{n}_j I_{jk} \mathbf{n}_k \tag{d}$$

This expression may be replaced with

$$\mathbf{I}_a = \mathbf{n}_a \cdot \sum_{j=1}^{3} \sum_{k=1}^{3} \mathbf{n}_j I_{jk} \mathbf{n}_k \tag{e}$$

if it is understood that the right-hand member of Equation (e) has the same meaning as that of (d). The double summation in Equation (e) is then called an *inertia dyadic* of S for O; and if the symbol $\mathbf{I}$ is used to denote this quantity, that is, if $\mathbf{I}$ is defined as

$$\mathbf{I} = \sum_{j=1}^{3} \sum_{k=1}^{3} \mathbf{n}_j I_{jk} \mathbf{n}_k \tag{f}$$

then $\mathbf{I}_a$ can be expressed as

$$\mathbf{I}_a \underset{(e),(f)}{=} \mathbf{n}_a \cdot \mathbf{I} \tag{g}$$

Together, Equations (d), (f), and (g) constitute a definition of dot multiplication of a vector ($\mathbf{n}_a$) with a dyadic ($\mathbf{I}$). The dyadic is said to be *premultiplied* by the vector, because $\mathbf{n}_a$ precedes $\mathbf{I}$ in Equation (g). In a similar way, *postmultiplication* of the dyadic $\mathbf{I}$ by $\mathbf{n}_a$ is defined by setting

$$\mathbf{I} \cdot \mathbf{n}_a = \sum_{j=1}^{3} \sum_{k=1}^{3} \mathbf{n}_j I_{jk} \mathbf{n}_k \cdot \mathbf{n}_a \tag{h}$$

Now,

$$\mathbf{n}_k \cdot \mathbf{n}_a \underset{(a)}{=} a_k \tag{i}$$

and

$$I_{jk} \underset{(3.25)}{=} I_{kj} \tag{j}$$

Consequently,

$$\mathbf{I} \cdot \mathbf{n}_a \underset{(h),(i),(j)}{=} \sum_{j=1}^{3} \sum_{k=1}^{3} \mathbf{n}_j I_{kj} a_k \tag{k}$$

or, since j and k may be interchanged,

$$\mathbf{I} \cdot \mathbf{n}_a \underset{(k)}{=} \sum_{j=1}^{3} \sum_{k=1}^{3} \mathbf{n}_k I_{jk} a_j \tag{l}$$

The right-hand members of Equations (l) and (b) are the same. Hence

$$\mathbf{I}_a \underset{(l),(b)}{=} \mathbf{I} \cdot \mathbf{n}_a \tag{m}$$

and

$$\mathbf{n}_a \cdot \mathbf{I} \underset{(g),(m)}{=} \mathbf{I} \cdot \mathbf{n}_a \tag{n}$$

It follows from Equation (n) that, for every vector $\mathbf{v}$,

$$\mathbf{v} \cdot \mathbf{I} = \mathbf{I} \cdot \mathbf{v}$$

This equation expresses the *symmetry* property of the inertia dyadic $\mathbf{I}$.

Moments of inertia and products of inertia (see Section 3.12) can be regarded as scalars obtained by simultaneous premultiplication and postmultiplication of the inertia dyadic with appropriate unit vectors. This follows from the fact that I_{ab} is given, on the one hand, by

$$I_{ab} \underset{(3.14)}{=} \mathbf{I}_a \cdot \mathbf{n}_b \underset{(g)}{=} (\mathbf{n}_a \cdot \mathbf{I}) \cdot \mathbf{n}_b \tag{o}$$

and, on the other hand, by

$$\begin{aligned} I_{ab} &\underset{(3.25)}{=} I_{ba} \underset{(3.14)}{=} \mathbf{I}_b \cdot \mathbf{n}_a \\ &\underset{(m)}{=} (\mathbf{I} \cdot \mathbf{n}_b) \cdot \mathbf{n}_a = \mathbf{n}_a \cdot (\mathbf{I} \cdot \mathbf{n}_b) \end{aligned} \tag{p}$$

so that

$$(\mathbf{n}_a \cdot \mathbf{I}) \cdot \mathbf{n}_b \underset{(o),(p)}{=} \mathbf{n}_a \cdot (\mathbf{I} \cdot \mathbf{n}_b)$$

which shows that the parentheses are unnecessary and that

$$I_{ab} \underset{(o)}{=} \mathbf{n}_a \cdot \mathbf{I} \cdot \mathbf{n}_b$$

In summary, the inertia dyadic $\mathbf{I}$ of a set S for a point O is defined as

$$\mathbf{I} = \sum_{j=1}^{3} \sum_{k=1}^{3} \mathbf{n}_j I_{jk} \mathbf{n}_k \tag{3.39}$$

and can be used to express the second moment $\mathbf{I}_a$ of S relative to O and the product of inertia I_{ab} of S relative to O (see Section 3.12) as

$$\mathbf{I}_a = \mathbf{I} \cdot \mathbf{n}_a \tag{3.40}$$

and

$$I_{ab} = \mathbf{n}_a \cdot \mathbf{I} \cdot \mathbf{n}_b \tag{3.41}$$

The inertia dyadic is symmetric; that is, for any vector $\mathbf{v}$,

$$\mathbf{v} \cdot \mathbf{I} = \mathbf{I} \cdot \mathbf{v} \tag{3.42}$$

3.22 Inertia Torque Expressed in Terms of the Inertia Dyadic. If $\mathbf{I}$ is the inertia dyadic of a rigid body B for the mass center B^* of B (see Section 3.21), and $\boldsymbol{\omega}$ and $\boldsymbol{\alpha}$ are the angular velocity and the angular acceleration of B in a reference frame R, the inertia torque $\mathbf{T}^*$ for B in R (see Section 3.10) can be expressed as

$$\boxed{\mathbf{T}^* = (\mathbf{I} \cdot \boldsymbol{\omega}) \times \boldsymbol{\omega} - \mathbf{I} \cdot \boldsymbol{\alpha}} \tag{3.43}$$

Proof:

$$\mathbf{T}^* \underset{(3.12)}{=} -\sum_{i=1}^{\bar{N}} m_i \mathbf{r}_i \times \mathbf{a}_i \tag{a}$$

Let $\mathbf{a}^*$ denote the acceleration of B^* in R. Then

$$\mathbf{a}_i \underset{(2.23)}{=} \mathbf{a}^* + \boldsymbol{\alpha} \times \mathbf{r}_i + \boldsymbol{\omega} \times (\boldsymbol{\omega} \times \mathbf{r}_i) \tag{b}$$

and

$$\mathbf{T}^* \underset{(a),(b)}{=} -\left(\sum_{i=1}^{\bar{N}} m_i \mathbf{r}_i\right) \times \mathbf{a}^* - \sum_{i=1}^{\bar{N}} m_i \mathbf{r}_i \times (\boldsymbol{\alpha} \times \mathbf{r}_i) - \sum_{i=1}^{\bar{N}} m_i \mathbf{r}_i \times [\boldsymbol{\omega} \times (\boldsymbol{\omega} \times \mathbf{r}_i)] \tag{c}$$

The first sum in the right-hand member of Equation (c) vanishes because $\mathbf{r}_i$ is the position vector of P_i relative to the mass center of B. To study the second sum, note that $\boldsymbol{\alpha}$ can be expressed as

$$\boldsymbol{\alpha} = \alpha \mathbf{n}_\alpha \tag{d}$$

where α is a certain scalar and $\mathbf{n}_\alpha$ is a unit vector parallel to $\boldsymbol{\alpha}$. Then

$$\sum_{i=1}^{\bar{N}} m_i \mathbf{r}_i \times (\boldsymbol{\alpha} \times \mathbf{r}_i) \underset{(d)}{=} \alpha \sum_{i=1}^{\bar{N}} m_i \mathbf{r}_i \times (\mathbf{n}_\alpha \times \mathbf{r}_i) \underset{(3.13)}{=} \alpha \mathbf{I}_\alpha \underset{(3.40)}{=} \alpha \mathbf{I} \cdot \mathbf{n}_\alpha \underset{(d)}{=} \mathbf{I} \cdot \boldsymbol{\alpha} \tag{e}$$

Next, note that

$$\mathbf{r}_i \times [\boldsymbol{\omega} \times (\boldsymbol{\omega} \times \mathbf{r}_i)] = \boldsymbol{\omega} \times [\mathbf{r}_i \times (\boldsymbol{\omega} \times \mathbf{r}_i)] \tag{f}$$

Consequently,

$$\sum_{i=1}^{\bar{N}} m_i \mathbf{r}_i \times [\boldsymbol{\omega} \times (\boldsymbol{\omega} \times \mathbf{r}_i)] \underset{(f)}{=} \boldsymbol{\omega} \times \sum_{i=1}^{\bar{N}} m_i \mathbf{r}_i \times (\boldsymbol{\omega} \times \mathbf{r}_i)$$

and, proceeding as in the derivation of Equation (e), one obtains

$$\sum_{i=1}^{\bar{N}} m_i \mathbf{r}_i \times [\boldsymbol{\omega} \times (\boldsymbol{\omega} \times \mathbf{r}_i)] = \boldsymbol{\omega} \times (\mathbf{I} \cdot \boldsymbol{\omega}) \tag{g}$$

Hence,

$$\mathbf{T}^* \underset{(c)}{=} -\mathbf{I} \underset{(e)}{\cdot} \boldsymbol{\alpha} - \boldsymbol{\omega} \times (\mathbf{I} \underset{(g)}{\cdot} \boldsymbol{\omega})$$

▪ EXAMPLE

In Figure 3.20, $\mathbf{n}_1$ and $\mathbf{n}_2$ are unit vectors parallel to edges of a uniform,

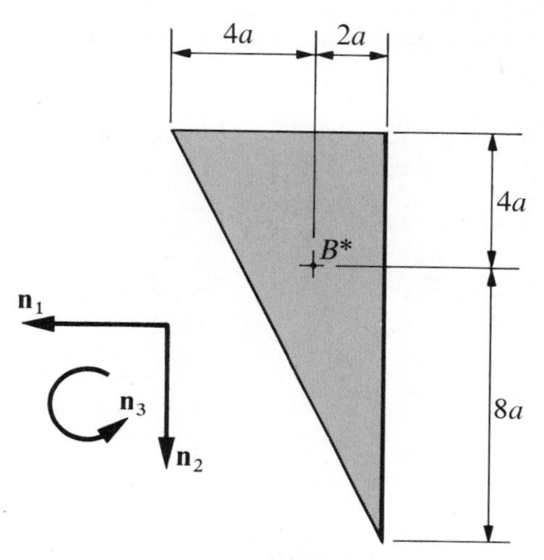

Figure 3.20

right-triangular plate B of mass m, and $\mathbf{n}_3$ designates the unit vector $\mathbf{n}_1 \times \mathbf{n}_2$. The inertia torque $\mathbf{T}^*$ for B in a reference frame R is to be determined for a state of motion during which the angular velocity $\boldsymbol{\omega}$ of B in R is given by

$$\boldsymbol{\omega} = \omega_1\mathbf{n}_1 + \omega_2\mathbf{n}_2 + \omega_3\mathbf{n}_3$$

where ω_1, ω_2, and ω_3 are certain functions of time t.

The moments and products of inertia of B relative to B^* for $\mathbf{n}_1$, $\mathbf{n}_2$, and $\mathbf{n}_3$ have the values listed in Table 3.1. [Symmetry considerations show

TABLE 3.1

	I_{jk}	k = 1	2	3
j	1	$8ma^2$	$2ma^2$	0
	2	$2ma^2$	$2ma^2$	0
	3	0	0	$10ma^2$

that $\mathbf{n}_3$ is parallel to a principal axis of B for B^*, and thus I_{31} and I_{32} vanish in accordance with Equation (3.31).] The inertia dyadic $\mathbf{I}$ of B for B^* is thus given by

$$\begin{aligned}\mathbf{I} \underset{(3.39)}{=} & \; 8ma^2\mathbf{n}_1\mathbf{n}_1 + 2ma^2\mathbf{n}_1\mathbf{n}_2 + 2ma^2\mathbf{n}_2\mathbf{n}_1 \\ & + 2ma^2\mathbf{n}_2\mathbf{n}_2 + 10ma^2\mathbf{n}_3\mathbf{n}_3 \\ = & \; 2ma^2(4\mathbf{n}_1\mathbf{n}_1 + \mathbf{n}_1\mathbf{n}_2 + \mathbf{n}_2\mathbf{n}_1 + \mathbf{n}_2\mathbf{n}_2 + 5\mathbf{n}_3\mathbf{n}_3)\end{aligned}$$

Hence,

$$\begin{aligned}\mathbf{I}\cdot\boldsymbol{\omega} = & \; 2ma^2(4\mathbf{n}_1\mathbf{n}_1\cdot\boldsymbol{\omega} + \mathbf{n}_1\mathbf{n}_2\cdot\boldsymbol{\omega} \\ & + \mathbf{n}_2\mathbf{n}_1\cdot\boldsymbol{\omega} + \mathbf{n}_2\mathbf{n}_2\cdot\boldsymbol{\omega} + 5\mathbf{n}_3\mathbf{n}_3\cdot\boldsymbol{\omega}) \\ = & \; 2ma^2[(4\omega_1 + \omega_2)\mathbf{n}_1 + (\omega_1 + \omega_2)\mathbf{n}_2 + 5\omega_3\mathbf{n}_3]\end{aligned}$$

and

$$\begin{aligned}(\mathbf{I}\cdot\boldsymbol{\omega})\times\boldsymbol{\omega} = & \; 2ma^2\,\{[(\omega_1 + \omega_2)\omega_3 - 5\omega_3\omega_2]\mathbf{n}_1 \\ & + [5\omega_3\omega_1 - (4\omega_1 + \omega_2)\omega_3]\mathbf{n}_2 \\ & + [(4\omega_1 + \omega_2)\omega_2 - (\omega_1 + \omega_2)\omega_1]\mathbf{n}_3\} \\ = & \; 2ma^2[(\omega_1\omega_3 - 4\omega_2\omega_3)\mathbf{n}_1 \\ & + (\omega_1\omega_3 - \omega_2\omega_3)\mathbf{n}_2 \\ & + (3\omega_1\omega_2 + \omega_2^2 - \omega_1^2)\mathbf{n}_3]\end{aligned}$$

Next, the angular acceleration of B in R is given by [see Problem 2(h)]

$$\boldsymbol{\alpha} = \dot{\omega}_1\mathbf{n}_1 + \dot{\omega}_2\mathbf{n}_2 + \dot{\omega}_3\mathbf{n}_3$$

Consequently,

$$\mathbf{I}\cdot\boldsymbol{\alpha} = 2ma^2[(4\dot{\omega}_1 + \dot{\omega}_2)\mathbf{n}_1 + (\dot{\omega}_1 + \dot{\omega}_2)\mathbf{n}_2 + 5\dot{\omega}_3\mathbf{n}_3]$$

and

$$\begin{aligned}\mathbf{T}^* \underset{(3.43)}{=} & \; (\mathbf{I}\cdot\boldsymbol{\omega})\times\boldsymbol{\omega} - \mathbf{I}\cdot\boldsymbol{\alpha} \\ = & \; 2ma^2[(\omega_1\omega_3 - 4\omega_2\omega_3 - 4\dot{\omega}_1 - \dot{\omega}_2)\mathbf{n}_1 \\ & + (\omega_1\omega_3 - \omega_2\omega_3 - \dot{\omega}_1 - \dot{\omega}_2)\mathbf{n}_2 \\ & + (3\omega_1\omega_2 + \omega_2^2 - \omega_1^2 - 5\dot{\omega}_3)\mathbf{n}_3]\end{aligned}$$

3.23 Mutually Perpendicular Principal Axes. Referring to Section 3.22, let $\mathbf{n}_1$, $\mathbf{n}_2$, $\mathbf{n}_3$ be a right-handed set of mutually perpendicular unit vectors, each parallel to a principal axis of B for B^*, but not necessarily fixed in B; let I_1, I_2, I_3 be the associated principal moments of inertia of B for B^*; and let $\omega_i = \boldsymbol{\omega}\cdot\mathbf{n}_i$ and $\alpha_i = \boldsymbol{\alpha}\cdot\mathbf{n}_i$, for $i = 1, 2, 3$. Then

$$\begin{aligned}\mathbf{T}^* = & \; [\omega_2\omega_3(I_2 - I_3) - \alpha_1 I_1]\mathbf{n}_1 \\ & + [\omega_3\omega_1(I_3 - I_1) - \alpha_2 I_2]\mathbf{n}_2 \\ & + [\omega_1\omega_2(I_1 - I_2) - \alpha_3 I_3]\mathbf{n}_3 \qquad (3.44)\end{aligned}$$

(If $\mathbf{n}_1$, $\mathbf{n}_2$, and $\mathbf{n}_3$ are fixed in B or in R, then α_i may be replaced with $\dot{\omega}_i$.)

Proof:

$$\mathbf{I} \underset{(3.39),(3.31)}{=} I_1\mathbf{n}_1\mathbf{n}_1 + I_2\mathbf{n}_2\mathbf{n}_2 + I_3\mathbf{n}_3\mathbf{n}_3$$

$$\mathbf{I}\cdot\boldsymbol{\omega} = I_1\mathbf{n}_1\mathbf{n}_1\cdot\boldsymbol{\omega} + I_2\mathbf{n}_2\mathbf{n}_2\cdot\boldsymbol{\omega} + I_3\mathbf{n}_3\mathbf{n}_3\cdot\boldsymbol{\omega}$$

$$= I_1\omega_1\mathbf{n}_1 + I_2\omega_2\mathbf{n}_2 + I_3\omega_3\mathbf{n}_3$$

$$\mathbf{I}\cdot\boldsymbol{\alpha} = I_1\alpha_1\mathbf{n}_1 + I_2\alpha_2\mathbf{n}_2 + I_3\alpha_3\mathbf{n}_3$$

Substitute into Equation (3.43).

▪ EXAMPLE

Figure 3.21 is a schematic representation of a vehicle V that carries a

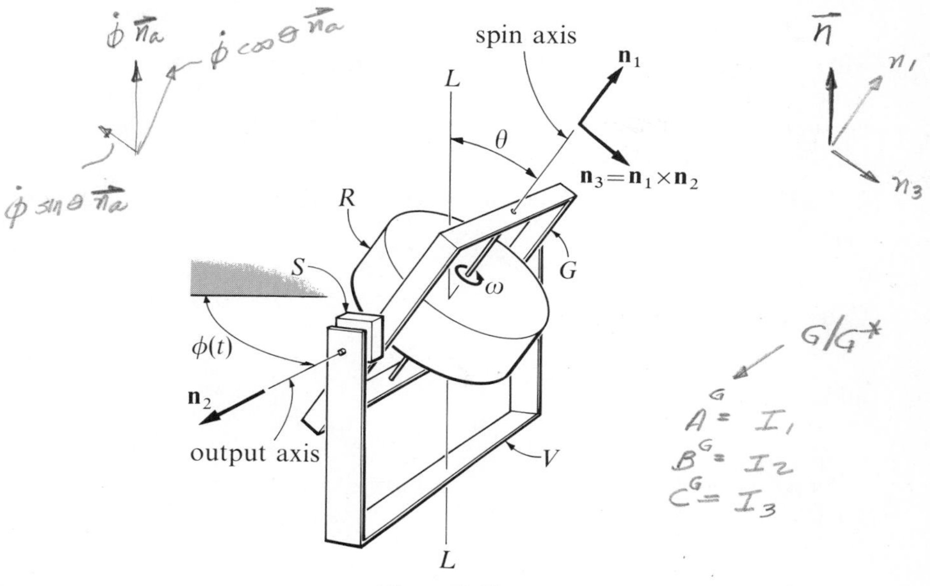

Figure 3.21

gyroscopic rate-of-turn indicator consisting of a rotor R, a gimbal ring G, and a spring-and-damper assembly S. The inertia properties of R and G are described as follows: The point of intersection of the spin axis and output axis is the mass center of R, and R has a moment of inertia J about the spin axis, and a moment of inertia I about any line passing through the mass center of R and perpendicular to the spin axis. The mass center of G coincides with that of R, and the spin axis, the output axis, and a line perpendicular to both of these and passing through their point of intersection are all principal axes of inertia of G, the corresponding moments of inertia being A, B, and C, respectively.

This system moves in such a way that line LL, which is fixed in V, remains parallel to its initial orientation; the angle ϕ is made to vary in a prescribed manner; and the rotor R is driven (by means of a motor that forms an integral part of G) in such a way that the angular speed ω of R in G remains constant.

With θ used as the (only) generalized coordinate of the system comprised of R and G, the associated generalized inertia force F^* is to be determined.

The velocity of the common mass center of R and G is independent of θ. Hence, the partial rate of change of the position of this point with respect to θ is equal to zero and

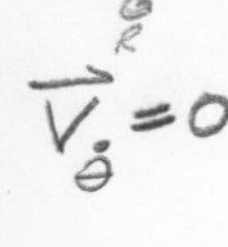

$$F^* \underset{(3.10)}{=} \boldsymbol{\omega}_{\dot{\theta}}{}^G \cdot \mathbf{T}_G{}^* + \boldsymbol{\omega}_{\dot{\theta}}{}^R \cdot \mathbf{T}_R{}^* \tag{a}$$

where $\mathbf{T}_G{}^*$ and $\mathbf{T}_R{}^*$ denote the inertia torques for G and R.

Let $\mathbf{n}_1$, $\mathbf{n}_2$, $\mathbf{n}_3$ be the right-handed set of mutually perpendicular unit vectors shown in Figure 3.21. Then the angular velocities of G and R are (see Sections 2.8 and 2.10)

$$\boldsymbol{\omega}^G = \dot{\phi} \cos \theta \mathbf{n}_1 - \dot{\theta} \mathbf{n}_2 - \dot{\phi} \sin \theta \mathbf{n}_3 \tag{b}$$

and

$$\boldsymbol{\omega}^R = \boldsymbol{\omega}^G + \omega \mathbf{n}_1 \tag{c}$$

Consequently (see Section 2.5),

$$\boldsymbol{\omega}_{\dot{\theta}}{}^G = \boldsymbol{\omega}_{\dot{\theta}}{}^R = -\mathbf{n}_2 \tag{d}$$

and

$$F^* \underset{(a),(d)}{=} -\mathbf{n}_2 \cdot \mathbf{T}_G{}^* - \mathbf{n}_2 \cdot \mathbf{T}_R{}^* \tag{e}$$

As $\mathbf{n}_1$, $\mathbf{n}_2$, and $\mathbf{n}_3$ are fixed in G and are parallel to principal axes of G for the mass center of G,

$$\begin{aligned} \mathbf{n}_2 \cdot \mathbf{T}_G{}^* &\underset{(3.44)}{=} \omega_3{}^G \omega_1{}^G (C - A) - \dot{\omega}_2{}^G B \\ &= \dot{\phi}^2 \sin \theta \cos \theta (A - C) + \ddot{\theta} B \end{aligned} \tag{f}$$

The unit vectors $\mathbf{n}_1$, $\mathbf{n}_2$, and $\mathbf{n}_3$ are parallel to principal axes of R for the mass center of R, although $\mathbf{n}_2$ and $\mathbf{n}_3$ are not fixed in R. Hence,

$$\mathbf{n}_2 \cdot \mathbf{T}_R{}^* \underset{(3.44)}{=} \omega_3{}^R \omega_1{}^R (I - J) - \alpha_2{}^R I \tag{g}$$

where

$$\omega_1{}^R \underset{(b),(c)}{=} \omega + \dot{\phi} \cos \theta \qquad \omega_3{}^R = -\dot{\phi} \sin \theta \tag{h}$$

and (see Section 2.11)

$$\alpha_2{}^R \underset{(b),(c)}{=} -(\ddot{\theta} + \omega \dot{\phi} \sin \theta) \tag{i}$$

Consequently,

$$F^* \underset{(e)}{=} -\dot{\phi}^2 \sin\theta \cos\theta(A - C) - \ddot{\theta}B$$
$$\underset{(f)}{} + (\omega + \dot{\phi}\cos\theta)\dot{\phi}\sin\theta(I - J) - (\ddot{\theta} + \omega\dot{\phi}\sin\theta)I$$
$$\underset{(g)}{} \underset{(h)}{} \underset{(i)}{} = -(I + B)\ddot{\theta} + (I - J + C - A)\dot{\phi}^2 \sin\theta\cos\theta - J\omega\dot{\phi}\sin\theta$$

3.24 Simple Angular Velocity. When a rigid body B has a simple angular velocity (see Section 2.8) parallel to a unit vector $\mathbf{n}_a$ in a reference frame R, the inertia torque $\mathbf{T}^*$ for B in R (see Section 3.10) can be expressed as

$$\mathbf{T}^* = \omega^2 \mathbf{I}_a \times \mathbf{n}_a - \dot{\omega}\mathbf{I}_a \tag{3.45}$$

where ω is the angular speed of B in R for $\mathbf{n}_a$, and $\mathbf{I}_a$ is the second moment of B relative to the mass center B^* of B for $\mathbf{n}_a$ (see Section 3.12). If I_a is the moment of inertia of B about the line passing through B^* and parallel to $\mathbf{n}_a$, then the following frequently useful relationship is an immediate consequence of Equations (3.45) and (3.15):

$$\mathbf{n}_a \cdot \mathbf{T}^* = -\dot{\omega} I_a \tag{3.46}$$

Proof: By hypothesis, the angular velocity $\boldsymbol{\omega}$ of B in R can be expressed as

$$\boldsymbol{\omega} \underset{(2.11)}{=} \omega \mathbf{n}_a$$

and the angular acceleration $\boldsymbol{\alpha}$ of B in R is given by

$$\boldsymbol{\alpha} \underset{(2.15)}{=} \dot{\omega}\mathbf{n}_a$$

Consequently,

$$\mathbf{T}^* \underset{(3.43)}{=} (\omega^2 \mathbf{I} \cdot \mathbf{n}_a) \times \mathbf{n}_a - \dot{\omega}\mathbf{I} \cdot \mathbf{n}_a$$
$$\underset{(3.40)}{=} \omega^2 \mathbf{I}_a \times \mathbf{n}_a - \dot{\omega}\mathbf{I}_a$$

▪ EXAMPLE

Figure 3.22 shows the system previously discussed in the examples in Sections 2.22 and 2.23. C represents a uniform circular disk of radius r and mass m mounted on a fixed horizontal shaft S and pressed into contact with a uniform rectangular plate R of length $2a$, width $2b$, and mass M. Plate R is free to slide on a horizontal support H, and C is free to rotate on S, but C's motion along S is prescribed as a function $s(t)$ of the time t; no slip is permitted to occur at the point of contact between C and R.

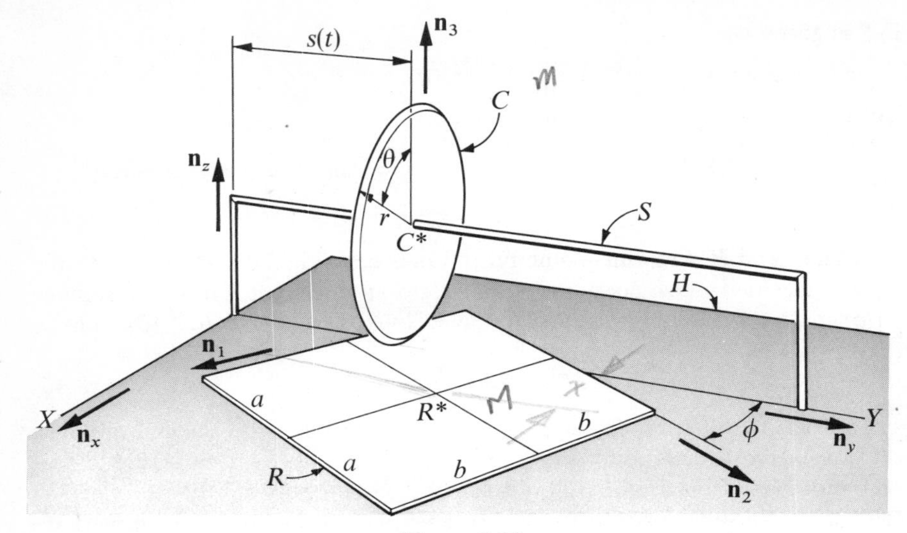

Figure 3.22

In the example in Section 2.22 it was pointed out that C and R comprise a simple nonholonomic system possessing two degrees of freedom. If generalized coordinates q_1 and q_2 are defined as $q_1 = x$ and $q_2 = \phi$, where x is the X coordinate of the mass center R^* of R and ϕ is the angle between an edge of R and the Y axis (see Figure 3.22), then the associated generalized inertia forces F_1^* and F_2^* are found as follows.

With self-explanatory notation,

$$F_r^* \underset{(3.10)}{=} \tilde{\mathbf{v}}_{\dot{q}_r}{}^{R^*} \cdot \mathbf{F}_R{}^* + \tilde{\boldsymbol{\omega}}_{\dot{q}_r}{}^{R} \cdot \mathbf{T}_R{}^* + \tilde{\mathbf{v}}_{\dot{q}_r}{}^{C^*} \cdot \mathbf{F}_C{}^* + \tilde{\boldsymbol{\omega}}_{\dot{q}_r}{}^{C} \cdot \mathbf{T}_C{}^* \tag{a}$$

where (see the example in Section 2.23)

$$\tilde{\mathbf{v}}_{\dot{q}_1}{}^{R^*} = \mathbf{n}_x \qquad \tilde{\mathbf{v}}_{\dot{q}_2}{}^{R^*} = -x\mathbf{n}_y \tag{b}$$

$$\tilde{\boldsymbol{\omega}}_{\dot{q}_1}{}^{R} = 0 \qquad \tilde{\boldsymbol{\omega}}_{\dot{q}_2}{}^{R} = -\mathbf{n}_z \tag{c}$$

$$\tilde{\mathbf{v}}_{\dot{q}_1}{}^{C^*} = 0 \qquad \tilde{\mathbf{v}}_{\dot{q}_2}{}^{C^*} = 0 \tag{d}$$

$$\tilde{\boldsymbol{\omega}}_{\dot{q}_1}{}^{C} = -\frac{1}{r}\mathbf{n}_y \qquad \tilde{\boldsymbol{\omega}}_{\dot{q}_2}{}^{C} = \frac{1}{r}[y - s(t)]\mathbf{n}_y \tag{e}$$

Hence,

$$F_1^* \underset{(a)}{=} \mathbf{n}_x \underset{(b)}{\cdot} \mathbf{F}_R{}^* + \underset{(c)}{0} + \underset{(d)}{0} - \frac{1}{r}\mathbf{n}_y \underset{(e)}{\cdot} \mathbf{T}_C{}^* \tag{f}$$

and, similarly,

$$F_2^* = -x\mathbf{n}_y \cdot \mathbf{F}_R{}^* - \mathbf{n}_z \cdot \mathbf{T}_R{}^* + \frac{1}{r}[y - s(t)]\mathbf{n}_y \cdot \mathbf{T}_C{}^* \tag{g}$$

$\mathbf{F}_R{}^*$ is given by

$$\mathbf{F}_R{}^* \underset{(3.11)}{=} -M(\ddot{x}\mathbf{n}_x + \ddot{y}\mathbf{n}_y)$$

Consequently,

$$\mathbf{n}_x \cdot \mathbf{F}_R{}^* = -M\ddot{x} \tag{h}$$

and

$$\mathbf{n}_y \cdot \mathbf{F}_R{}^* = -M\ddot{y} \tag{i}$$

Both C and R have simple angular velocities:

$$\boldsymbol{\omega}^C = \omega^C\mathbf{n}_y \qquad \boldsymbol{\omega}^R = \omega^R\mathbf{n}_z$$

where

$$\omega^C = \dot{\theta}$$

and

$$\omega^R = -\dot{\phi}$$

The moments of inertia of C about the line passing through C^* and parallel to $\mathbf{n}_y$, and of R about the line passing through R^* and parallel to $\mathbf{n}_z$ have the values $mr^2/2$ and $M(a^2 + b^2)/3$, respectively. Hence,

$$\mathbf{n}_y \cdot \mathbf{T}_C{}^* \underset{(3.46)}{=} -\frac{\ddot{\theta}mr^2}{2} \tag{j}$$

and

$$\mathbf{n}_z \cdot \mathbf{T}_R{}^* \underset{(3.46)}{=} \frac{\ddot{\phi}M(a^2 + b^2)}{3} \tag{k}$$

Thus

$$F_1{}^* \underset{(f)}{=} \underset{(h)}{-M\ddot{x}} + \underset{(j)}{\frac{\ddot{\theta}mr}{2}}$$

and

$$F_2{}^* \underset{(g)}{=} \underset{(i)}{Mx\ddot{y}} - \underset{(k)}{\frac{\ddot{\phi}M(a^2 + b^2)}{3}} - \underset{(j)}{\frac{\ddot{\theta}mr[y - s(t)]}{2}}$$

3.25 Alternative Expressions for Generalized Inertia Forces. In the solution of certain problems, it is convenient to replace Equations (3.8) with

$$F_r{}^* = \sum_{i=1}^{N} \tilde{\mathbf{v}}_{u_r}{}^{P_i} \cdot \mathbf{F}_i{}^* \qquad r = 1, \cdot \cdot \cdot , n - m \tag{3.47}$$

and, correspondingly, to replace Equations (3.10) with

$$(F_r{}^*)_B = \tilde{\mathbf{v}}_{u_r} \cdot \mathbf{F}^* + \tilde{\boldsymbol{\omega}}_{u_r} \cdot \mathbf{T}^* \qquad r = 1, \cdot \cdot \cdot , n - m \tag{3.48}$$

where $\tilde{\mathbf{v}}_{u_r}$ and $\tilde{\boldsymbol{\omega}}_{u_r}$ are the quantities previously discussed in Section 2.24.

▪ EXAMPLE

In Figure 3.23, S represents a uniform sphere of radius r and mass m

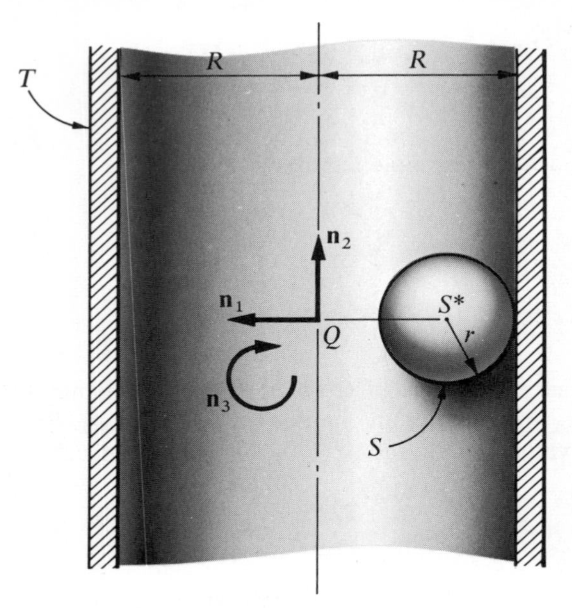

Figure 3.23

that rolls on the interior surface of a tube T of inner radius R. If $\mathbf{n}_1$, $\mathbf{n}_2$, and $\mathbf{n}_3$ are unit vectors directed as shown, the angular velocity $\boldsymbol{\omega}$ of S and the velocity $\mathbf{v}$ of the center S^* of S can be expressed as

$$\boldsymbol{\omega} = \omega_1\mathbf{n}_1 + \omega_2\mathbf{n}_2 + \omega_3\mathbf{n}_3 \tag{a}$$

and

$$\mathbf{v} = \boldsymbol{\omega} \times (r\mathbf{n}_1) = r(\omega_3\mathbf{n}_2 - \omega_2\mathbf{n}_3) \tag{b}$$

In the sequel it becomes necessary to know the angular velocity $\boldsymbol{\Omega}$ of a reference frame in which $\mathbf{n}_1$, $\mathbf{n}_2$, and $\mathbf{n}_3$ are fixed. This angular velocity is the same as that of the plane P in which S^* is fixed and which contains the axis of T. It may be found as follows.

The velocity $\mathbf{v}$ of S^* can be expressed as

$$\mathbf{v} \underset{(2.20)}{=} \mathbf{v}^Q + \boldsymbol{\Omega} \times [(R - r)(-\mathbf{n}_1)]$$

where Q is the point of P that lies at the foot of the perpendicular dropped on the axis of T from S^*. Both $\mathbf{v}^Q$ and $\boldsymbol{\Omega}$ are known to be parallel to $\mathbf{n}_2$. Hence they can be expressed as

$$\mathbf{v}^Q = v^Q\,\mathbf{n}_2 \qquad \boldsymbol{\Omega} = \Omega\mathbf{n}_2$$

and it follows that

$$v^Q \mathbf{n}_2 + (R - r)\Omega \mathbf{n}_3 \underset{\text{(b)}}{=} r(\omega_3 \mathbf{n}_2 - \omega_2 \mathbf{n}_3)$$

Consequently,

$$\Omega = \frac{r\omega_2}{r - R}$$

and

$$\boldsymbol{\Omega} = \frac{r\omega_2}{r - R} \mathbf{n}_2 \tag{c}$$

Hence, if u_1, u_2, and u_3 are defined as

$$u_r = \omega_r \qquad r = 1, 2, 3 \tag{d}$$

then

$$\boldsymbol{\omega} \underset{\text{(a),(d)}}{=} u_1\mathbf{n}_1 + u_2\mathbf{n}_2 + u_3\mathbf{n}_3 \tag{e}$$

$$\mathbf{v} \underset{\text{(b),(d)}}{=} r(u_3\mathbf{n}_2 - u_2\mathbf{n}_3) \tag{f}$$

and

$$\boldsymbol{\Omega} \underset{\text{(c),(d)}}{=} \frac{ru_2}{r - R} \mathbf{n}_2 \tag{g}$$

The partial rates of change of the position of S^* and of the orientation of S are now seen to be

$$\tilde{\mathbf{v}}_{u_1} = 0 \qquad \tilde{\mathbf{v}}_{u_2} = -r\mathbf{n}_3 \qquad \tilde{\mathbf{v}}_{u_3} = r\mathbf{n}_2 \tag{h}$$

and

$$\tilde{\boldsymbol{\omega}}_{u_1} = \mathbf{n}_1 \qquad \tilde{\boldsymbol{\omega}}_{u_2} = \mathbf{n}_2 \qquad \tilde{\boldsymbol{\omega}}_{u_3} = \mathbf{n}_3 \tag{i}$$

and the generalized inertia forces F_1^*, F_2^*, and F_3^* are thus given by

$$F_1^* \underset{(3.48)}{=} \mathbf{n}_1 \cdot \mathbf{T}^* \tag{j}$$

$$F_2^* \underset{(3.48)}{=} -r\mathbf{n}_3 \cdot \mathbf{F}^* + \mathbf{n}_2 \cdot \mathbf{T}^* \tag{k}$$

$$F_3^* \underset{(3.48)}{=} r\mathbf{n}_2 \cdot \mathbf{F}^* + \mathbf{n}_3 \cdot \mathbf{T}^* \tag{l}$$

where $\mathbf{F}^*$ and $\mathbf{T}^*$ are the inertia force and the inertia torque for S. To find $\mathbf{F}^*$ and $\mathbf{T}^*$, note that the acceleration $\mathbf{a}$ of S^* and the angular acceleration $\boldsymbol{\alpha}$ of S are

$$\mathbf{a} = \frac{d\mathbf{v}}{dt} \underset{(2.13)}{=} r(\dot{u}_3\mathbf{n}_2 \underset{\text{(f)}}{-} \dot{u}_2\mathbf{n}_3) + \boldsymbol{\Omega} \times \mathbf{v}$$

$$\underset{\text{(f),(g)}}{=} \frac{(ru_2)^2}{R - r} \mathbf{n}_1 + r\dot{u}_3\mathbf{n}_2 - r\dot{u}_2\mathbf{n}_3$$

and

$$\boldsymbol{\alpha} \underset{(2.15)}{=} \frac{d\boldsymbol{\omega}}{dt} \underset{(2.13)}{=} \dot{u}_1\mathbf{n}_1 + \dot{u}_2\mathbf{n}_2 \underset{(e)}{+} \dot{u}_3\mathbf{n}_3 + \boldsymbol{\Omega} \times \boldsymbol{\omega}$$

$$\underset{(g),(e)}{=} \left(\dot{u}_1 - \frac{ru_2u_3}{R - r}\right)\mathbf{n}_1 + \dot{u}_2\mathbf{n}_2 + \left(\dot{u}_3 + \frac{ru_1u_2}{R - r}\right)\mathbf{n}_3$$

Hence,

$$\mathbf{F}^* \underset{(3.11)}{=} -m\left[\frac{(ru_2)^2}{R - r}\mathbf{n}_1 + r\dot{u}_3\mathbf{n}_2 - r\dot{u}_2\mathbf{n}_3\right]$$

and

$$\mathbf{T}^* \underset{(3.44)}{=} -\tfrac{2}{5}mr^2\left[\left(\dot{u}_1 - \frac{ru_2u_3}{R - r}\right)\mathbf{n}_1 + \dot{u}_2\mathbf{n}_2 + \left(\dot{u}_3 + \frac{ru_1u_2}{R - r}\right)\mathbf{n}_3\right]$$

Consequently,

$$F_1^* \underset{(j)}{=} -\tfrac{2}{5}mr^2\left(\dot{u}_1 - \frac{ru_2u_3}{R - r}\right)$$

$$F_2^* \underset{(k)}{=} -mr^2\dot{u}_2 - \tfrac{2}{5}mr^2\dot{u}_2 = -\tfrac{7}{5}mr^2\dot{u}_2$$

$$F_3^* \underset{(l)}{=} -\tfrac{2}{5}mr^2\left(\tfrac{7}{2}\dot{u}_3 + \frac{ru_1u_2}{R - r}\right)$$

Chapter **4**

Energy Functions

4.1 Potential Functions. If S is a holonomic system with generalized coordinates $q_1, \cdots, q_n$ in a reference frame R, and $F_1', \cdots, F_n'$ are the contributions to the generalized active forces $F_1, \cdots, F_n$, respectively, of certain contact and/or body forces acting on particles of S, there may exist a function P of $q_1, \cdots, q_n$, and t, such that

$$F_r' = -\frac{\partial P}{\partial q_r} \qquad r = 1, \cdots, n \tag{4.1}$$

P is called a *potential function* for the contact and/or body forces in question.

▪ EXAMPLE

Referring to the example in Section 3.2, let F_1' and F_2' be the contributions to the generalized active forces F_1 and F_2, respectively, of forces exerted on the particles P_1 and P_2 by the springs σ_1 and σ_2; that is, let

$$F_1' = -k_1q_1 + k_2(q_2 - q_1) \tag{a}$$

and

$$F_2' = -k_2(q_2 - q_1) \tag{b}$$

Then a potential function $P(q_1, q_2, t)$ for the forces exerted on the particles

P_1 and P_2 by the springs σ_1 and σ_2 can be found by solving the differential equations

$$\frac{\partial P}{\partial q_1} = -F_1' \underset{(a)}{=} k_1 q_1 - k_2(q_2 - q_1) \tag{c}$$

and

$$\frac{\partial P}{\partial q_2} = -F_2' \underset{(b)}{=} k_2(q_2 - q_1) \tag{d}$$

Equation (c) is satisfied if

$$P \underset{(c)}{=} \tfrac{1}{2}k_1 q_1^2 - k_2(q_2 q_1 - \tfrac{1}{2}q_1^2) + f(q_2, t) \tag{e}$$

where $f(q_2, t)$ is any function whatsoever of q_2 and t; but $f(q_2, t)$ must be such that (d) is satisfied; that is,

$$-k_2 q_1 + \frac{\partial f}{\partial q_2} \underset{(d),(e)}{=} k_2(q_2 - q_1)$$

or

$$\frac{\partial f}{\partial q_2} = k_2 q_2$$

A function f that meets this requirement is

$$f = \tfrac{1}{2}k_2 q_2^2 + w(t) \tag{f}$$

where $w(t)$ is any function whatsoever of t. Consequently, the following is a potential function for the spring forces:

$$P \underset{(e),(f)}{=} \tfrac{1}{2}k_1 q_1^2 - k_2(q_2 q_1 - \tfrac{1}{2}q_1^2) + \tfrac{1}{2}k_2 q_2^2 + w(t)$$

There also exists a potential function for the gravitational forces acting on P_1 and P_2; namely, the (time-dependent) function $-g \cos \theta(m_1 q_1 + m_2 q_2)$. The negatives of the partial derivatives of this function with respect to q_1 and q_2 are $m_1 g \cos \theta$ and $m_2 g \cos \theta$, respectively, and these are the contributions to F_1 and F_2 of gravitational forces acting on P_1 and P_2.

The example in Section 3.7 shows that one may encounter forces for which there exists no potential function. The contribution of damper forces to one of the generalized forces was there found to be $-\delta \dot{q}_2$; and partial differentiation with respect to q_1 and q_2 of a function of q_1, q_2, and t cannot lead to such a term.

4.2 Gravitational Forces. A potential function P (see Section 4.1) for the gravitational forces $\mathbf{G}_1, \cdots, \mathbf{G}_N$ discussed in Section 3.5 is given by

$$P = mgh \tag{4.2}$$

where h is a function of $q_1, \cdots, q_n$, and t whose absolute value is equal to the distance between the mass center P^* of S and any fixed, horizontal plane H, and h is positive or negative according to whether P^* lies above or below H.

Proof: The contribution $(F_r)_G$ of $\mathbf{G}_1, \cdots, \mathbf{G}_N$ to the generalized active force F_r is

$$(F_r)_G \underset{(3.7)}{=} mg\mathbf{k} \cdot \mathbf{v}_{\dot{q}_r}{}^{P^*} \tag{a}$$

With h as defined, the velocity $\mathbf{v}^{P^*}$ of P^* can be expressed as

$$\mathbf{v}^{P^*} = -\frac{dh}{dt}\mathbf{k} + \cdots = -\left(\sum_{r=1}^{n} \frac{\partial h}{\partial q_r}\dot{q}_r + \frac{\partial h}{\partial t}\right)\mathbf{k} + \cdots$$

where the dots represent a vector perpendicular to $\mathbf{k}$. Consequently,

$$\mathbf{k} \cdot \mathbf{v}_{\dot{q}_r}{}^{P^*} \underset{(\text{Sec. } 2.13)}{=} -\frac{\partial h}{\partial q_r} \tag{b}$$

and

$$(F_r)_G \underset{(a),(b)}{=} -mg\frac{\partial h}{\partial q_r} \tag{c}$$

Let

$$P = mgh \tag{d}$$

Then

$$\frac{\partial P}{\partial q_r} \underset{(d)}{=} mg\frac{\partial h}{\partial q_r} \underset{(c)}{=} -(F_r)_G$$

and [see Equations (4.1)] P is a potential function for the gravitational forces under consideration.

▪ EXAMPLE

To obtain a potential function for the gravitational forces acting on the links discussed in the example in Section 3.5, note that the total mass m of the linkage is

$$m = 9M$$

and let

$$h = -\tfrac{3}{2}L \sin \theta$$

Then the absolute value of h is equal to the distance between the mass center B and the horizontal plane H passing through A; h is negative when B lies below H; and

$$P = mgh = -13.5MgL \sin \theta$$

4.3 Springs. If a light spring σ is attached to a particle of a system S, or if σ connects two particles of S to each other, a potential function P (see Section 4.1) for the forces exerted by the spring is given by

$$P = \int_0^x f(\xi)\, d\xi \tag{4.3}$$

where x is a function of $q_1, \cdots, q_n$, and t that measures the extension of σ, and $f(x)$ defines the character of σ. For example, if σ is a linear spring with spring constant k, then

$$f(x) = kx \tag{4.4}$$

and

$$P = \tfrac{1}{2}kx^2 \tag{4.5}$$

Proof: In Figure 4.1, $\mathbf{n}$ is a unit vector parallel to the line connecting

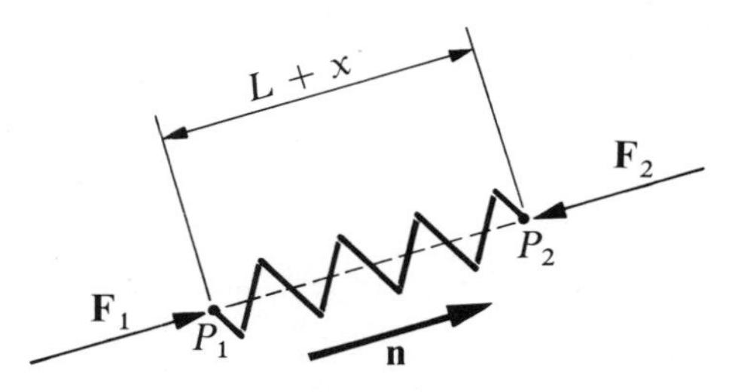

Figure 4.1

the end points of σ, L designates the "natural" length of σ, and $\mathbf{F}_1$ and $\mathbf{F}_2$ are the forces exerted by σ on the particles P_1 and P_2 to which σ is attached. These forces have the same magnitude, $|f(x)|$, and both are parallel to $\mathbf{n}$. Hence,

$$\mathbf{F}_1 = f(x)\mathbf{n} \qquad \mathbf{F}_2 = -f(x)\mathbf{n} \tag{a}$$

If P_1 and P_2 are particles of a system S, then F_r', the contribution of $\mathbf{F}_1$ and $\mathbf{F}_2$ to the generalized active force F_r, is given by

$$\begin{aligned} F_{r'} &\underset{(3.3)}{=} \mathbf{v}_{\dot{q}_r}{}^{P_1} \cdot \mathbf{F}_1 + \mathbf{v}_{\dot{q}_r}{}^{P_2} \cdot \mathbf{F}_2 \\ &\underset{(a)}{=} -(\mathbf{v}_{\dot{q}_r}{}^{P_2} - \mathbf{v}_{\dot{q}_r}{}^{P_1}) \cdot \mathbf{n}f(x) \end{aligned} \tag{b}$$

Now

$$\begin{aligned} \mathbf{v}_{\dot{q}_r}{}^{P_2} - \mathbf{v}_{\dot{q}_r}{}^{P_1} &\underset{(2.19)}{=} \frac{\partial}{\partial \dot{q}_r}(\mathbf{v}^{P_2} - \mathbf{v}^{P_1}) \\ &= \frac{\partial}{\partial \dot{q}_r}\frac{d}{dt}[(L + x)\mathbf{n}] \\ &= \frac{\partial}{\partial \dot{q}_r}\left[\frac{dx}{dt}\mathbf{n} + (L + x)\frac{d\mathbf{n}}{dt}\right] \\ &= \frac{\partial}{\partial \dot{q}_r}\left\{\sum_{s=1}^{n}\left[\frac{\partial x}{\partial q_s}\mathbf{n} + (L + x)\frac{\partial \mathbf{n}}{\partial q_s}\right]\dot{q}_s + \frac{\partial x}{\partial t}\mathbf{n} + (L + x)\frac{\partial \mathbf{n}}{\partial t}\right\} \\ &= \frac{\partial x}{\partial q_r}\mathbf{n} + (L + x)\frac{\partial \mathbf{n}}{\partial q_r} \end{aligned} \tag{c}$$

Consequently,

$$F_r' \underset{(b),(c)}{=} -f(x) \frac{\partial x}{\partial q_r} \tag{d}$$

Let

$$P = \int_0^x f(\xi)\, d\xi \tag{e}$$

Then

$$\frac{\partial P}{\partial q_r} = \frac{dP}{dx} \frac{\partial x}{\partial q_r} \underset{(e)}{=} f(x) \frac{\partial x}{\partial q_r} \underset{(d)}{=} -F_r'$$

and [see Equation (4.1)] P is a potential function for the spring forces. A similar proof shows that this result applies also when only one end of σ is attached to a particle of the system S.

▪ EXAMPLE

In the example in Section 3.2, one end of the spring σ_1 and both ends of the spring σ_2 are attached to particles of the system under consideration. The extensions x_1 and x_2 of the springs are

$$x_1 = q_1 \qquad x_2 = q_2 - q_1$$

Hence a potential function P for the forces exerted on the particles P_1 and P_2 by the springs σ_1 and σ_2 is

$$P \underset{(4.5)}{=} \tfrac{1}{2}k_1 x_1^2 + \tfrac{1}{2}k_2 x_2^2$$
$$= \tfrac{1}{2}k_1 q_1^2 + \tfrac{1}{2}k_2(q_2 - q_1)^2$$

(Compare this result with that obtained in the example in Section 4.1.)

4.4 Potential Energy. If S is a holonomic system with generalized coordinates $q_1, \cdots, q_n$ in a reference frame R, and $F_1, \cdots, F_n$ are the associated generalized active forces for S in R (see Section 3.2), there may exist potential functions (see Section 4.1) for *all* contact and body forces that contribute to the generalized active forces. The sum of these potential functions is then a function P of $q_1, \cdots, q_n$, and t such that

$$F_r = -\frac{\partial P}{\partial q_r} \qquad r = 1, \cdots, n \tag{4.6}$$

A function P that satisfies Equations (4.6) is called a *potential energy* of S in R, and S is said to be a *conservative* system whenever such a function exists. (This terminology may be misleading for the following reason: There exist conservative systems whose total mechanical energy is not conserved; that is, the sum of the potential and kinetic energy of such a system does not necessarily remain constant. See the example in Section 6.5.)

▪ EXAMPLE

In the example in Section 3.2, the only contact forces that contribute to the generalized forces F_1 and F_2 are the forces exerted on P_1 and P_2 by the springs σ_1 and σ_2, and

$$\tfrac{1}{2}k_1q_1^2 - k_2(q_2q_1 - \tfrac{1}{2}q_1^2) + \tfrac{1}{2}k_2q_2^2 + w(t)$$

is a potential function for these forces, as was shown in the example in Section 4.1, where it was also observed that

$$-g \cos \theta(m_1q_1 + m_2q_2)$$

is a potential function for the gravitational forces that contribute to F_1 and F_2. Hence the function P defined as

$$P = \tfrac{1}{2}k_1q_1^2 - k_2(q_2q_1 - \tfrac{1}{2}q_1^2) + \tfrac{1}{2}k_2q_2^2 + w(t) - g \cos \theta(m_1q_1 + m_2q_2)$$

is a potential energy of P_1 and P_2, and it may be verified by reference to the example in Section 4.1 that

$$F_1 = -\frac{\partial P}{\partial q_1} \qquad F_2 = -\frac{\partial P}{\partial q_2}$$

4.5 Determination of Generalized Forces. When an expression for a potential energy of a system (see Section 4.4) is readily available, generalized forces can usually be found more expeditiously by differentiating this potential energy [see Equation (4.6)] than by using any other method. By the same token, when knowledge of the generalized forces is required for the construction of a potential energy expression, the use of this expression clearly cannot facilitate the determination of generalized forces.

4.6 Dissipation Functions. If S is a holonomic system with generalized coordinates $q_1, \cdots, q_n$ in a reference frame R, and $F_1', \cdots, F_n'$ are the contributions to the generalized active forces $F_1, \cdots, F_n$, respectively, of certain contact forces acting on particles of S, there may exist a function $\mathfrak{F}$ of $q_1, \cdots, q_n, \dot{q}_1, \cdots, \dot{q}_n$, and t, such that

$$F_r' = -\frac{\partial \mathfrak{F}}{\partial \dot{q}_r} \qquad r = 1, \cdots, n \tag{4.7}$$

$\mathfrak{F}$ is called a *dissipation function* for the contact forces in question.

▪ EXAMPLE

Referring to the example in Section 3.7, let F_1' and F_2' be the contributions to the generalized active forces F_1 and F_2, respectively, of damping forces exerted on the rods A and B by the viscous fluid; that is, let

$$F_1' = 0 \qquad F_2' = -\delta\dot{q}_2 \tag{a}$$

Then the function $\mathfrak{F}$ defined as

$$\mathfrak{F} = \tfrac{1}{2}\delta\dot{q}_2^2 \tag{b}$$

is a dissipation function for the damper forces, because

$$\frac{\partial\mathfrak{F}}{\partial\dot{q}_1} \underset{\text{(b)}}{=} 0 \underset{\text{(a)}}{=} -F_1'$$

and

$$\frac{\partial\mathfrak{F}}{\partial\dot{q}_2} \underset{\text{(b)}}{=} \delta\dot{q}_2 \underset{\text{(a)}}{=} -F_2'$$

4.7 Forces Proportional to Velocity. If $\mathbf{F}_1, \cdots, \mathbf{F}_N$ are contact forces acting on the particles $P_1, \cdots, P_N$ of a holonomic system, and each of these forces is a "resisting" force proportional to the velocity of the particle on which it acts—that is,

$$\mathbf{F}_i = -c\mathbf{v}_i \qquad i = 1, \cdots, N \tag{4.8}$$

where c is a positive constant—then a dissipation function $\mathfrak{F}$ for these forces (see Section 4.6) is given by

$$\mathfrak{F} = \frac{c}{2}\sum_{i=1}^{N} \mathbf{v}_i^2 \tag{4.9}$$

Proof: The contribution F_r' to the generalized force F_r of the forces $\mathbf{F}_1, \cdots, \mathbf{F}_N$ is given by

$$\begin{aligned} F_r' &\underset{(3.3)}{=} \sum_{i=1}^{N} \mathbf{v}_{\dot{q}_r}^{P_i} \cdot \mathbf{F}_i \\ &\underset{(4.8)}{=} -\sum_{i=1}^{N} c\mathbf{v}_{\dot{q}_r}^{P_i} \cdot \mathbf{v}_i \\ &\underset{(2.19)}{=} -c\sum_{i=1}^{N} \frac{\partial\mathbf{v}_i}{\partial\dot{q}_r} \cdot \mathbf{v}_i \\ &= -\frac{\partial}{\partial\dot{q}_r}\left(\frac{c}{2}\sum_{i=1}^{N} \mathbf{v}_i^2\right) \end{aligned}$$

Hence, if $\mathfrak{F}$ is defined as in Equation (4.9), F_r' can be expressed as

$$F_r' = -\frac{\partial \mathfrak{F}}{\partial \dot{q}_r}$$

4.8 Kinetic Energy and Generalized Inertia Forces. The *kinetic energy* K of a set of N particles $P_1, \cdots, P_N$ in a reference frame R is defined as

$$K = \tfrac{1}{2} \sum_{i=1}^{N} m_i \mathbf{v}_i^2 \tag{4.10}$$

where m_i is the mass of P_i and $\mathbf{v}_i$ is the velocity of P_i in R. If these particles comprise a holonomic system S possessing n generalized coordinates $q_1, \cdots, q_n$ in R (see Section 2.2), K can be regarded as a function of the $2n + 1$ independent variables $q_1, \cdots, q_n, \dot{q}_1, \cdots, \dot{q}_n$, and t; and the generalized inertia forces $F_1^*, \cdots, F_n^*$ (see Section 3.9) can then be expressed as

$$F_r^* = \frac{\partial K}{\partial q_r} - \frac{d}{dt}\frac{\partial K}{\partial \dot{q}_r} \qquad r = 1, \cdots, n \tag{4.11}$$

Proof: By definition,

$$\begin{aligned} F_r^* &\underset{(3.8),(3.9)}{=} -\sum_{i=1}^{N} m_i \mathbf{v}_{\dot{q}_r}{}^{P_i} \cdot \mathbf{a}_i \\ &\underset{(2.25)}{=} -\sum_{i=1}^{N} \frac{m_i}{2}\left(\frac{d}{dt}\frac{\partial \mathbf{v}_i^2}{\partial \dot{q}_r} - \frac{\partial \mathbf{v}_i^2}{\partial q_r}\right) \\ &= \frac{\partial}{\partial q_r}\sum_{i=1}^{N}\frac{m_i}{2}\mathbf{v}_i^2 - \frac{d}{dt}\frac{\partial}{\partial \dot{q}_r}\sum_{i=1}^{N}\frac{m_i}{2}\mathbf{v}_i^2 \\ &\underset{(4.10)}{=} \frac{\partial K}{\partial q_r} - \frac{d}{dt}\frac{\partial K}{\partial \dot{q}_r} \end{aligned}$$

p 37

▪ EXAMPLE

The kinetic energy K of the system described in the example in Section 3.2 is given by

$$\begin{aligned} K &= \tfrac{1}{2}m_1(\mathbf{v}^{P_1})^2 + \tfrac{1}{2}m_2(\mathbf{v}^{P_2})^2 \\ &= \tfrac{1}{2}m_1[\dot{q}_1^2 + (L_1 + q_1)^2\dot{\theta}^2] \\ &\qquad + \tfrac{1}{2}m_2[\dot{q}_2^2 + (L_1 + L_2 + q_2)^2\dot{\theta}^2] \end{aligned}$$

The generalized forces F_1^* and F_2^* are thus

$$F_1^* = \frac{\partial K}{\partial q_1} - \frac{d}{dt}\frac{\partial K}{\partial \dot{q}_1}$$
$$= m_1(L_1 + q_1)\dot{\theta}^2 - \frac{d}{dt}(m_1\dot{q}_1)$$
$$= m_1[(L_1 + q_1)\dot{\theta}^2 - \ddot{q}_1]$$

and

$$F_2^* = \frac{\partial K}{\partial q_2} - \frac{d}{dt}\frac{\partial K}{\partial \dot{q}_2}$$
$$= m_2(L_1 + L_2 + q_2)\dot{\theta}^2 - \frac{d}{dt}(m_2\dot{q}_2)$$
$$= m_2[(L_1 + L_2 + q_2)\dot{\theta}^2 - \ddot{q}_2]$$

The method used here for the determination of F_1^* and F_2^* should be compared with that employed in the example in Section 3.9.

4.9 Kinetic Energies of Translation and Rotation. The *kinetic energy of translation* K_v and the *kinetic energy of rotation* K_ω of a rigid body B in a reference frame R are defined as follows:

$$K_v = \tfrac{1}{2}m\mathbf{v}^2 \tag{4.12}$$

where m is the mass of B, and $\mathbf{v}$ is the velocity of the mass center B^* of B in R; and

$$K_\omega = \tfrac{1}{2}\boldsymbol{\omega} \cdot \mathbf{I} \cdot \boldsymbol{\omega} \tag{4.13}$$

where $\boldsymbol{\omega}$ is the angular velocity of B in R, and $\mathbf{I}$ is the inertia dyadic of B for B^* (see Section 3.21).

In terms of K_v and K_ω, the kinetic energy K of B in R is given by

$$K = K_v + K_\omega \tag{4.14}$$

Proof: Let m_i and $\mathbf{v}_i$ be the mass and the velocity in R of a typical particle P_i of B. Then

$$\mathbf{v}_i \underset{(2.20)}{=} \mathbf{v} + \boldsymbol{\omega} \times \mathbf{r}_i \tag{a}$$

where $\mathbf{r}_i$ is the position vector of P_i relative to B^*; and, if $\bar{N}$ is the number of particles comprising B,

$$K \underset{(4.10)}{=} \tfrac{1}{2}\sum_{i=1}^{\bar{N}} m_i\mathbf{v}_i^2$$
$$\underset{(a)}{=} \tfrac{1}{2}\sum_{i=1}^{\bar{N}} m_i[\mathbf{v}^2 + 2\mathbf{v} \cdot \boldsymbol{\omega} \times \mathbf{r}_i + (\boldsymbol{\omega} \times \mathbf{r}_i)^2]$$

or

$$K = \tfrac{1}{2}\left(\sum_{i=1}^{\bar{N}} m_i\right)\mathbf{v}^2 + \mathbf{v}\cdot\boldsymbol{\omega}\times\left(\sum_{i=1}^{N} m_i\mathbf{r}_i\right) + \tfrac{1}{2}\sum_{i=1}^{\bar{N}} m_i(\boldsymbol{\omega}\times\mathbf{r}_i)^2 \qquad \text{(b)}$$

The first sum in the right-hand member of Equation (b) is equal to the mass m of B, and the second sum is equal to zero because $\mathbf{r}_i$ is the position vector of P_i relative to the mass center of B. To show that the third sum is equal to $\boldsymbol{\omega}\cdot\mathbf{I}\cdot\boldsymbol{\omega}$, express $\boldsymbol{\omega}$ as

$$\boldsymbol{\omega} = \omega\mathbf{n}_\omega \qquad \text{(c)}$$

where ω is a certain scalar and $\mathbf{n}_\omega$ is a unit vector parallel to $\boldsymbol{\omega}$, and note that

$$\begin{aligned}(\boldsymbol{\omega}\times\mathbf{r}_i)^2 &= \boldsymbol{\omega}\cdot\mathbf{r}_i\times(\boldsymbol{\omega}\times\mathbf{r}_i)\\ &\underset{\text{(c)}}{=} \boldsymbol{\omega}\cdot\mathbf{r}_i\times(\mathbf{n}_\omega\times\mathbf{r}_i)\omega\end{aligned} \qquad \text{(d)}$$

and, therefore,

$$\begin{aligned}\sum_{i=1}^{\bar{N}} m_i(\boldsymbol{\omega}\times\mathbf{r}_i)^2 &\underset{\text{(d)}}{=} \boldsymbol{\omega}\cdot\sum_{i=1}^{\bar{N}} m_i\mathbf{r}_i\times(\mathbf{n}_\omega\times\mathbf{r}_i)\omega\\ &\underset{(3.13)}{=} \boldsymbol{\omega}\cdot\mathbf{I}_\omega\omega \underset{(3.40)}{=} \boldsymbol{\omega}\cdot\mathbf{I}\cdot\mathbf{n}_\omega\omega \underset{\text{(c)}}{=} \boldsymbol{\omega}\cdot\mathbf{I}\cdot\boldsymbol{\omega}\end{aligned}$$

4.10 Kinetic Energy of Rotation. The kinetic energy of rotation K_ω of a rigid body B in a reference frame R (see Section 4.9) can be expressed in the following alternative forms:

$$K_\omega = \tfrac{1}{2}I\omega^2 \qquad (4.15)$$

where I is the moment of inertia of B about the line that passes through the mass center B^* of B and is parallel to $\boldsymbol{\omega}$;

$$K_\omega = \tfrac{1}{2}(I_1\omega_1^2 + I_2\omega_2^2 + I_3\omega_3^2) \qquad (4.16)$$

where I_1, I_2, I_3 are moments of inertia of B about mutually perpendicular principal axes of B for B^*, and ω_1, ω_2, ω_3 are the associated measure numbers of the angular velocity of B in R; and

$$K_\omega = \tfrac{1}{2}\sum_{j=1}^{3}\sum_{k=1}^{3}\omega_j I_{jk}\omega_k \qquad (4.17)$$

where the subscripts refer to any three mutually perpendicular axes, and I_{jk} is a product of inertia of B for two such axes passing through B^*.

Proof: The inertia dyadic **I** of B for the mass center of B is given by

$$\mathbf{I} \underset{(3.39)}{=} \sum_{j=1}^{3} \sum_{k=1}^{3} \mathbf{n}_j I_{jk} \mathbf{n}_k \tag{a}$$

and the angular velocity $\boldsymbol{\omega}$ can be expressed both as

$$\boldsymbol{\omega} = \omega \mathbf{n}_\omega \tag{b}$$

where $\mathbf{n}_\omega$ is a unit vector parallel to $\boldsymbol{\omega}$, and as

$$\boldsymbol{\omega} = \sum_{i=1}^{3} \omega_i \mathbf{n}_i \tag{c}$$

Consequently,

$$K_\omega \underset{(4.13)}{=} \tfrac{1}{2} \underset{(b)}{\omega \mathbf{n}_\omega} \cdot \mathbf{I} \cdot \underset{(b)}{\mathbf{n}_\omega \omega}$$

$$\underset{(3.41)}{=} \tfrac{1}{2} I \omega^2 \underset{(b)}{=} \tfrac{1}{2} I \boldsymbol{\omega}^2$$

Alternatively,

$$K_\omega \underset{(4.13)}{=} \tfrac{1}{2} \underset{(c)}{\sum_{i=1}^{3} \omega_i \mathbf{n}_i} \cdot \underset{(a)}{\sum_{j=1}^{3} \sum_{k=1}^{3} \mathbf{n}_j I_{jk} \mathbf{n}_k} \cdot \underset{(c)}{\sum_{i=1}^{3} \omega_i \mathbf{n}_i}$$

$$= \tfrac{1}{2} \left(\sum_{i=1}^{3} \sum_{j=1}^{3} \sum_{k=1}^{3} \omega_i \mathbf{n}_i \cdot \mathbf{n}_j I_{jk} \mathbf{n}_k \right) \cdot \sum_{i=1}^{3} \omega_i \mathbf{n}_i$$

Now, $\mathbf{n}_i \cdot \mathbf{n}_j$ vanishes unless $i = j$. Hence,

$$K_\omega = \tfrac{1}{2} \left(\sum_{j=1}^{3} \sum_{k=1}^{3} \omega_j I_{jk} \mathbf{n}_k \right) \cdot \sum_{i=1}^{3} \omega_i \mathbf{n}_i$$

$$= \tfrac{1}{2} \sum_{j=1}^{3} \sum_{k=1}^{3} \sum_{i=1}^{3} \omega_j I_{jk} \omega_i \mathbf{n}_k \cdot \mathbf{n}_i$$

and, since $\mathbf{n}_k \cdot \mathbf{n}_i$ vanishes unless $i = k$,

$$K_\omega = \tfrac{1}{2} \sum_{j=1}^{3} \sum_{k=1}^{3} \omega_j I_{jk} \omega_k$$

This concludes the proof of Equation (4.17). To obtain Equation (4.16), it is necessary only to note that, when the subscripts refer to principal axes, I_{jk} vanishes unless $j = k$.

▪ EXAMPLE

The vertex V of a uniform, right-circular cone C is made to move with a speed $v(t)$ on a horizontal line. The base of C has a radius r, and C has a height h and mass M. The generalized inertia force F^* corresponding to the coordinate ϕ (see Figure 4.2) is to be determined.

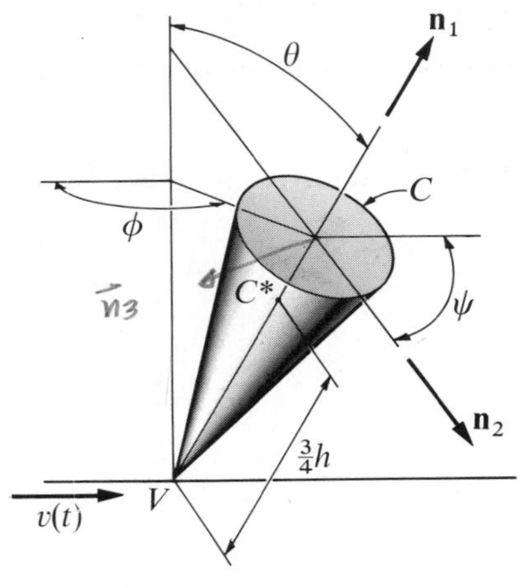

Figure 4.2

If $\mathbf{n}_1$ and $\mathbf{n}_2$ are unit vectors directed as shown in Figure 4.2, and $\mathbf{n}_3 = \mathbf{n}_1 \times \mathbf{n}_2$, the angular velocity of C is given by

$$\boldsymbol{\omega} = \omega_1\mathbf{n}_1 + \omega_2\mathbf{n}_2 + \omega_3\mathbf{n}_3$$

where

$$\omega_1 = \dot{\psi} + \dot{\phi}\cos\theta \qquad \omega_2 = -\dot{\phi}\sin\theta \qquad \omega_3 = \dot{\theta} \tag{a}$$

and the velocity $\mathbf{v}$ of V can be expressed as

$$\mathbf{v} = v_1\mathbf{n}_1 + v_2\mathbf{n}_2 + v_3\mathbf{n}_3$$

where

$$\begin{aligned} v_1 &= -v\sin\theta\cos\phi \\ v_2 &= -v\cos\theta\cos\phi \\ v_3 &= v\sin\phi \end{aligned}$$

The velocity $\mathbf{v}^*$ of the mass center C^* of C is then obtained by noting that

$$\mathbf{v}^* \underset{(2.20)}{=} \mathbf{v} + \tfrac{3}{4}h\boldsymbol{\omega} \times \mathbf{n}_1$$

so that

$$\mathbf{v}^* = v_1{}^*\mathbf{n}_1 + v_2{}^*\mathbf{n}_2 + v_3{}^*\mathbf{n}_3$$

where

$$\left.\begin{aligned} v_1{}^* &= -v\sin\theta\cos\phi \\ v_2{}^* &= -v\cos\theta\cos\phi + \tfrac{3}{4}h\dot{\theta} \\ v_3{}^* &= v\sin\phi + \tfrac{3}{4}h\dot{\phi}\sin\theta \end{aligned}\right\} \tag{b}$$

The unit vectors $\mathbf{n}_1$ and $\mathbf{n}_2$ are parallel to principal axes of inertia of C for C^*, and the associated moments of inertia are

$$I_1 = \tfrac{3}{10}Mr^2 \qquad I_2 = \tfrac{3}{80}M(4r^2 + h^2) \tag{c}$$

Let K_v and K_ω denote the kinetic energy of translation and rotation of C, respectively. Then

$$K_v \underset{(4.12)}{=} \tfrac{1}{2}M(v_1{}^{*2} + v_2{}^{*2} + v_3{}^{*2})$$

and

$$K_\omega \underset{(4.16)}{=} \tfrac{1}{2}(I_1\omega_1{}^2 + I_2\omega_2{}^2 + I_3\omega_3{}^2) \tag{d}$$

Consequently,

$$\begin{aligned}
F^* &\underset{(4.11),(4.14)}{=} \frac{\partial}{\partial \phi}(K_v + K_\omega) - \frac{d}{dt}\frac{\partial}{\partial \dot{\phi}}(K_v + K_\omega) \\
&\underset{(d)}{=} M\left(v_1{}^*\frac{\partial v_1{}^*}{\partial \phi} + v_2{}^*\frac{\partial v_2{}^*}{\partial \phi} + v_3{}^*\frac{\partial v_3{}^*}{\partial \phi}\right) \\
&\quad + I_1\omega_1\frac{\partial \omega_1}{\partial \phi} + I_2\omega_2\frac{\partial \omega_2}{\partial \phi} + I_3\omega_3\frac{\partial \omega_3}{\partial \phi} \\
&\quad - \frac{d}{dt}\left(Mv_3{}^*\frac{\partial v_3{}^*}{\partial \dot{\phi}} + I_1\omega_1\frac{\partial \omega_1}{\partial \dot{\phi}} + I_2\omega_2\frac{\partial \omega_2}{\partial \dot{\phi}}\right) \\
&\underset{(a),(b)}{=} \tfrac{3}{4}Mhv(\dot{\theta}\cos\theta\sin\phi + \dot{\phi}\sin\theta\cos\phi) \\
&\quad - \frac{d}{dt}[\tfrac{3}{4}Mh\sin\theta(v\sin\phi + \tfrac{3}{4}h\dot{\phi}\sin\theta) \\
&\quad + I_1(\dot{\psi} + \dot{\phi}\cos\theta)\cos\theta + I_2\dot{\phi}\sin^2\theta] \\
&= -\tfrac{3}{4}Mh\dot{v}\sin\theta\sin\phi \\
&\quad - \frac{d}{dt}[(\tfrac{9}{16}Mh^2 + I_2)\dot{\phi}\sin^2\theta + I_1(\dot{\psi} + \dot{\phi}\cos\theta)\cos\theta]
\end{aligned}$$

or

$$\begin{aligned}
F^* \underset{(c)}{=} -\tfrac{3}{4}M\Big\{&h\dot{v}\sin\theta\sin\phi \\
&+ \frac{1}{10}\frac{d}{dt}[\tfrac{1}{2}(4r^2 + 15h^2)\dot{\phi}\sin^2\theta + 4r^2(\dot{\psi} + \dot{\phi}\cos^2\theta)]\Big\}
\end{aligned}$$

4.11 Rigid Body with One Point Fixed. If one point P of a rigid body B remains fixed in a reference frame R, the kinetic energy K of B in R can be expressed in the following alternative forms:

$$K = \tfrac{1}{2}\boldsymbol{\omega}\cdot\mathbf{I}\cdot\boldsymbol{\omega} \tag{4.18}$$

where $\boldsymbol{\omega}$ is the angular velocity of B in R, and $\mathbf{I}$ is the inertia dyadic of B for P (see Section 3.21);

$$K = \tfrac{1}{2}I\omega^2 \tag{4.19}$$

where I is the moment of inertia of B about the line that passes through P and is parallel to $\boldsymbol{\omega}$;

$$K = \tfrac{1}{2}(I_1\omega_1^2 + I_2\omega_2^2 + I_3\omega_3^3) \tag{4.20}$$

where I_1, I_2, I_3 are moments of inertia of B about mutually perpendicular principal axes of B for P, and ω_1, ω_2, ω_3 are the associated measure numbers of the angular velocity of B in R; and

$$K = \tfrac{1}{2}\sum_{j=1}^{3}\sum_{k=1}^{3} \omega_j I_{jk}\omega_k \tag{4.21}$$

where the subscripts refer to any three mutually perpendicular axes, and I_{jk} is a product of inertia of B for two such axes passing through P. To prove these statements, proceed as in Sections 4.9 and 4.10.

4.12 Homogeneous Kinetic Energy Functions, Inertia Coefficients. The kinetic energy K of a holonomic system S that possesses n generalized coordinates $q_1, \cdots, q_n$ in a reference frame R can be expressed in the form

$$K = K_0 + K_1 + K_2 \tag{4.22}$$

where K_k is homogeneous and of degree k in the variables $\dot{q}_1, \cdots, \dot{q}_n$. In particular,

$$K_2 = \tfrac{1}{2}\sum_{r=1}^{n}\sum_{s=1}^{n} m_{rs}\dot{q}_r\dot{q}_s \tag{4.23}$$

where m_{rs}, called an *inertia coefficient* of S in R, is a function of $q_1, \cdots, q_n$, and t. In terms of the masses and partial rates of change of position of the particles of S, m_{rs} is given by

$$m_{rs} = \sum_{i=1}^{N} m_i \mathbf{v}_{\dot{q}_r}{}^{P_i} \cdot \mathbf{v}_{\dot{q}_s}{}^{P_i} \tag{4.24}$$

which shows that

$$m_{rs} = m_{sr} \tag{4.25}$$

Furthermore, if B is a rigid body belonging to S, the contribution $(m_{rs})_B$ of B to m_{rs} is

$$(m_{rs})_B = m\mathbf{v}_{\dot{q}_r} \cdot \mathbf{v}_{\dot{q}_s} + \boldsymbol{\omega}_{\dot{q}_r} \cdot \mathbf{I} \cdot \boldsymbol{\omega}_{\dot{q}_s} \tag{4.26}$$

where m is the mass of B, $\mathbf{v}_{\dot{q}_r}$ is the partial rate of change with respect to q_r of the position of the mass center B^* of B, $\boldsymbol{\omega}_{\dot{q}_r}$ is the partial rate of change with respect to q_r of the orientation of B in R, and $\mathbf{I}$ is the inertia dyadic of B for B^* (see Section 3.21).

Proofs: The velocity $\mathbf{v}_i$ of a typical particle P_i of S can be expressed as

$$\mathbf{v}_i \underset{(2.16)}{=} \sum_{r=1}^{n} \mathbf{v}_{\dot{q}_r}{}^{P_i}\dot{q}_r + \mathbf{v}_t{}^{P_i}$$

Hence,

$$\mathbf{v}_i{}^2 = \sum_{r=1}^{n}\sum_{s=1}^{n} \mathbf{v}_{\dot{q}_r}{}^{P_i} \cdot \mathbf{v}_{\dot{q}_s}{}^{P_i}\dot{q}_r\dot{q}_s + 2\sum_{r=1}^{n} \mathbf{v}_{\dot{q}_r}{}^{P_i} \cdot \mathbf{v}_t{}^{P_i}\dot{q}_r + (\mathbf{v}_t{}^{P_i})^2 \tag{a}$$

and

$$K = \tfrac{1}{2}\sum_{i=1}^{N} m_i\mathbf{v}_i{}^2 \underset{(a)}{=} \tfrac{1}{2}\sum_{i=1}^{N}\sum_{r=1}^{n}\sum_{s=1}^{n} m_i\mathbf{v}_{\dot{q}_r}{}^{P_i} \cdot \mathbf{v}_{\dot{q}_s}{}^{P_i}\dot{q}_r\dot{q}_s + \sum_{i=1}^{N}\sum_{r=1}^{n} m_i\mathbf{v}_{\dot{q}_r}{}^{P_i} \cdot \mathbf{v}_t{}^{P_i}\dot{q}_r + \tfrac{1}{2}\sum_{i=1}^{N} m_i(\mathbf{v}_t{}^{P_i})^2 \tag{b}$$

or, if K_0, K_1, and K_2 are defined as

$$K_0 = \tfrac{1}{2}\sum_{i=1}^{N} m_i(\mathbf{v}_t{}^{P_i})^2 \tag{c}$$

$$K_1 = \sum_{i=1}^{N}\sum_{r=1}^{n} m_i\mathbf{v}_{\dot{q}_r}{}^{P_i} \cdot \mathbf{v}_t{}^{P_i}\dot{q}_r \tag{d}$$

and

$$K_2 = \tfrac{1}{2}\sum_{i=1}^{N}\sum_{r=1}^{n}\sum_{s=1}^{n} m_i\mathbf{v}_{\dot{q}_r}{}^{P_i} \cdot \mathbf{v}_{\dot{q}_s}{}^{P_i}\dot{q}_r\dot{q}_s \tag{e}$$

then

$$K = K_0 + K_1 + K_2$$

As the order in which the summations in (e) are performed is immaterial, K_2 can be expressed as

$$K_2 \underset{(e)}{=} \tfrac{1}{2}\sum_{r=1}^{n}\sum_{s=1}^{n}\sum_{i=1}^{N} m_i\mathbf{v}_{\dot{q}_r}{}^{P_i} \cdot \mathbf{v}_{\dot{q}_s}{}^{P_i}\,\dot{q}_r\dot{q}_s$$

from which Equation (4.23) follows if (4.24) is regarded as a definition; and Equation (4.26) is obtained from (4.24) by proceeding as in Section

4.9 after noting that

$$\mathbf{v}_{\dot{q}_r}{}^{P_i} \underset{(2.22)}{=} \mathbf{v}_{\dot{q}_r} + \boldsymbol{\omega}_{\dot{q}_r} \times \mathbf{r}_i$$

where $\mathbf{r}_i$ is the position vector of P_i relative to B^*.

▪ EXAMPLE

Figure 4.3 shows a double pendulum consisting of light strings of lengths

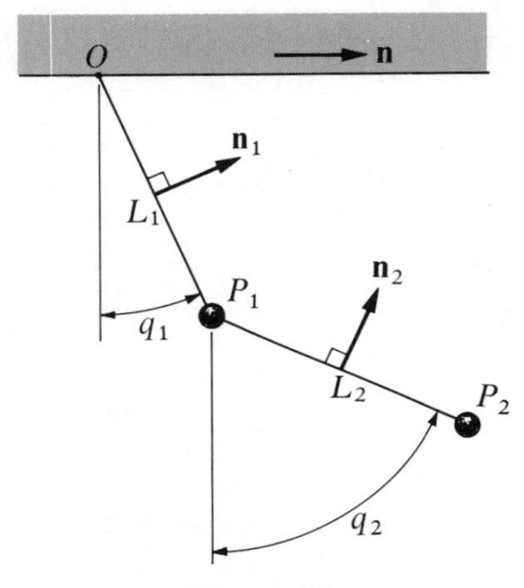

Figure 4.3

L_1 and L_2 attached to particles P_1 and P_2 of masses m_1 and m_2. If the point O is made to move with a speed v parallel to a unit vector $\mathbf{n}$, the velocities $\mathbf{v}_1$ and $\mathbf{v}_2$ of P_1 and P_2 are

$$\mathbf{v}_1 = v\mathbf{n} + L_1\dot{q}_1\mathbf{n}_1$$
$$\mathbf{v}_2 = v\mathbf{n} + L_1\dot{q}_1\mathbf{n}_1 + L_2\dot{q}_2\mathbf{n}_2$$

where $\mathbf{n}_1$ and $\mathbf{n}_2$ are unit vectors and q_1 and q_2 measure angles, as indicated in Figure 4.3. The kinetic energy K of the system is thus given by

$$\begin{aligned} K &= \tfrac{1}{2}m_1\mathbf{v}_1{}^2 + \tfrac{1}{2}m_2\mathbf{v}_2{}^2 \\ &= \tfrac{1}{2}m_1(v^2 + 2vL_1\dot{q}_1 \cos q_1 + L_1{}^2\dot{q}_1{}^2) \\ &\quad + \tfrac{1}{2}m_2[v^2 + 2v(L_1\dot{q}_1 \cos q_1 + L_2\dot{q}_2 \cos q_2) \\ &\qquad + 2L_1L_2\dot{q}_1\dot{q}_2 \cos (q_2 - q_1) + L_1{}^2\dot{q}_1{}^2 + L_2{}^2\dot{q}_2{}^2] \end{aligned}$$

If K_0, K_1, and K_2 are now defined as

$$\begin{aligned} K_0 &= \tfrac{1}{2}(m_1 + m_2)v^2 \\ K_1 &= (m_1 + m_2)L_1 \cos q_1 v\dot{q}_1 + m_2L_2 \cos q_2 v\dot{q}_2 \\ K_2 &= \tfrac{1}{2}[(m_1 + m_2)L_1{}^2\dot{q}_1{}^2 + 2m_2L_1L_2 \cos (q_2 - q_1)\dot{q}_1\dot{q}_2 + m_2L_2{}^2\dot{q}_2{}^2] \end{aligned}$$

then K_0, K_1, and K_2 are homogeneous of degree zero, one, and two, respectively, in the variables $\dot{q}_1$ and $\dot{q}_2$; and K_2 can be expressed as

$$K_2 = \tfrac{1}{2}(m_{11}\dot{q}_1^2 + m_{12}\dot{q}_1\dot{q}_2 + m_{21}\dot{q}_2\dot{q}_1 + m_{22}\dot{q}_2^2)$$

where

$$m_{11} = (m_1 + m_2)L_1^2 \quad m_{12} = m_2L_1L_2\cos(q_2 - q_1)$$
$$m_{21} = m_{12} \quad m_{22} = m_2L_2^2$$

The inertia coefficients can also be found by using Equation (4.24). For example,

$$\begin{aligned} m_{12} &= m_1\mathbf{v}_{\dot{q}_1}^{P_1} \cdot \mathbf{v}_{\dot{q}_2}^{P_1} + m_2\mathbf{v}_{\dot{q}_1}^{P_2} \cdot \mathbf{v}_{\dot{q}_2}^{P_2} \\ &= m_1(L_1\mathbf{n}_1) \cdot 0 + m_2(L_1\mathbf{n}_1) \cdot (L_2\mathbf{n}_2) \\ &= m_2L_1L_2\cos(q_2 - q_1) \end{aligned}$$

4.13 Quadratic Kinetic Energy Function. The following observation will be found useful in the sequel: When the kinetic energy K of a holonomic system that possesses n generalized coordinates $q_1, \cdots, q_n$ in a reference frame R is homogeneous and of second degree in the variables $\dot{q}_1, \cdots, \dot{q}_n$, then

$$\sum_{r=1}^{n} \frac{\partial K}{\partial \dot{q}_r}\dot{q}_r = 2K \tag{4.27}$$

Proof: Use Euler's theorem for homogeneous functions[1] or note that (see Section 4.12)

$$K = \tfrac{1}{2}\sum_{r=1}^{n}\sum_{s=1}^{n} m_{rs}\dot{q}_r\dot{q}_s \tag{a}$$

so that

$$\begin{aligned} \frac{\partial K}{\partial \dot{q}_r} \underset{(a)}{=} & \tfrac{1}{2}\frac{\partial}{\partial \dot{q}_r}\sum_{\alpha=1}^{n}\sum_{\beta=1}^{n} m_{\alpha\beta}\dot{q}_\alpha\dot{q}_\beta \\ &= \tfrac{1}{2}\sum_{\alpha=1}^{n}\sum_{\beta=1}^{n} m_{\alpha\beta}\left(\frac{\partial \dot{q}_\alpha}{\partial \dot{q}_r}\dot{q}_\beta + \dot{q}_\alpha\frac{\partial \dot{q}_\beta}{\partial \dot{q}_r}\right) \\ &= \tfrac{1}{2}\left(\sum_{\beta=1}^{n} m_{r\beta}\dot{q}_\beta + \sum_{\alpha=1}^{n} m_{\alpha r}\dot{q}_\alpha\right) \\ &= \tfrac{1}{2}\left(\sum_{s=1}^{n} m_{rs}\dot{q}_s + \sum_{s=1}^{n} m_{sr}\dot{q}_s\right) \\ &\underset{(4.25)}{=} \sum_{s=1}^{n} m_{rs}\dot{q}_s \end{aligned}$$

[1] If $f(\lambda x_1, \cdots, \lambda x_n) = \lambda^m f(x_1, \cdots, x_n)$ for all λ, then $\sum_{r=1}^{n}\frac{\partial f}{\partial x_r}x_r = mf$.

from which it follows that

$$\sum_{r=1}^{n} \frac{\partial K}{\partial \dot{q}_r} \dot{q}_r = \sum_{r=1}^{n} \sum_{s=1}^{n} m_{rs} \dot{q}_s \dot{q}_r \underset{\text{(a)}}{=} 2K$$

▪ EXAMPLE

If v in the example in Section 4.12 is taken equal to zero, the kinetic energy of the system becomes

$$K = \tfrac{1}{2} m_1 L_1^2 \dot{q}_1^2 + \tfrac{1}{2} m_2 [L_1^2 \dot{q}_1^2 + L_2^2 \dot{q}_2^2 + 2L_1 L_2 \dot{q}_1 \dot{q}_2 \cos (q_2 - q_1)]$$

Hence,

$$\frac{\partial K}{\partial \dot{q}_1} = (m_1 + m_2) L_1^2 \dot{q}_1 + m_2 L_1 L_2 \dot{q}_2 \cos (q_2 - q_1)$$

$$\frac{\partial K}{\partial \dot{q}_2} = m_2 L_2^2 \dot{q}_2 + m_2 L_1 L_2 \dot{q}_1 \cos (q_2 - q_1)$$

and

$$\begin{aligned} \frac{\partial K}{\partial \dot{q}_1} \dot{q}_1 + \frac{\partial K}{\partial \dot{q}_2} \dot{q}_2 &= (m_1 + m_2) L_1^2 \dot{q}_1^2 + m_2 L_2^2 \dot{q}_2^2 \\ &\quad + 2 m_2 L_1 L_2 \dot{q}_1 \dot{q}_2 \cos (q_2 - q_1) \\ &= 2K \end{aligned}$$

4.14 Virtual Work. If $P_1, \cdots, P_N$ are the particles of a simple nonholonomic system S possessing $n - m$ degrees of freedom in a reference frame R (see Section 2.22), $\delta \mathbf{p}_1, \cdots, \delta \mathbf{p}_N$ are mutually compatible virtual displacements of $P_1, \cdots, P_N$ in R (see Section 2.28), and $\mathbf{F}_i$ is the resultant of all contact and body forces acting on P_i, the quantity δW defined as

$$\delta W = \sum_{i=1}^{N} \mathbf{F}_i \cdot \delta \mathbf{p}_i \tag{4.28}$$

is called the *virtual work of the active forces* for the virtual displacements under consideration. Similarly, δW^*, the *virtual work of the inertia forces* for these virtual displacements, is defined as

$$\delta W^* = \sum_{i=1}^{N} \mathbf{F}_i^* \cdot \delta \mathbf{p}_i \tag{4.29}$$

where $\mathbf{F}_i^*$ is the inertia force for P_i in R (see Section 3.9).

■ EXAMPLE

Figure 4.4 shows once again the system described in the example in

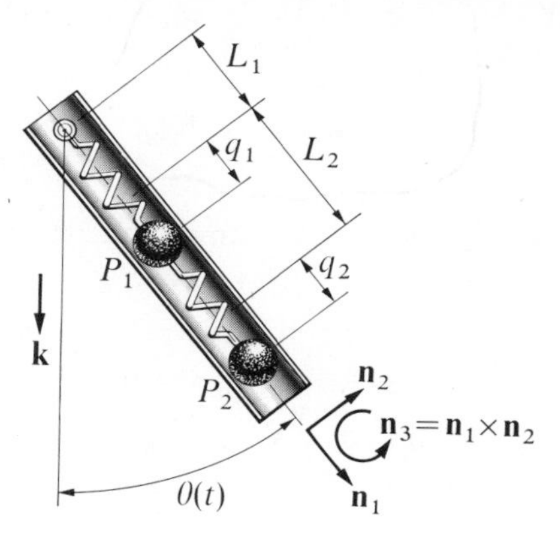

Figure 4.4

Section 3.2. The resultants $\mathbf{F}_1$ and $\mathbf{F}_2$ of all contact and body forces acting on P_1 and P_2, respectively, were there found to be

$$\mathbf{F}_1 = [-k_1q_1 + k_2(q_2 - q_1)]\mathbf{n}_1 + R_{12}\mathbf{n}_2 + R_{13}\mathbf{n}_3 + m_1g\mathbf{k}$$

and

$$\mathbf{F}_2 = -k_2(q_2 - q_1)\mathbf{n}_1 + R_{22}\mathbf{n}_2 + R_{23}\mathbf{n}_3 + m_2g\mathbf{k}$$

The most general mutually compatible virtual displacements $\delta\mathbf{p}_1$ and $\delta\mathbf{p}_2$ of P_1 and P_2 are (see Section 2.28)

$$\delta\mathbf{p}_1 = \delta q_1\mathbf{n}_1 \qquad \delta\mathbf{p}_2 = \delta q_2\mathbf{n}_1$$

Consequently, the most general expression for the virtual work δW of the active forces is

$$\delta W \underset{(4.28)}{=} [-k_1q_1 + k_2(q_2 - q_1) + m_1g\cos\theta]\,\delta q_1 + [-k_2(q_2 - q_1) + m_2g\cos\theta]\,\delta q_2$$

and the corresponding virtual work δW^* of the inertia forces is

$$\delta W^* \underset{(4.29)}{=} -m_1\mathbf{a}_1\cdot\mathbf{n}_1\,\delta q_1 - m_2\mathbf{a}_2\cdot\mathbf{n}_2\,\delta q_2$$

where $\mathbf{a}_1$ and $\mathbf{a}_2$ are the accelerations given in the example in Section 3.9. Thus

$$\delta W^* = -m_1[\ddot{q}_1 - (L_1 + q_1)\dot{\theta}^2]\,\delta q_1 - m_2[\ddot{q}_2 - (L_1 + L_2 + q_2)\dot{\theta}^2]\,\delta q_2$$

4.15 Virtual Work and Generalized Forces. Virtual work (see Section 4.14) and generalized forces (see Sections 3.2 and 3.9) are related to each other as follows:

$$\delta W = \sum_{r=1}^{n-m} F_r \, \delta q_r \tag{4.30}$$

$$\delta W^* = \sum_{r=1}^{n-m} F_r^* \, \delta q_r \tag{4.31}$$

where δq_r is any quantity having the same dimensions as the generalized coordinate q_r.

Proof: By definition,

$$\delta \mathbf{p}_i \underset{(2.38)}{=} \sum_{r=1}^{n-m} \tilde{\mathbf{v}}_{\dot{q}_r}^{P_i} \, \delta q_r$$

Hence,

$$\begin{aligned} \delta W &\underset{(4.28)}{=} \sum_{i=1}^{N} \mathbf{F}_i \cdot \sum_{r=1}^{n-m} \tilde{\mathbf{v}}_{\dot{q}_r}^{P_i} \, \delta q_r \\ &= \sum_{r=1}^{n-m} \sum_{i=1}^{N} \tilde{\mathbf{v}}_{\dot{q}_r}^{P_i} \cdot \mathbf{F}_i \, \delta q_r \\ &\underset{(3.3)}{=} \sum_{r=1}^{n-m} F_r \, \delta q_r \end{aligned}$$

and

$$\begin{aligned} \delta W^* &\underset{(4.29)}{=} \sum_{i=1}^{N} \mathbf{F}_i^* \cdot \sum_{r=1}^{n-m} \tilde{\mathbf{v}}_{\dot{q}_r}^{P_i} \, \delta q_r \\ &= \sum_{r=1}^{n-m} \sum_{i=1}^{N} \tilde{\mathbf{v}}_{\dot{q}_r}^{P_i} \cdot \mathbf{F}_i^* \, \delta q_r \\ &\underset{(3.8)}{=} \sum_{r=1}^{n-m} F_r^* \, \delta q_r \end{aligned}$$

▪ EXAMPLE

The expression for δW obtained in the example in Section 4.14 can be seen to be equivalent to $F_1 \, \delta q_1 + F_2 \, \delta q_2$ if F_1 and F_2 are the generalized forces found in the example in Section 3.2; and, with F_1^* and F_2^* as found in the example in Section 3.9, $F_1^* \, \delta q_1 + F_2^* \, \delta q_2$ is equal to δW^* as given in the example in Section 4.14.

4.16 Virtual Work Theorems. It follows from Equations (4.30) and (4.31) that a number of statements made previously in connection with generalized forces apply in slightly modified form to virtual work. For example (see Section 3.3), if $\mathbf{F}$ is a contact force exerted on a particle P of S by a smooth rigid body B whose motion in R is prescribed as a function of time, then $\mathbf{F}$ contributes nothing to δW. Or (see Section 3.4) if B is a rigid body belonging to S, the particles comprising B may exert contact and body forces on each other; the total contribution of these forces to δW is equal to zero; and if $\mathbf{F}_1, \cdots, \mathbf{F}_{N'}$ are N' forces acting respectively on particles $P_1, \cdots, P_{N'}$ of B, and this set of forces is equivalent (see Section 3.1) to a couple with torque $\mathbf{T}$ together with a force $\mathbf{F}$ applied at a point Q of B, then the contribution to δW of the forces $\mathbf{F}_1, \cdots, \mathbf{F}_{N'}$ is given by

$$(\delta W)_B = \mathbf{F} \cdot \delta\mathbf{q} + \mathbf{T} \cdot \delta\boldsymbol{\alpha} \tag{4.32}$$

where $\delta\mathbf{q}$ and $\delta\boldsymbol{\alpha}$ are a virtual displacement of Q and a virtual rotation of B, respectively. As a final example (see Section 3.10), the contribution to δW^* of all inertia forces for the particles of B is

$$(\delta W^*)_B = \mathbf{F}^* \cdot \delta\mathbf{p} + \mathbf{T}^* \cdot \delta\boldsymbol{\alpha} \tag{4.33}$$

where $\mathbf{F}^*$ and $\mathbf{T}^*$ are the inertia force and the inertia torque for B in R (see Section 3.10), and $\delta\mathbf{p}$ is a virtual displacement of the mass center of B in R. Similarly, Sections 3.5 through 3.8 have counterparts applicable to virtual work.

▪ EXAMPLE

In Figure 4.5, P, Q, R, and S designate the centers of four identical

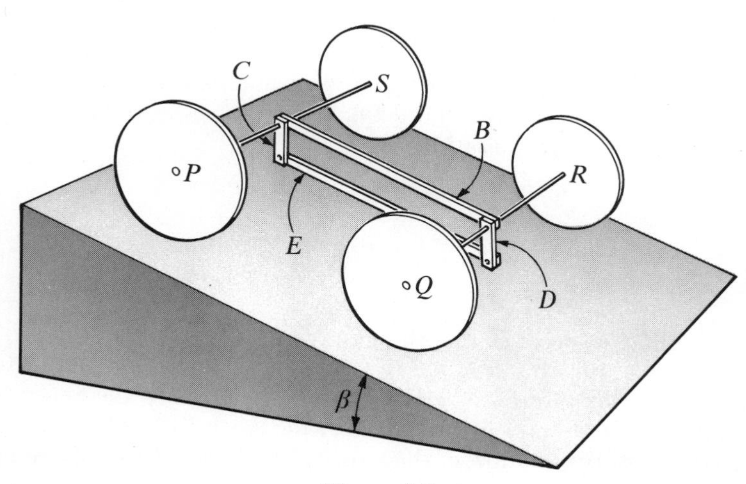

Figure 4.5

wheels, each of radius r, mass M, and moment of inertia I about its axis of symmetry. The axles of the wheels (which are comparatively light) are connected to each other with a bar B of length $4r$ and mass M', and a pin-connected linkage consisting of bars C, D, and E is attached as shown. Link C has a length L and mass m, and links D and E are identical with C and B, respectively.

Assuming that this system can move in such a way that the wheels roll upward or downward on a plane inclined to the horizontal at an angle β while the angle θ (see Figure 4.6) remains constant, the total virtual

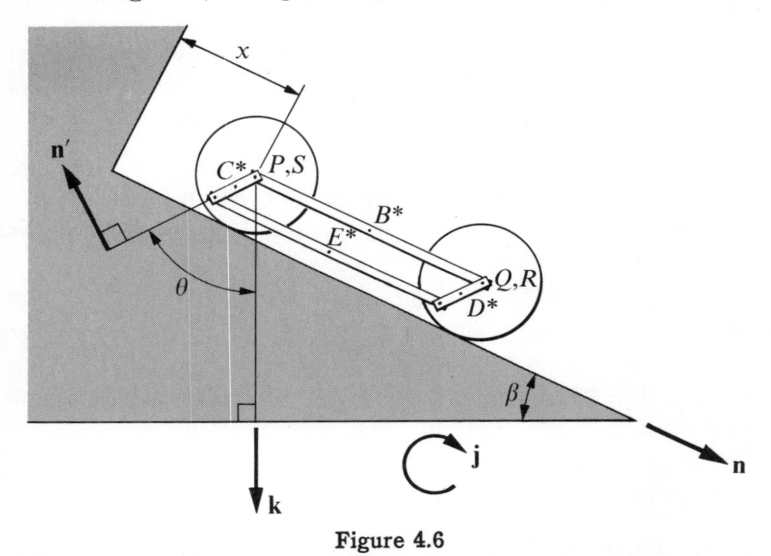

Figure 4.6

work of all active and inertia forces is to be determined for this motion, for the most general virtual displacement of the system.

Expressed in terms of the generalized coordinates x and θ and the unit vectors $\mathbf{n}$, $\mathbf{n}'$, and $\mathbf{j}$ shown in Figure 4.6, the velocities of the mass centers of the bodies under consideration are

$$\mathbf{v}^P = \mathbf{v}^Q = \mathbf{v}^R = \mathbf{v}^S = \mathbf{v}^{B^*} = \dot{x}\mathbf{n}$$

$$\mathbf{v}^{C^*} = \mathbf{v}^{D^*} = \dot{x}\mathbf{n} + \frac{L}{2}\,\dot{\theta}\mathbf{n}'$$

$$\mathbf{v}^{E^*} = \dot{x}\mathbf{n} + L\dot{\theta}\mathbf{n}'$$

and the angular velocity $\boldsymbol{\omega}$ of each of the four wheels is

$$\boldsymbol{\omega} = \frac{\dot{x}}{r}\,\mathbf{j}$$

With self-explanatory notation, the most general set of mutually compatible virtual displacements of the mass centers is thus given by

$$\left.\begin{aligned}\delta\mathbf{p} &= \delta\mathbf{q} = \delta\mathbf{r} = \delta\mathbf{s} = \delta\mathbf{b} = \delta x\mathbf{n}\\ \delta\mathbf{c} &= \delta\mathbf{d} = \delta x\mathbf{n} + \frac{L}{2}\,\delta\theta\mathbf{n}'\\ \delta\mathbf{e} &= \delta x\mathbf{n} + L\,\delta\theta\mathbf{n}'\end{aligned}\right\} \tag{a}$$

and the associated virtual rotation $\delta\boldsymbol{\alpha}$ of each of the wheels is

$$\delta\boldsymbol{\alpha} = \frac{\delta x}{r}\mathbf{j} \tag{b}$$

The only forces contributing to the virtual work of the active forces are the gravitational forces acting on the various bodies. Hence

$$\begin{aligned}\delta W &\underset{(4.32)}{=} Mg\mathbf{k}\cdot\delta\mathbf{p} + Mg\mathbf{k}\cdot\delta\mathbf{q} + Mg\mathbf{k}\cdot\delta\mathbf{r} + Mg\mathbf{k}\cdot\delta\mathbf{s}\\ &\qquad + M'g\mathbf{k}\cdot\delta\mathbf{b} + mg\mathbf{k}\cdot\delta\mathbf{c} + mg\mathbf{k}\cdot\delta\mathbf{d} + M'g\mathbf{k}\cdot\delta\mathbf{e}\\ &\underset{(a)}{=} g[(4M + 2M' + 2m)\,\delta x\mathbf{k}\cdot\mathbf{n} + (m + M')L\,\delta\theta\mathbf{k}\cdot\mathbf{n}']\\ &= g[2(2M + M' + m)\,\delta x\sin\beta - (m + M')L\,\delta\theta\sin\theta]\end{aligned}$$

For the motion under consideration, the acceleration of the mass center of each of the bodies is $\ddot{x}\mathbf{n}$; the angular acceleration of each of the wheels is $(\ddot{x}/r)\mathbf{j}$; and both the angular velocity and the angular acceleration of each of the bars is equal to zero. The total virtual work of the inertia forces is thus

$$\begin{aligned}\delta W^* &\underset{(4.33)}{=} -M\ddot{x}\mathbf{n}\cdot\delta\mathbf{p} - M\ddot{x}\mathbf{n}\cdot\delta\mathbf{q} - M\ddot{x}\mathbf{n}\cdot\delta\mathbf{r} - M\ddot{x}\mathbf{n}\cdot\delta\mathbf{s}\\ &\qquad -M'\ddot{x}\mathbf{n}\cdot\delta\mathbf{b} - m\ddot{x}\mathbf{n}\cdot\delta\mathbf{c} - m\ddot{x}\mathbf{n}\cdot\delta\mathbf{d} - M'\ddot{x}\mathbf{n}\cdot\delta\mathbf{e}\\ &\qquad -I\frac{\ddot{x}}{r}\mathbf{j}\cdot\delta\boldsymbol{\alpha} - I\frac{\ddot{x}}{r}\mathbf{j}\cdot\delta\boldsymbol{\alpha} - I\frac{\ddot{x}}{r}\mathbf{j}\cdot\delta\boldsymbol{\alpha} - I\frac{\ddot{x}}{r}\mathbf{j}\cdot\delta\boldsymbol{\alpha}\\ &\underset{(a),(b)}{=} -\ddot{x}\Big[(4M + 2M' + 2m)\,\delta x\\ &\qquad + (m + M')L\,\delta\theta\mathbf{n}\cdot\mathbf{n}' + 4\frac{I}{r^2}\,\delta x\Big]\\ &= -\ddot{x}\left[2\left(2M + M' + m + 2\frac{I}{r^2}\right)\delta x - (m + M')L\,\delta\theta\cos(\beta - \theta)\right]\end{aligned}$$

and the total virtual work is given by

$$\begin{aligned}\delta W + \delta W^* = 2\Big[g(2M + M' + m)\sin\beta - \ddot{x}\Big(2M + M'\\ + m + 2\frac{I}{r^2}\Big)\Big]\,\delta x + (m + M')[\ddot{x}\cos(\beta - \theta) - g\sin\theta]\,L\,\delta\theta\end{aligned}$$

Problems

▪ PROBLEM SET 4

(Section 3.1)

4(a) The lines L_1, L_2, and L_3 in Figure 4(a) are mutually perpendicular.

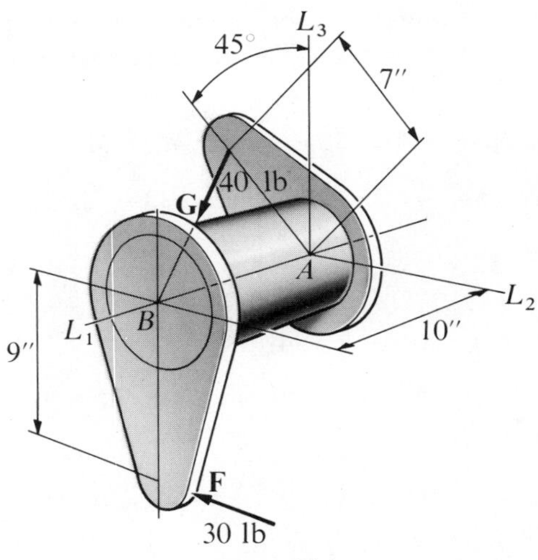

Figure 4(a)

Determine the magnitudes of the moments of the forces **F** and **G** about point A, and find the cosine of the angle between each of these moment vectors and line L_3.

Results: 403.5 in.-lb, 229.3 in.-lb, -0.743, 0.707.

4(b) A force **F** is applied at a point whose rectangular Cartesian coordinates are x, y, z. In terms of unit vectors $\mathbf{n}_x, \mathbf{n}_y, \mathbf{n}_z$ respectively parallel to the coordinate axes, **F** can be expressed as

$$\mathbf{F} = F_x\mathbf{n}_x + F_y\mathbf{n}_y + F_z\mathbf{n}_z$$

and the moment **M** of **F** about the origin of the coordinate system is given by

$$\mathbf{M} = M_x\mathbf{n}_x + M_y\mathbf{n}_y + M_z\mathbf{n}_z$$

Determine M_y, assuming (a) that the coordinate system is right-handed and (b) that it is left-handed.

Result: $zF_x - xF_z$, $xF_z - zF_x$.

4(c) Referring to Figure 4(a), and letting **H** be a force such that the set of three forces **F**, **G**, **H** is a couple and the line of action of **H** passes through point B, determine the magnitude of the torque **T** of this couple, and find the angle between **T** and **H**.

Results: 270 in.-lb, 33 degrees.

4(d) Show that the moments of a set S of bound vectors about all points of any line that is parallel to the resultant of S are equal to each other.

4(e) If the resultant **R** of a set S of bound vectors is not equal to zero, the points about which S has a minimum moment $\mathbf{M}^*$ lie on a line L^* that is parallel to **R**. L^* is called the *central axis* of S.

Letting **M** be the moment of S about an arbitrarily selected reference point O, show that $\mathbf{M}^*$ is given by

$$\mathbf{M}^* = \frac{\mathbf{R} \cdot \mathbf{M}}{\mathbf{R}^2} \mathbf{R}$$

and that L^* passes through the point whose position vector $\mathbf{p}^*$ relative to O is given by

$$\mathbf{p}^* = \frac{\mathbf{R} \times \mathbf{M}}{\mathbf{R}^2}$$

SUGGESTION: Make use of the fact that, when **M** is resolved into two components, one parallel to R, the other perpendicular to R, the magnitude of the first of these is the same for all points O.

4(f) When a set of bound vectors consists of a couple of torque **T** and a single vector parallel to **T**, it is called a *wrench*.

Show that when the resultant of a set S of bound vectors is not equal to zero, S can be replaced with a wrench consisting of a couple whose torque is equal to the minimum moment of S [see Problem 4(e)], together with the resultant of S applied at any point on the central axis of S.

4(g) The set of two forces **F** and **G** shown in Figure 4(a) is to be replaced with a single force and a couple, and the torque of the couple is to be as small as possible.

Determine the magnitude of the torque of the couple, and find the shortest distance from point A to the line of action of the force.

Results: 226 in.-lb, 6.57 in.

4(h) Show that a set of vectors whose lines of action intersect at a point O can be replaced with the resultant of the vectors, applied at O (Varignon's theorem).

4(i) Letting S be a set of bound vectors whose lines of action are coplanar and whose resultant is not equal to zero, show that S can be replaced with a single vector, this vector being the resultant of S applied at any point on the central axis of S [see Problem 4(e)].

4(j) The vectors $\mathbf{v}_1, \cdots, \mathbf{v}_n$ of a set S of bound vectors are all parallel to a unit vector $\mathbf{n}$. Show that (a) S can be replaced with its resultant, applied at any point on the central axis of S [see Problem 4(e)], and (b) the central axis of S passes through the mass center of a set of (fictitious) particles $P_1, \cdots, P_n$ of masses $m_1, \cdots, m_n$, particle P_i being situated at any point on the line of action of $\mathbf{v}_i$ and mass m_i being chosen such that

$$\mathbf{v}_i = km_i\mathbf{n}$$

where k is a constant of proportionality.

4(k) S' is a set of two forces, $\mathbf{F}'$ and $\mathbf{G}'$, and the line of action of $\mathbf{F}'$ is the line L_1 shown in Figure 4(a). Assuming that S' is equivalent to the set S of two forces **F** and **G** shown in Figure 4(a), determine the magnitudes of $\mathbf{F}'$ and $\mathbf{G}'$, and find the distance from B to the line of action of $\mathbf{G}'$.

Results: 32.8 lb, 21.3 lb, 12.7 in.

4(l) In Figure 4(l), Q designates a particle of mass m, and B represents a uniform, thin rod of length $2L$ and mass M. Both $\mathbf{n}_1$ and $\mathbf{n}_2$ are unit vectors, and R denotes the distance between Q and the mass center B^* of B.

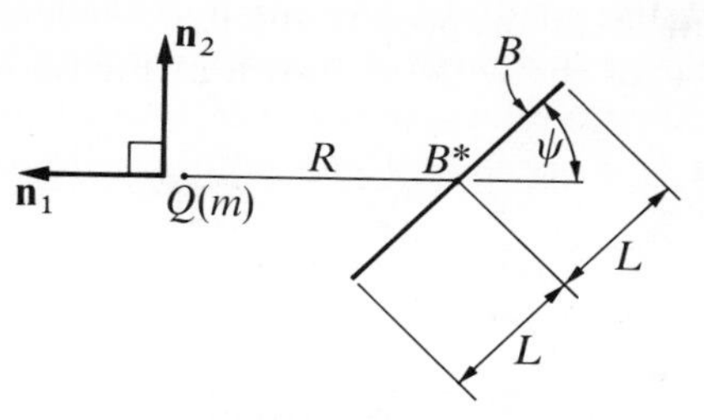

Figure 4(1)

Replacing the set of gravitational forces exerted by Q on B with a couple of torque **T** and a force **F** applied at B^*, (a) determine **T** and **F**, and (b) verify that **T** and **F** approach

$$\frac{GmML^2}{2R^3} \sin 2\psi \, \mathbf{n}_1 \times \mathbf{n}_2$$

and

$$\frac{GmM}{R^2} \mathbf{n}_1$$

respectively, as L/R approaches zero.

Results:

$$\mathbf{T} = \frac{GmM}{2L} \left\{ \left[1 + \frac{L}{R}\cos\psi\right] \left[1 + \frac{2L}{R}\cos\psi + \left(\frac{L}{R}\right)^2\right]^{-1/2} - \left[1 - \frac{L}{R}\cos\psi\right] \left[1 - \frac{2L}{R}\cos\psi + \left(\frac{L}{R}\right)^2\right]^{-1/2} \right\} \csc\psi \, \mathbf{n}_1 \times \mathbf{n}_2$$

$$\mathbf{F} = \frac{GmM}{2LR} \left[\left\{ \left[1 + \frac{2L}{R}\cos\psi + \left(\frac{L}{R}\right)^2\right]^{-1/2} + \left[1 - \frac{2L}{R}\cos\psi + \left(\frac{L}{R}\right)^2\right]^{-1/2} \right\} \frac{L\mathbf{n}_1}{R} + \left\{ \left[1 + \frac{L}{R}\cos\psi\right] \left[1 + \frac{2L}{R}\cos\psi + \left(\frac{L}{R}\right)^2\right]^{-1/2} - \left[1 - \frac{L}{R}\cos\psi\right] \left[1 - \frac{2L}{R}\cos\psi + \left(\frac{L}{R}\right)^2\right]^{-1/2} \right\} \csc\psi \, \mathbf{n}_2 \right]$$

▪ PROBLEM SET 5

(Sections 3.2–3.8)

5(a) The ends A and B of a uniform rod of weight W and length $2L$ are constrained to remain on a smooth circular wire that is made to rotate with constant angular speed Ω about a fixed vertical axis passing through the center C of the wire, as shown in Figure 5(a).

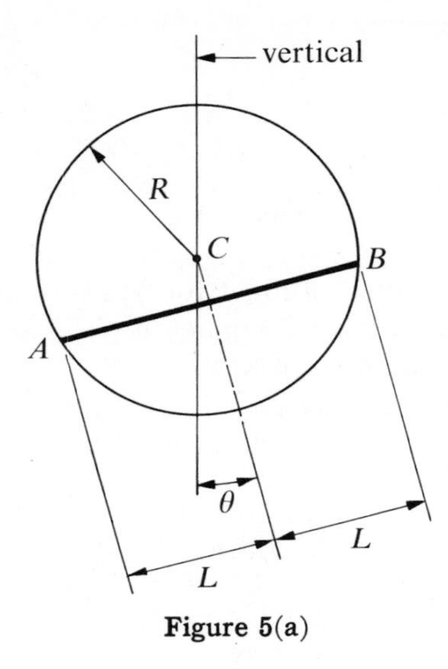

Figure 5(a)

Using the angle θ as a generalized coordinate, determine the associated generalized active force.

Result: $-W(R^2 - L^2)^{1/2} \sin \theta$.

5(b) Figure 5(b) shows 33 pin-connected uniform rods, each of weight

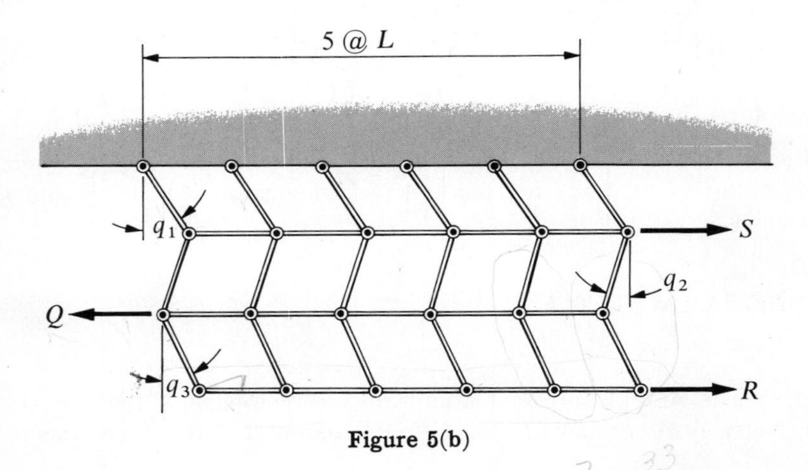

Figure 5(b)

W and length L, suspended from a horizontal support. Contact forces of magnitude Q, R, and S are applied to this system, as shown in the sketch.

Letting the angles q_1, q_2, and q_3 be generalized coordinates of this system, determine the generalized active forces F_1, F_2, and F_3.

Results:

$$F_1 = (-Q + R + S)L \cos q_1 - 30\, WL \sin q_1$$
$$F_2 = (Q - R)L \cos q_2 - 19\, WL \sin q_2$$
$$F_3 = RL \cos q_3 - 8WL \sin q_3$$

5(c) A system comprised of two particles of masses m_1 and m_2 moves in such a way that the distance r between the particles varies with time. Letting r be one of the generalized coordinates of the system, determine the contribution to the associated generalized active force of the gravitational forces exerted by the particles on each other.

Result: $-Gm_1m_2/r^2$.

5(d) A rigid block B of width $2w$ and weight W is supported by two elastic beams, each of length L and flexural rigidity EI. The beams are "built in" both at their supports and at P_1 and P_2 [see Figure 5(d)].

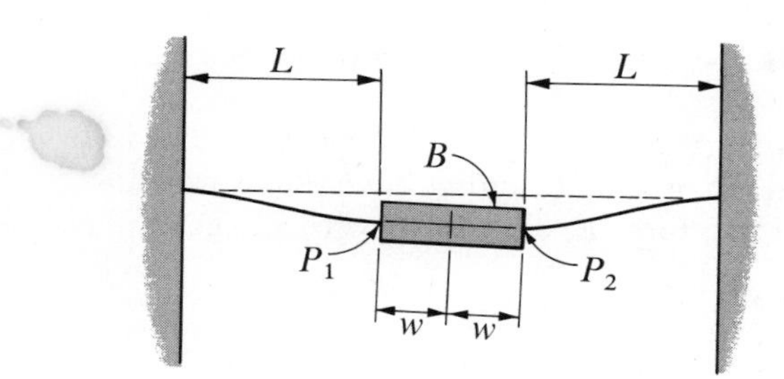

Figure 5(d)

Using the deflections q_1 and q_2 of the beams at P_1 and P_2 as generalized coordinates of B, and treating the beams as massless, determine the generalized active forces F_1 and F_2.

Results:

$$F_1 = \frac{12EI}{L^3}\left[-\left(1 + \frac{L}{2w} + \frac{1}{6}\frac{L^2}{w^2}\right)q_1 + \frac{1}{2}\frac{L}{w}\left(1 + \frac{1}{3}\frac{L}{w}\right)q_2\right] + \frac{W}{2}$$
$$F_2 = \frac{12EI}{L^3}\left[\frac{1}{2}\frac{L}{w}\left(1 + \frac{1}{3}\frac{L}{w}\right)q_1 - \left(1 + \frac{L}{2w} + \frac{1}{6}\frac{L^2}{w^2}\right)q_2\right] + \frac{W}{2}$$

5(e) Figure 5(e) shows two sharp-edged circular disks, D_1 and D_2, each

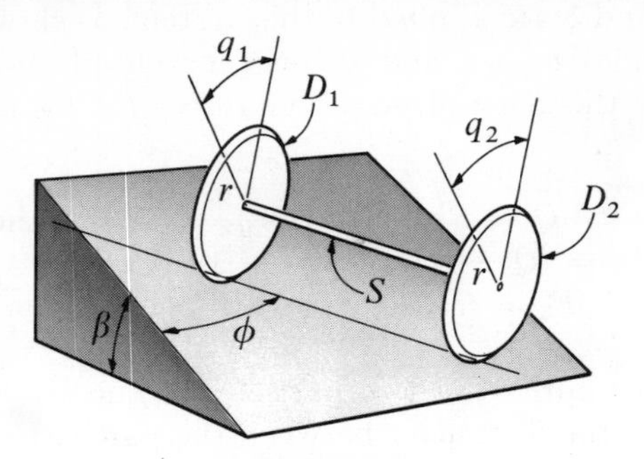

Figure 5(e)

of radius r and weight w, mounted at the extremities of a shaft S of length $2L$ and weight $2W$, the axis of S coinciding with those of D_1 and D_2. The disks are supported by a plane that is inclined at an angle β to the horizontal, and no slip is permitted to occur at their points of contact with this plane. Furthermore, each disk is free to rotate on S.

Letting q_1 and q_2 measure angles between a line normal to the supporting surface and lines fixed in the disks, as shown in Figure 5(e), determine the generalized active forces F_1 and F_2.

Result: $F_1 = F_2 = (w + W)r \sin \beta \sin \phi.$

5(f) A couple of torque $T\mathbf{i}_1 \times \mathbf{j}_1$ is applied to the rod of length $3L$ of Problem 3(j). Assuming that P and Q have equal weights W, determine the generalized active force F_1 associated with the generalized coordinate θ_1.

Result:

$$\frac{(T/3) \sin (\theta_2 - \theta_1) + WL[\sin \theta_1 \sin (\theta_2 - \theta_3) - \sin \theta_2 \sin (\theta_3 - \theta_1)]}{\sin (\theta_3 - \theta_2)}$$

5(g) Figure 5(g) is a schematic representation of a "reduction gear" consisting of a fixed bevel gear A, moving bevel gears B, C, C', D, D', and E, and an "arm" F. C and D are rigidly connected to each other, as are C' and D'. The number of teeth of each gear is listed in Figure 5(g).

Couples of torques $T_b\mathbf{n}$ and $T_e\mathbf{n}$ are applied to the shafts carrying B and E, respectively, $\mathbf{n}$ being a unit vector directed as shown in Figure 5(g). Letting q_1 be the angle of rotation of gear B, and neglecting all gravitational effects, determine the generalized active force F_1.

Result: $T_b + 244T_e.$

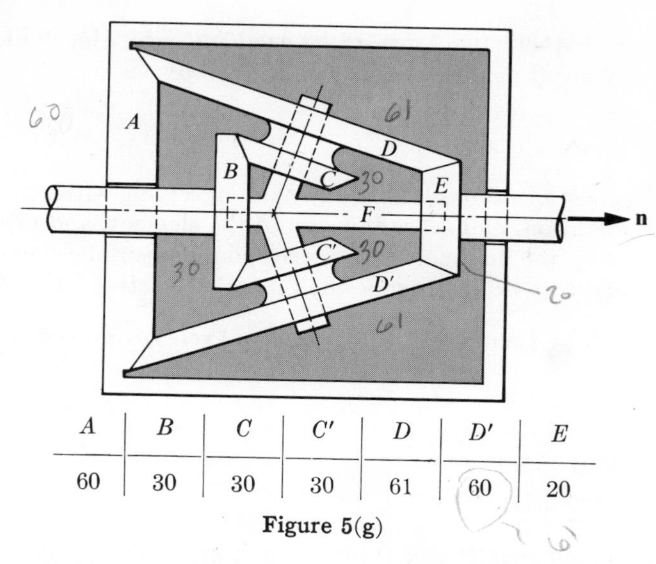

A	B	C	C'	D	D'	E
60	30	30	30	61	60	20

Figure 5(g)

5(h) A uniform sphere S of radius R is placed in a spherical cavity in a rigid body B, as indicated in Figure 5(h). The radius of the cavity is

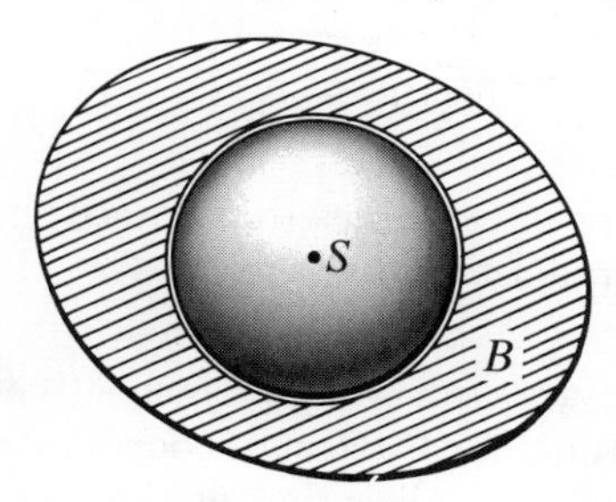

Figure 5(h)

only slightly larger than that of the sphere; the space between S and B is filled with a viscous fluid; and the center of S coincides with the mass center of B.

This system has nine degrees of freedom. Accordingly, nine quantities of the kind discussed in Section 2.24 can be introduced by expressing the angular velocities of B and S and the velocity of the center of S as

$$\boldsymbol{\omega}^B = u_1\mathbf{n}_1 + u_2\mathbf{n}_2 + u_3\mathbf{n}_3$$
$$\boldsymbol{\omega}^S = u_4\mathbf{n}_1 + u_5\mathbf{n}_2 + u_6\mathbf{n}_3$$

and

$$\mathbf{v} = u_7\mathbf{n}_1 + u_8\mathbf{n}_2 + u_9\mathbf{n}_3$$

respectively, where $\mathbf{n}_1$, $\mathbf{n}_2$, and $\mathbf{n}_3$ are mutually perpendicular unit vectors fixed in B.

Assuming that (1) the force $d\mathbf{F}$ exerted on S by the fluid across a differential element of the surface of S is given by

$$d\mathbf{F} = -c^B\mathbf{v}^P\, dA$$

where c is a constant, P designates any point of S lying within the element under consideration, and dA is the area of the element, and (2) the force exerted on B by the fluid across the corresponding surface element of the cavity is equal to $-d\mathbf{F}$, determine the generalized active forces F_2 and F_4.

Results: $F_2 = \frac{8}{3}\pi c R^4(u_5 - u_2), \qquad F_4 = \frac{8}{3}\pi c R^4(u_1 - u_4).$

▪ PROBLEM SET 6

(Sections 3.9–3.15)

6(a) Assuming that P and Q in Problem 3(j) have equal weights W, determine the generalized inertia force $F_1{}^*$ associated with θ_1.

Result: $$\frac{WL^2}{g}\left[\ddot{\theta}_1 + 2\ddot{\theta}_2 \frac{\sin(\theta_3 - \theta_1)}{\cos(\theta_3 - \theta_2)}\right]$$

6(b) Show by means of examples that $\mathbf{I}_a$ (see Section 3.12) may be, but is not necessarily, parallel to $\mathbf{n}_a$.

6(c) $\mathbf{n}_x$, $\mathbf{n}_y$, and $\mathbf{n}_z$ are unit vectors respectively parallel to the axes OX, OY, OZ of a rectangular Cartesian coordinate system, and each unit vector points in the positive direction of the axis to which it is parallel. Letting S be a set of N particles, m_i the mass of the particle P_i, and x_i, y_i, z_i the coordinates of P_i, express I_x, the moment of inertia of S about the X axis, and I_{yz}, the product of inertia of S with respect to O for $\mathbf{n}_y$ and $\mathbf{n}_z$, each in terms of the masses and coordinates of the particles. Then answer the following questions:

Does it matter whether $\mathbf{n}_x$, $\mathbf{n}_y$, $\mathbf{n}_z$ is a right-handed or left-handed set of unit vectors? Would the results be altered if OX, OY, OZ were not mutually perpendicular? Is the minus sign in the expression for I_{yz} the result of a "sign convention"?

Results:

$$I_x = \sum_{i=1}^{N} m_i(y_i^2 + z_i^2) \qquad I_{yz} = -\sum_{i=1}^{N} m_i y_i z_i$$

6(d) Show by means of examples that products of inertia can be positive, negative, or zero, and that radii of gyration can be equal to zero.

6(e) The moments of inertia of a set S of particles with respect to the axes OX, OY, OZ of a rectangular Cartesian coordinate system have the values A, B, C, respectively. Letting OX', OY', OZ' be a second set of mutually perpendicular axes, and A', B', C' the associated moments of inertia of S, show that

$$A' + B' + C' = A + B + C$$

6(f) A line L_a passes through a point O and is perpendicular to the unit vector $\mathbf{n}_3$ of a set of mutually perpendicular unit vectors $\mathbf{n}_1$, $\mathbf{n}_2$, $\mathbf{n}_3$. Under these circumstances, I_a, the moment of inertia of a body B about line L_a, depends on the orientation of L_a in the plane that passes through O and is perpendicular to $\mathbf{n}_3$.

Express the maximum value of I_a in terms of I_1, I_2, and I_{12}.

Result:

$$\frac{I_1 + I_2}{2} + \left[\left(\frac{I_1 - I_2}{2}\right)^2 + I_{12}{}^2\right]^{1/2}$$

6(g) In Figure 6(g), $\mathbf{n}_1$, $\mathbf{n}_2$, $\mathbf{n}_3$ are mutually perpendicular unit vectors; Q designates the mass center of a body B having a mass of 12 slugs; and products of inertia of B with respect to point O for $\mathbf{n}_1$, $\mathbf{n}_2$, $\mathbf{n}_3$ are tabulated in units of slug-ft².

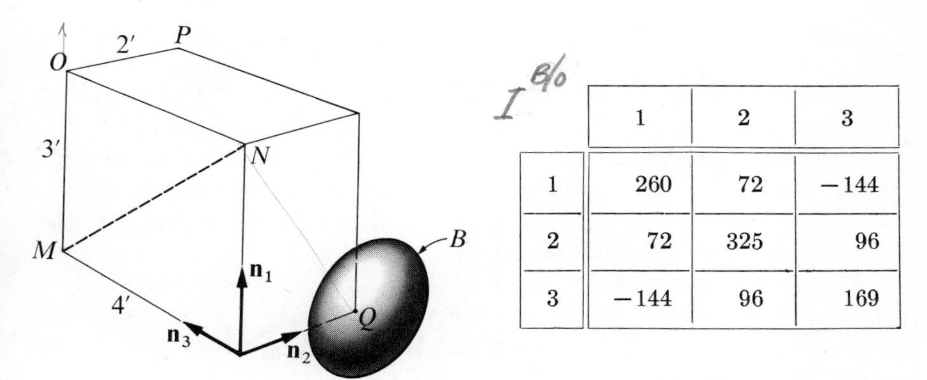

	1	2	3
1	260	72	−144
2	72	325	96
3	−144	96	169

Figure 6(g)

Determine the moment of inertia of B with respect to line MN.

Result: 3316/25 slug-ft².

6(h) Figure 6(h) shows a body consisting of a uniform, right-circular

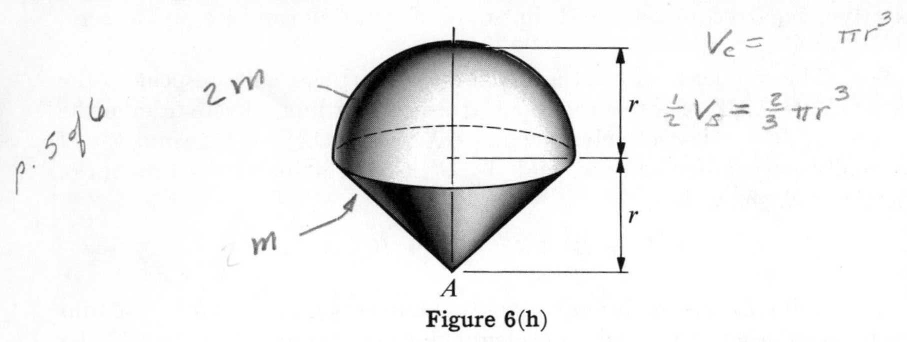

Figure 6(h)

cone of mass m and a uniform hemisphere of mass $2m$.

Determine the moment of inertia of this body about a line that passes through point A and is perpendicular to the axis of the cone.

Result: $(101/20)mr^2$.

6(i) The particles of a set S lie in a plane. Show that the moment of inertia of S about a line L that is normal to the plane is equal to the sum of the moments of inertia of S about any two lines that lie in the plane, are perpendicular to each other, and intersect L.

▪ PROBLEM SET 7

(Sections 3.16–3.21)

7(a) Referring to Figure 7(a), suppose that P is the mass center of a

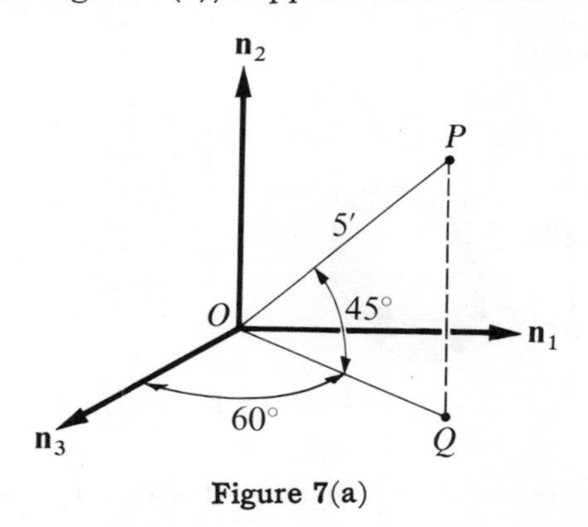

Figure 7(a)

body B of mass of 2 slugs, that $\mathbf{n}_1$, $\mathbf{n}_2$, $\mathbf{n}_3$ are parallel to principal axes of B for P, and that the corresponding principal moments of inertia have values 20, 30, 40 slug-ft^2.

Determine the moment of inertia of B about line OQ.

Result: 50 slug-ft^2.

7(b) Four identical particles are placed at points M, N, O, and P of Figure 6(g). Determine the minimum radius of gyration of this set of particles.

Result: 1.4 ft.

7(c) Verify each of the following statements:

A principal axis for the mass center of a body is a principal axis for each point of the axis.

If a principal axis for a point other than the mass center passes through the mass center, it is a principal axis for the mass center.

A line that is a principal axis for two of its points is a principal axis for the mass center.

The three principal axes for any point on a principal axis for the mass center are parallel to principal axes for the mass center.

If two principal moments of inertia for a given point are equal to each other, the moments of inertia with respect to all lines passing through this point and lying in the plane determined by the associated principal axes are equal to each other.

7(d) Determine the smallest angle between line AB and any principal axis for point A of the thin, uniform plate represented by the shaded portion of Figure 7(d).

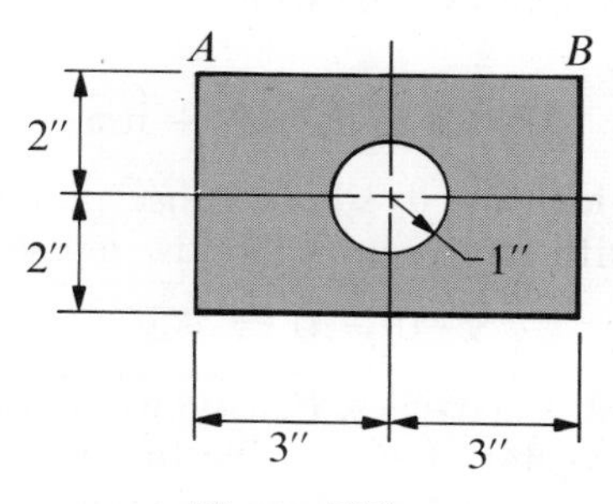

Figure 7(d)

Result: 30 degrees.

7(e) Two identical, thin, uniform, right-triangular plates are attached to each other as shown in Figure 7(e). When a/b is given, k, the minimum

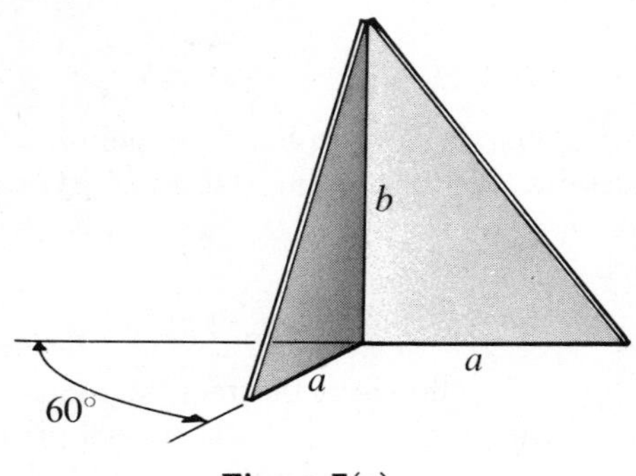

Figure 7(e)

radius of gyration of this assembly, can be expressed in the form $k = nb$. Determine n for $a/b = 2$ and for $a/b = \frac{1}{2}$.

Results: $\frac{1}{3}$, $[35 - (241)^{1/2}]^{1/2}/24$.

7(f) Referring to Problem 6(g), determine the inertia dyadic of the body B for point N.

Result: $68\mathbf{n}_1\mathbf{n}_1 + 72\mathbf{n}_1\mathbf{n}_2 + 72\mathbf{n}_2\mathbf{n}_1 + 133\mathbf{n}_2\mathbf{n}_2 + 169\mathbf{n}_3\mathbf{n}_3$.

7(g) Evaluate $\mathbf{n} \cdot \mathbf{I} \cdot \mathbf{n}$, where $\mathbf{n}$ is a unit vector parallel to line MN in Figure 6(g) and $\mathbf{I}$ is the inertia dyadic found in Problem 7(f). Compare the result obtained with that of Problem 6(g).

7(h) If $\mathbf{n}_1$, $\mathbf{n}_2$, $\mathbf{n}_3$ are mutually perpendicular unit vectors, the quantity $\mathbf{U}$ defined as

$$\mathbf{U} = \mathbf{n}_1\mathbf{n}_1 + \mathbf{n}_2\mathbf{n}_2 + \mathbf{n}_3\mathbf{n}_3$$

is called the unit or identity dyadic because premultiplication or postmultiplication of $\mathbf{U}$ with any vector $\mathbf{v}$ yields that vector:

$$\mathbf{v} \cdot \mathbf{U} = \mathbf{U} \cdot \mathbf{v} = \mathbf{v}$$

Letting S be a set of N particles, P_i a typical particle of S, m_i the mass of P_i, $\mathbf{p}_i$ the position vector of P_i relative to a point O, and $\mathbf{I}$ the inertia dyadic of S for O, show that $\mathbf{I}$ can be expressed as

$$\mathbf{I} = \sum_{i=1}^{N} m_i(\mathbf{U}\mathbf{p}_i^2 - \mathbf{p}_i\mathbf{p}_i)$$

7(i) If S is a set of N particles $P_1, \cdots, P_N$ moving with velocities $\mathbf{v}_1, \cdots, \mathbf{v}_N$ in a reference frame R, the angular momentum $\mathbf{H}$ of S in R relative to the mass center S^* of S is given by

$$\mathbf{H} = \sum_{i=1}^{N} m_i \mathbf{r}_i \times \mathbf{v}_i$$

where m_i and $\mathbf{r}_i$ are the mass of P_i and the position vector of P_i relative to S^*, respectively. Show that, if the particles of S form a rigid body B, $\mathbf{H}$ can be expressed as

$$\mathbf{H} = \mathbf{I} \cdot \boldsymbol{\omega}$$

where $\mathbf{I}$ is the inertia dyadic of B for the mass center of B and $\boldsymbol{\omega}$ is the angular velocity of B in R.

7(j) The moments of inertia of a rigid body B about mutually perpendicular principal axes X_1, X_2, X_3 of B for the mass center B^* of B have the values I_1, I_2, I_3; and $\mathbf{n}_1$, $\mathbf{n}_2$, $\mathbf{n}_3$ are unit vectors parallel to X_1, X_2, X_3, respectively.

Letting $\omega_i = \boldsymbol{\omega} \cdot \mathbf{n}_i$, with $i = 1, 2, 3$, where $\boldsymbol{\omega}$ is the angular velocity of B in a reference frame R, determine the angular momentum $\mathbf{H}$ of B in R relative to B^*. Express the result in terms of I_i, ω_i, and $\mathbf{n}_i$, for $i = 1, 2, 3$.

Result: $\mathbf{H} = I_1\omega_1\mathbf{n}_1 + I_2\omega_2\mathbf{n}_2 + I_3\omega_3\mathbf{n}_3$.

7(k) Show that the angular momentum of a rigid body B in a reference frame R relative to the mass center B^* of B is parallel to the angular velocity $\boldsymbol{\omega}$ of B in R if and only if $\boldsymbol{\omega}$ is parallel to a principal axis of inertia of B for B^*.

7(l) If S is a set of particles $P_1, \cdots, P_N$ moving with velocities $\mathbf{v}_1, \cdots, \mathbf{v}_N$ in a reference frame R, the angular momentum $\mathbf{H}^{S/O}$ of S in R relative to a point O that is fixed in R is given by

$$\mathbf{H}^{S/O} = \sum_{i=1}^{N} m_i \mathbf{p}_i \times \mathbf{v}_i$$

where m_i and $\mathbf{p}_i$ are the mass of P_i and the position vector of P_i relative to O, respectively.

Letting $\mathbf{H}^{S/S^*}$ denote the angular momentum of S in R relative to the mass center S^* of S [see Problem 7(i)], show that

$$\mathbf{H}^{S/O} = \mathbf{H}^{S/S^*} + \mathbf{H}^{S^*/O}$$

where $\mathbf{H}^{S^*/O}$ is the angular momentum in R relative to O of a fictitious particle whose motion is identical to that of S^* and whose mass is equal to $m_1 + \cdots + m_N$.

7(m) Letting K_ω be the kinetic energy of rotation of a rigid body B, $\boldsymbol{\omega}$ the angular velocity of B, and $\mathbf{H}$ the angular momentum of B relative to the mass center B^* of B, and considering all motions of B such that the magnitude of $\mathbf{H}$ remains constant, show that K_ω attains maximum and minimum values only when $\boldsymbol{\omega}$ is parallel to a principal axis of B for B^*.

▪ PROBLEM SET 8

(Sections 3.22–3.25)

8(a) In Figure 8(a), ABC represents a uniform right-triangular plate

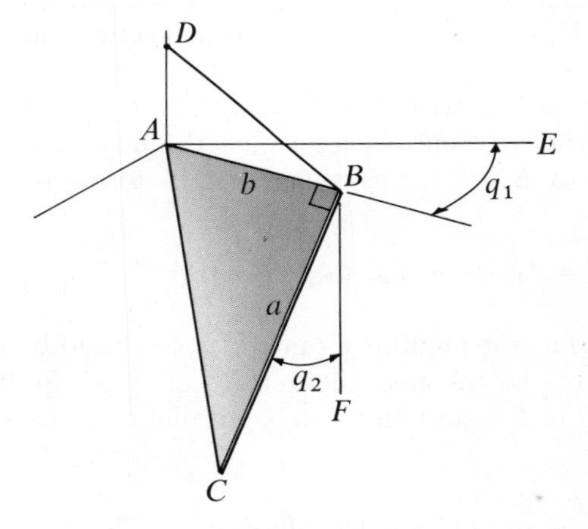

Figure 8(a)

of mass m, supported as follows: Vertex A is fixed and vertex B is attached to a string that is fastened at D, the length of the string being such that line AB is horizontal. (Line AD is vertical.)

Letting q_1 be the angle between AB and a fixed horizontal line AE, and q_2 the angle between BC and the vertical line BF, determine the generalized inertia force F_2^*.

Result:

$$\frac{ma}{12}\,[2a(\sin q_2 \cos q_2 \dot{q}_1^2 - \ddot{q}_2) - 3b \cos q_2 \ddot{q}_1]$$

3

8(b) Show that the inertia torque $\mathbf{T}^*$ for a rigid body B in a reference frame R can be expressed as

$$\mathbf{T}^* = -\dot{\mathbf{H}}$$

where $\mathbf{H}$ is the angular momentum of B in R relative to the mass center of B and the dot denotes differentiation with respect to time in reference frame R.

8(c) Referring to Problem 5(a), determine the generalized inertia force associated with θ.

Result:

$$\frac{-W}{3g}[(3R^2 - 2L^2)\ddot{\theta} + \Omega^2(4L^2 - 3R^2)\sin\theta\cos\theta]$$

8(d) Determine the generalized inertia forces $F_1{}^*$ and $F_2{}^*$ for the system described in Problem 5(e).

Results:

$$F_1{}^* = -(J_1\ddot{q}_1 + J_2\ddot{q}_2) \qquad F_2{}^* = -(J_2\ddot{q}_1 + J_1\ddot{q}_2)$$

where

$$J_1 = \frac{wr^2}{g}\left(\frac{3}{2} + \frac{2}{3}\frac{W}{w} + \frac{1}{8}\frac{r^2}{L^2}\right) \qquad J_2 = \frac{wr^2}{g}\left(\frac{1}{3}\frac{W}{w} - \frac{1}{8}\frac{r^2}{L^2}\right)$$

8(e) A uniform circular disk of radius R and mass m is rigidly attached to a shaft whose axis passes through the center of the disk. The shaft revolves about its axis, which is fixed, with a uniform angular speed ω. Due to misalignment, the normal to the plane of the disk makes an angle θ with the shaft axis.

Determine the magnitude of the inertia torque for the disk.

Result: $(mR^2\omega^2/8)\sin 2\theta$.

8(f) Referring to Problem 5(d), and assuming that the height of B is negligible in comparison with the width $2w$, determine the generalized inertia forces $F_1{}^*$ and $F_2{}^*$.

Results: $-(W/6g)(2\ddot{q}_1 + \ddot{q}_2)$, $-(W/6g)(\ddot{q}_1 + 2\ddot{q}_2)$.

8(g) Three uniform bars, A, B, C, each of mass m and length L, are pin-connected at the midpoints of B and C; and light, inextensible strings join the end points of B and C, as shown in Figure 8(g).

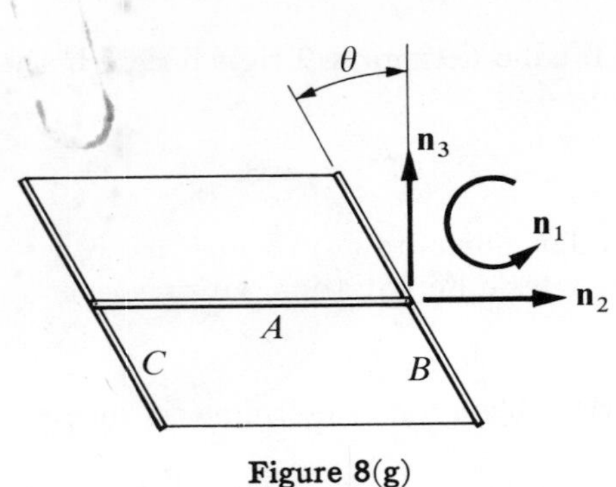

Figure 8(g)

Noting that this system possesses seven degrees of freedom, let $\mathbf{n}_1$, $\mathbf{n}_2$, $\mathbf{n}_3$ be mutually perpendicular unit vectors directed as shown, and define $u_1, \cdots, u_7$ as

$$u_i = \begin{cases} \boldsymbol{\omega} \cdot \mathbf{n}_i & i = 1, 2, 3 \\ \dot{\theta} & i = 4 \\ \mathbf{v} \cdot \mathbf{n}_{i-4} & i = 5, 6, 7 \end{cases}$$

where $\boldsymbol{\omega}$ is the angular velocity of bar A, θ is the angle shown in Figure 8(g), and $\mathbf{v}$ is the velocity of the midpoint of A.

Dropping all terms of second or higher degree in θ or derivatives of θ, determine the generalized inertia forces associated with u_1, u_4, and u_7.

Results:

$$\frac{-mL^2}{12}[5u_2u_3 + 9\dot{u}_1 + 2\dot{u}_4 + 2(u_2{}^2 - u_3{}^2)\theta]$$

$$\frac{mL^2}{6}[u_2u_3 - \dot{u}_1 - \dot{u}_4 + (u_3{}^2 - u_2{}^2)\theta]$$

$$-3m(\dot{u}_7 + u_1u_6 - u_2u_5)$$

▪ PROBLEM SET 9

(Sections 4.1–4.7)

9(a) A system comprised of two particles of masses m_1 and m_2 moves in such a way that the distance r between the particles varies with time. Find a potential function for the gravitational forces exerted by the particles on each other.

Result: $-Gm_1m_2/r$.

9(b) In Figure 9(b), B designates a rigid body, B^* the mass center of B,

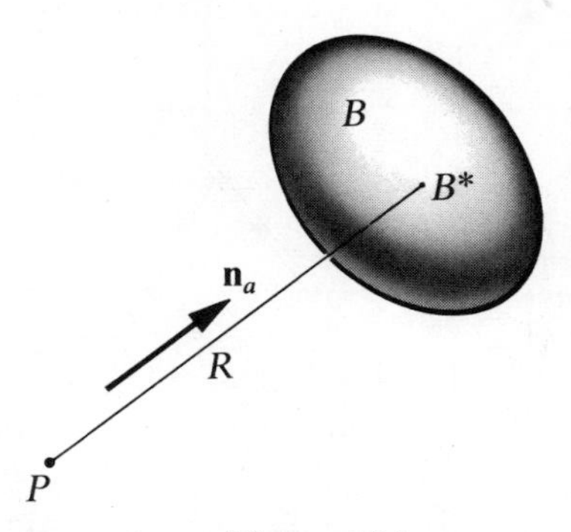

Figure 9(b)

P a fixed particle of mass M, R the distance between P and B^*, and $\mathbf{n}_a$ a unit vector parallel to line PB^*.

Letting I be the moment of inertia of B about line PB^*, and assuming that (a) B^* is constrained to move with constant speed on a circle of radius R centered at P and (b) the system of gravitational forces exerted on B by P is approximately equivalent to a force applied at B^* and directed from B^* toward P, together with a couple having a torque $\mathbf{T}$ given by

$$\mathbf{T} = \frac{3GM\mathbf{n}_a}{R^3} \times \mathbf{I}_a$$

where $\mathbf{I}_a$ is the second moment of B relative to B^* for $\mathbf{n}_a$, show that $3GMI/2R^3$ is a potential function for the gravitational forces under consideration.

SUGGESTION: Letting q_1, q_2, q_3 be generalized coordinates that determine the orientation of B, note that the contribution of the given force system to the generalized active force F_r can be expressed as [see Equation (3.5)]

$$\frac{3GM}{R^3} \boldsymbol{\omega}_{\dot{q}_r} \times \mathbf{n}_a \cdot \mathbf{I}_a$$

and that [see Equation (2.5)]

$$\boldsymbol{\omega}_{\dot{q}_r} \times \mathbf{n}_a = -\frac{{}^B\partial \mathbf{n}_a}{\partial q_r}$$

9(c) A uniform thin rod of cross-sectional area A is partly immersed in a fluid of weight density w. The lowest point of the rod is at a depth y, and the rod makes an angle θ with the vertical, as indicated in Figure 9(c).

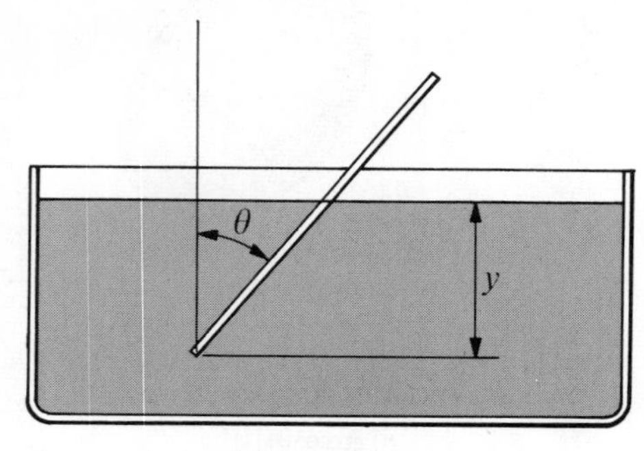

Figure 9(c)

Find a potential function for the forces exerted on the rod by the fluid.

Result: $(wAy^2/2) \sec \theta$.

9(d) Find a potential function for the gravitational forces acting on the system described in Problem 5(b).

Result: $-WL(30 \cos q_1 + 19 \cos q_2 + 8 \cos q_3)$.

9(e) Determine a potential energy of the system described in Problem 5(a).

Result: $-W(R^2 - L^2)^{1/2} \cos \theta$.

9(f) Determine a potential energy of the system described in Problem 5(d).

Result:

$$\frac{6EI}{L^3}\left[\left(1 + \frac{L}{2w} + \frac{1}{6}\frac{L^2}{w^2}\right)(q_1^2 + q_2^2) - \frac{L}{w}\left(1 + \frac{1}{3}\frac{L}{w}\right) q_1 q_2\right] - \frac{W}{2}(q_1 + q_2)$$

9(g) Three corners of a uniform cube are attached to fixed supports by means of identical, linear springs of spring constant k. When the springs are undeformed, their axes coincide with edges of the cube, as indicated in Figure 9(g).

To bring the cube into a general position, the center is displaced to a point whose coordinates in the fixed-axes system X_1, X_2, X_3 [see Figure

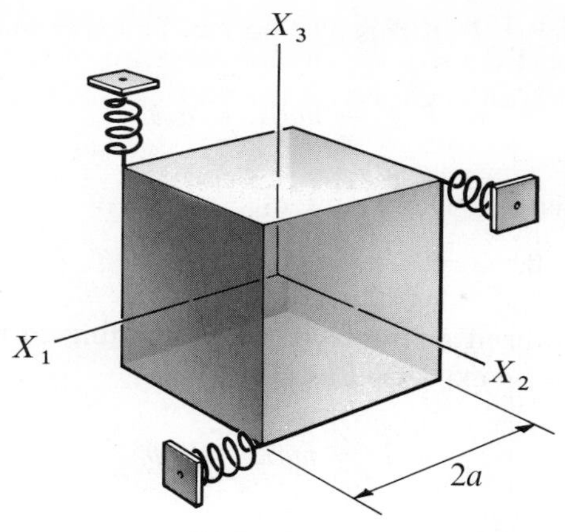

Figure 9(g)

9(g)] are x_1, x_2, x_3; and three axes A_1, A_2, A_3, which are fixed in the cube and initially coincident with X_1, X_2, X_3, respectively, are brought into new positions by means of successive rotations of amount θ_1 about A_1, θ_2 about A_2, and θ_3 about A_3.

Dropping all terms of third or higher degree in x_i and θ_i, for $i = 1, 2, 3$, find a potential function for the forces exerted on the cube by the springs.

Result:

$$\tfrac{1}{2}ka^2\left[\left(\theta_1+\theta_2-\frac{x_3}{a}\right)^2+\left(\theta_2+\theta_3-\frac{x_1}{a}\right)^2+\left(\theta_3+\theta_1-\frac{x_2}{a}\right)^2\right]$$

9(h) A simple pendulum of mass m and length L is attached to a linear spring of natural length L' and spring constant $k = 5mg/L$, as shown in Figure 9(h).

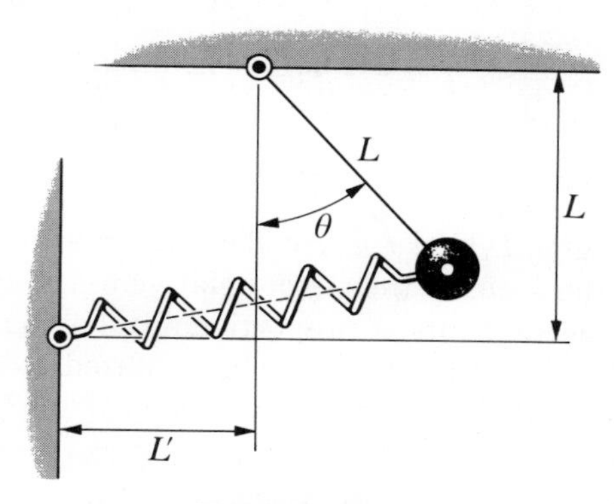

Figure 9(h)

A potential energy P of this system can be expressed as

$$P = mgL \sum_{n=1}^{\infty} a_n \theta^n$$

where a_n is a constant. Determine $a_1, \cdots, a_4$.

Results: $a_1 = 0,\ a_2 = 3,\ a_3 = 0,\ a_4 = -\frac{7}{8}$.

9(i) The generalized active force corresponding to the coordinate θ in Problem 9(h) can be expressed as

$$F = mgL \sum_{n=1}^{\infty} b_n \theta^n$$

where b_n is a constant.

Find b_1, b_2, and b_3 (1) by using the results of Problem 9(h) and (2) without reference to potential energy, and comment briefly on the relative merits of the two methods used.

Results: $b_1 = -6,\ b_2 = 0,\ b_3 = 3.5$.

9(j) Two blocks are connected to each other and to a fixed support by means of springs and dashpots, as indicated in Figure 9(j). The springs

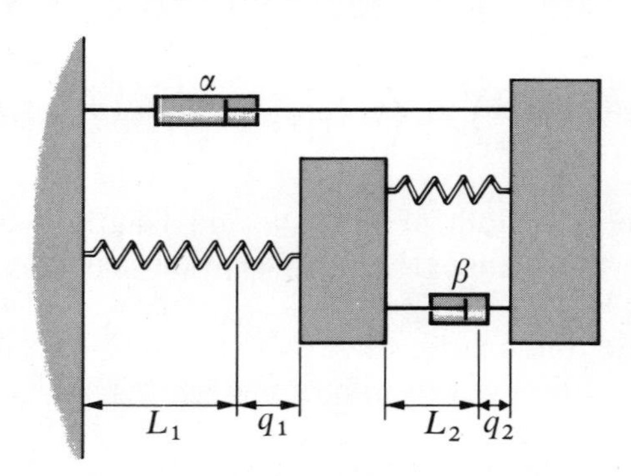

Figure 9(j)

have natural lengths L_1 and L_2, and the force exerted by each dashpot is proportional to the speed of the piston relative to the cylinder in which the piston moves, the constants of proportionality having the values α and β.

Using q_1 and q_2 [see Figure 9(j)] as generalized coordinates, determine a dissipation function for the dashpot forces.

Result:

$$\tfrac{1}{2}[\alpha \dot{q}_1^2 + 2\alpha \dot{q}_1 \dot{q}_2 + (\alpha + \beta)\dot{q}_2^2]$$

9(k) One point of a rigid body B is fixed, and the system of forces exerted on B by a fluid that surrounds B is equivalent to a force applied at the fixed point together with a couple having a torque $-c\boldsymbol{\omega}$, where c is a constant and $\boldsymbol{\omega}$ is the angular velocity of B.

Show that $c\boldsymbol{\omega}^2/2$ is a dissipation function for the forces exerted on B by the fluid.

SUGGESTION: Make use of the fact that $\boldsymbol{\omega}$ can be expressed as

$$\boldsymbol{\omega} = \boldsymbol{\omega}_{\dot{q}_1}\dot{q}_1 + \boldsymbol{\omega}_{\dot{q}_2}\dot{q}_2 + \boldsymbol{\omega}_{\dot{q}_3}\dot{q}_3 + \boldsymbol{\omega}_t$$

and that the contributions of the fluid forces to the generalized forces are then given by $-c\boldsymbol{\omega} \cdot \boldsymbol{\omega}_{\dot{q}_r}$, with $r = 1, 2, 3$.

▪ PROBLEM SET 10

(Sections 4.8–4.16)

10(a) A particle P of mass m moves on the line L_a described in Problem 2(g). The disk D has a radius r and, at a certain instant, P is at a distance $r/2$ from D^*; the distance between P and D^* is decreasing at a rate v; $\theta = 0$; and $\psi = \pi/2$ radian.

Determine the kinetic energy of P for this instant.

Result:

$$\tfrac{1}{2}m[r^2(\tfrac{5}{4}\dot{\psi}^2 + \tfrac{1}{4}\dot{\phi}^2 + \dot{\theta}^2 + \dot{\phi}\dot{\theta}) + v(v + 2r\dot{\psi})]$$

10(b) Referring to Problem 5(a), use the kinetic energy of the rod to determine the generalized inertia force associated with θ, and compare this method with that employed in the solution of Problem 8(c).

10(c) Determine the kinetic energy of the plate described in Problem 8(a), and use it to find the generalized inertia force $F_2{}^*$.

10(d) Determine the generalized inertia force $F_1{}^*$ for the system described in Problem 5(b).

Result:

$$\frac{WL^2}{g}[-29\ddot{q}_1 + 19(c_1c_2 - s_1s_2)\ddot{q}_2 - 8(s_3s_1 + c_3c_1)\ddot{q}_3 - 19(s_2c_1 + c_2s_1)\dot{q}_2{}^2 - 8(s_1c_3 - c_1s_3)\dot{q}_3{}^2]$$

10(e) Explain briefly why kinetic energy functions cannot be used to solve Problems 8(d) and 8(g).

10(f) Referring to Problem 3(d), determine the kinetic energy of the system comprised of the shaft and the four spheres for an instant at which the shaft has an angular speed ω, assuming that the spheres, each of mass m, are uniform and solid; the shaft has a moment of inertia J about its axis; $\theta = 30$ degrees; and the contact between C and S is one of pure rolling.

Result:

$$\tfrac{1}{2}[J + 18mr^2(2 + \sqrt{3})/5]\omega^2$$

10(g) Two square plates are rigidly connected to each other along one edge, as shown in Figure 10(g), and this assembly rotates about a vertical

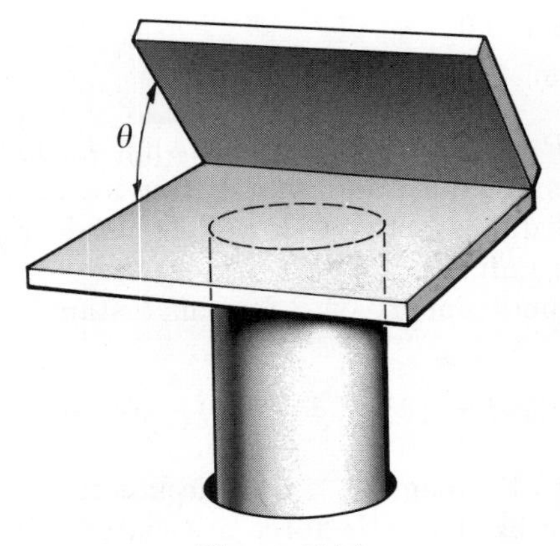

Figure 10(g)

axis that is normal to one of the plates and passes through its center.

Determine the value of θ for which the kinetic energy of the system has the smallest possible value.

Result: arc cos (0.75).

10(h) Two generalized coordinates of a holonomic system are said to be coupled dynamically when the associated inertia coefficient is not equal to zero. For example, as $m_{12} \neq 0$ in the example in Section 4.12, q_1 and q_2 are coupled dynamically.

The mass center B^* of a rigid body B is fixed. Principal axes A_1, A_2, and A_3 of B for B^* are initially aligned with fixed axes X_1, X_2, and X_3, respectively, and successive rotations of amounts q_1, q_2, and q_3 are then performed about A_1, A_2, and A_3 to bring B into a general position.

Determine which of the generalized coordinates q_1, q_2, and q_3 are coupled dynamically.

Result: q_1 and q_2; q_1 and q_3.

10(i) Referring to Problem 10(g), suppose that the two plates are connected by means of a smooth hinge, so that θ can vary freely, and consider a motion during which θ remains constant at 60 degrees and $L\omega^2 = Ng$, where L is the length of one side of either plate, ω is the angular speed of the shaft, N is a positive number, and g is the acceleration of gravity.

Determine the value of N for which the total virtual work of all active and inertia forces is equal to zero.

Result: $2\sqrt{3}$.

10(j) Figure 10(j) shows three pin-connected uniform bars, each of

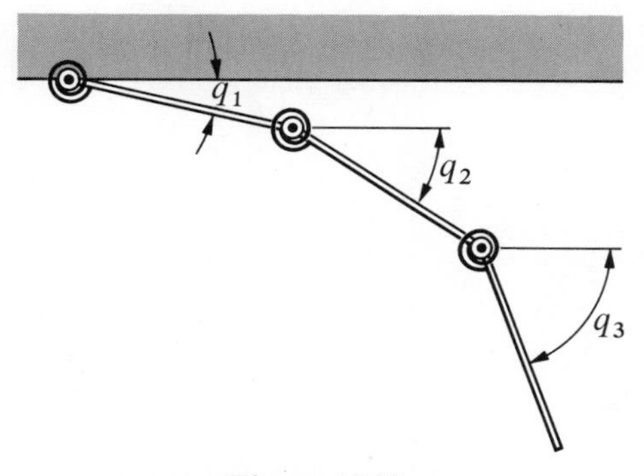

Figure 10(j)

weight W and length L. Identical linear torsion springs of modulus c are attached as indicated, and these springs are undeformed when the angles q_1, q_2, and q_3 between the bars and the horizontal are equal to zero.

Determine the virtual work of the active forces for this system for a set of virtual displacements such that $\delta q_2 = \delta q_3 = 0$.

Result: $(5WL/2) \cos q_1 + c(q_2 - 2q_1)$.

10(k) Referring to Problem 5(d), and letting P_1 and P_2 have equal virtual displacements of magnitude δq_1, determine the virtual work of the active forces.

Result: $[-(12\ EI/L^3)(q_1 + q_2) + W]\ \delta q_1$.

10(l) Evaluate the virtual work of the active forces for the system described in Problem 5(e) for a set of virtual displacements such that the virtual displacement of the center of D_2 is equal to zero.

Result: $(w + W)r \sin \beta \sin \phi\ \delta q_1$.

PART III

LAWS OF MOTION

THE MATHEMATICAL ANALYSIS of problems of mechanics generally proceeds in two stages. The first deals with the formulation of equations, a process that may be based on any one of a number of fundamental propositions, such as Newton's laws of motion, D'Alembert's principle, Hamilton's principle, and so forth. The equations are then solved, in the second stage, by analytical, numerical, or experimental (for example, analog) methods.

In Chapter 5, Lagrange's form of D'Alembert's principle is stated as a fundamental proposition, and is used as a point of departure for the development of various other methods of formulating equations of motion. The integration of these equations is discussed in Chapter 6.

Chapter **5**

Formulation of Equations of Motion

5.1 Lagrange's Form of D'Alembert's Principle. There exist reference frames R, called *inertial reference frames*, such that, if S is any simple nonholonomic system possessing $n - m$ degrees of freedom in R, and F_r and F_r^* are generalized active and generalized inertia forces for S in R (see Sections 3.2, 3.9, and 3.25), then the $n - m$ equations

$$F_r + F_r^* = 0 \qquad r = 1, \cdots, n - m \tag{5.1}$$

are satisfied during every motion of S. This proposition is known as *Lagrange's form of D'Alembert's principle.* Equations (5.1), or any equations derived from them, are called *dynamical equations.*

Ultimately, the justification for regarding a reference frame as inertial can come only from experiments. One such experiment, first performed by Foucault in 1851, is discussed in the example that follows.

▪ EXAMPLE

In Figure 5.1, E represents the earth, regarded as a sphere whose center is the point E^*, and P designates a particle suspended by means of a light string of length L from a point Q that is fixed relative to E. Point O is the point of intersection of line QE^* and the surface of E; h is the distance from O to Q; ϕ measures the angle between line OE^* and the earth's polar axis, line NS; $\mathbf{k}$ is a unit vector parallel to NS; and $\mathbf{s}$, $\mathbf{e}$, and $\mathbf{u}$ are unit vectors pointing southward, eastward, and upward at O, respectively.

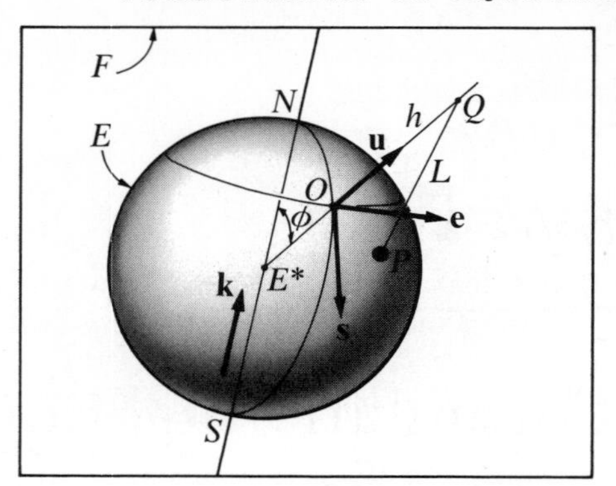

Figure 5.1

Finally, F represents a reference frame in which the line NS is fixed and in which E rotates about this axis once per sidereal day, so that the angular velocity $\boldsymbol{\omega}$ of E in F is given by

$$\boldsymbol{\omega} = \omega \mathbf{k} \tag{a}$$

where $\omega = 2\pi/86{,}164$ rad/s. (F differs from the so-called "astronomical" reference frame primarily because it shares the motion of point E^*. E^* is thus fixed in F, whereas it moves in the astronomical reference frame.)

The string and the particle P comprise a pendulum whose motion relative to the earth can be described in terms of the two angles q_1 and q_2 shown in Figure 5.2. Dynamical equations governing q_1 and q_2 will be

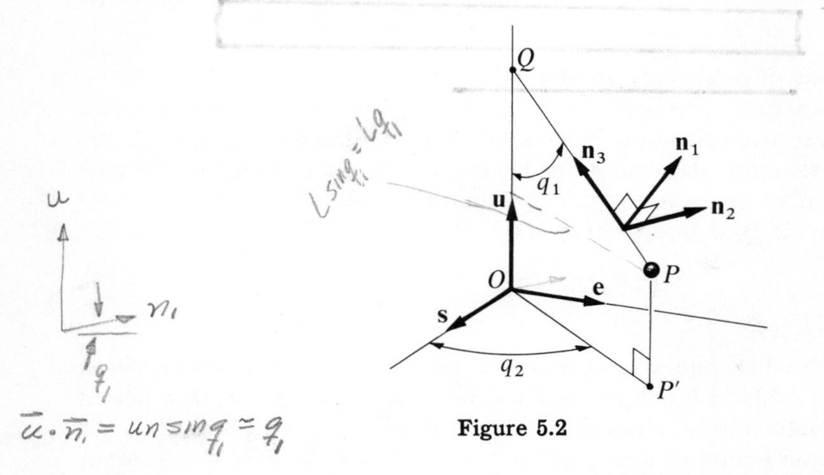

Figure 5.2

obtained by using Equation (5.1) and assuming that (a) F and (b) E is an inertial reference frame.

The analysis will be confined to small oscillations of the pendulum;

that is, all quantities of second or higher degree in q_1 and/or derivatives of q_1 will be regarded as negligible.

The velocity ${}^F\mathbf{v}^P$ of P in F can be expressed as

$$ {}^F\mathbf{v}^P \underset{(2.21)}{=} {}^F\mathbf{v}^{\bar{E}} + {}^E\mathbf{v}^P \tag{b} $$

where $\bar{E}$ denotes that point of E (or E extended) with which P coincides. The velocity ${}^F\mathbf{v}^{\bar{E}}$ is independent of $\dot{q}_1$ and $\dot{q}_2$; and (with $\sin q_1$ replaced by q_1)

$$ {}^E\mathbf{v}^P = L(\dot{q}_1\mathbf{n}_1 + \dot{q}_2 q_1 \mathbf{n}_2) \tag{c} $$

where $\mathbf{n}_1$ and $\mathbf{n}_2$ are unit vectors perpendicular to line PQ and to each other, and $\mathbf{n}_2$ is perpendicular to the plane determined by O, Q, and P, as indicated in Figure 5.2. Consequently, the partial rates of change of the position of P in F with respect to q_1 and q_2 (see Section 2.13) are

$$ {}^F\mathbf{v}_{\dot{q}_1}{}^P \underset{(b),(c)}{=} L\mathbf{n}_1 \qquad {}^F\mathbf{v}_{\dot{q}_2}{}^P = Lq_1\mathbf{n}_2 \tag{d} $$

The gravitational force $\mathbf{G}$ exerted on P by E is given with sufficient accuracy by

$$ \mathbf{G} = -mg\mathbf{u} \tag{e} $$

where m is the mass of P and g is the gravitational acceleration at point O. The force $\mathbf{S}$ exerted on P by the string can be expressed as

$$ \mathbf{S} = S\mathbf{n}_3 \tag{f} $$

where S is an unknown scalar and $\mathbf{n}_3$ is a unit vector parallel to the string ($\mathbf{n}_3 = \mathbf{n}_1 \times \mathbf{n}_2$), as shown in Figure 5.2.

If not only gravitational forces exerted on P by the sun, the moon, and so forth, but also contact forces exerted on P by the surrounding air are left out of account, then the generalized active forces F_1 and F_2 for P in F are given by

$$ \left.\begin{aligned} F_1 &\underset{(3.3)}{=} {}^E\mathbf{v}_{\dot{q}_1}{}^P \cdot (\mathbf{G} + \mathbf{S}) \underset{(d),(e),(f)}{=} -mgLq_1 \\ F_2 &\underset{(3.3)}{=} {}^F\mathbf{v}_{\dot{q}_2}{}^P \cdot (\mathbf{G} + \mathbf{S}) \underset{(d),(e),(f)}{=} 0 \end{aligned}\right\} \tag{g} $$

(The fact that $\mathbf{S}$ contributes nothing to F_1 and F_2 could have been anticipated by reference to Section 3.3, because the string may be replaced with a smooth sphere of radius L, center at Q.)

To find the generalized inertia forces $F_1{}^*$ and $F_2{}^*$ for P in F, note that the acceleration of P in F is given by

$$ {}^F\mathbf{a}^P \underset{(2.24)}{=} {}^E\mathbf{a}^P + {}^F\mathbf{a}^{\bar{E}} + 2\boldsymbol{\omega} \times {}^E\mathbf{v}^P \tag{h} $$

together with

$$
\left.\begin{aligned}
{}^E\mathbf{a}^P &= L[(\ddot{q}_1 - \dot{q}_2{}^2 q_1)\mathbf{n}_1 + (\ddot{q}_2 q_1 + 2\dot{q}_1\dot{q}_2)\mathbf{n}_2 + \cdots] \\
{}^F\mathbf{a}^{\bar{E}} &\underset{(2.23)}{=} \underset{(a)}{\omega^2}\mathbf{k} \times (\mathbf{k} \times \mathbf{p})
\end{aligned}\right\}
\quad \text{(i)}
$$

and

$$
2\boldsymbol{\omega} \times {}^E\mathbf{v}^P \underset{(a),(c)}{=} 2L\omega(\dot{q}_1\mathbf{k} \times \mathbf{n}_1 + \dot{q}_2 q_1 \mathbf{k} \times \mathbf{n}_2)
$$

where the three dots represent a vector parallel to $\mathbf{n}_3$ and where $\mathbf{p}$ is the position vector of P relative to E^* (see Figure 5.1).

The acceleration ${}^F\mathbf{a}^{\bar{E}}$, being proportional to ω^2, is relatively unimportant. When ${}^F\mathbf{a}^{\bar{E}}$ is omitted, $F_1{}^*$ is given by

$$
F_1{}^* \underset{(3.8),(3.9)}{=} -m{}^F\mathbf{v}_{\dot{q}_1}{}^P \cdot {}^F\mathbf{a}^P \underset{(d),(h),(i)}{=} -mL^2(\ddot{q}_1 - \dot{q}_2{}^2 q_1 + 2\omega\dot{q}_2 q_1 \mathbf{k} \times \mathbf{n}_2 \cdot \mathbf{n}_1) \quad \text{(j)}
$$

and, similarly,

$$
F_2{}^* = -mL^2 q_1[\ddot{q}_2 q_1 + 2\dot{q}_1(\dot{q}_2 + \omega\mathbf{k} \times \mathbf{n}_1 \cdot \mathbf{n}_2)] \quad \text{(k)}
$$

Next, introduce the abbreviation

$$
p^2 = \frac{g}{L}
$$

and note (see Figures 5.1 and 5.2) that

$$
\mathbf{k} \times \mathbf{n}_1 \cdot \mathbf{n}_2 = \mathbf{k} \cdot \mathbf{n}_3 = \cos\phi + q_1 \cos q_2 \sin\phi
$$

Substitution from Equations (g), (j), and (k) into Equation (5.1) then shows that, if F is an inertial reference frame, q_1 and q_2 are governed by the differential equations

$$
\left.\begin{aligned}
\ddot{q}_1 + (p^2 - \dot{q}_2{}^2 - 2\omega\dot{q}_2\cos\phi)q_1 &= 0 \\
\ddot{q}_2 q_1 + 2\dot{q}_1(\dot{q}_2 + \omega\cos\phi) &= 0
\end{aligned}\right\}
\quad \text{(l)}
$$

To obtain the equations that replace Equations (l) when E, rather than F, is assumed to be an inertial reference frame, one can proceed in a similar way, but use ${}^E\mathbf{v}_{\dot{q}_r}{}^P$ in place of ${}^F\mathbf{v}_{\dot{q}_r}{}^P$, and ${}^E\mathbf{a}^P$ instead of ${}^F\mathbf{a}^P$. However, the velocities ${}^F\mathbf{v}^P$ and ${}^E\mathbf{v}^P$ differ from each other only by a term that is independent of $\dot{q}_1$ and $\dot{q}_2$ [see Equation (b)], so that

$$
{}^E\mathbf{v}_{\dot{q}_r}{}^P = {}^F\mathbf{v}_{\dot{q}_r}{}^P \qquad r = 1, 2
$$

and ${}^E\mathbf{a}^P$ differs from ${}^F\mathbf{a}^P$ only by terms depending on ω [see Equation (h) and (i)]. Consequently, one needs only to omit the ω-dependent terms from Equations (l) in order to arrive at the differential equations that govern q_1 and q_2 if E is an inertial reference frame; and every solution of Equations (l) will be a solution of the equations thus obtained, once all ω-dependent terms have been deleted from such a solution.

To solve Equation (l), and thus to arrive at a description of the motion of the pendulum, it is convenient to introduce a vector $\mathbf{r} = Lq_1\mathbf{n}$, where $\mathbf{n}$ is a unit vector directed as shown in Figure 5.3. As may be seen by

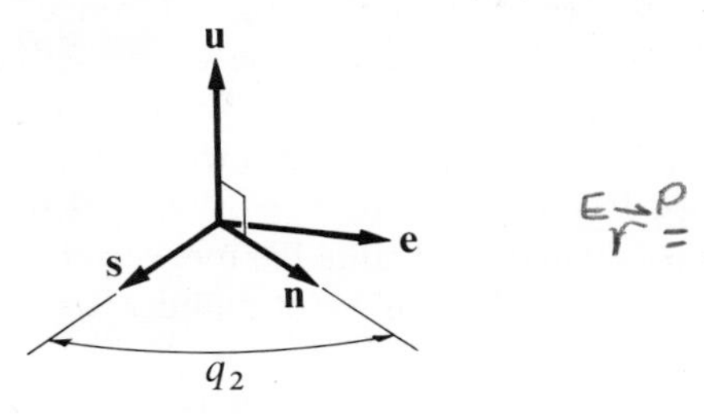

Figure 5.3

reference to Figure 5.2, this vector is very nearly equal to the position vector relative to point O of the orthogonal projection P' of point P on the horizontal plane passing through point 0, and it therefore characterizes the motion of P'. Furthermore, it may be verified by carrying out the indicated differentiations that the two nonlinear scalar differential equations (l) are together equivalent to the single linear vector differential equation

$$\ddot{\mathbf{r}} + 2\omega \cos \phi \mathbf{u} \times \dot{\mathbf{r}} + p^2\mathbf{r} = 0 \qquad \text{(m)}$$

where dots denote differentiation with respect to time in E; and this equation can be replaced by an even simpler one through the introduction of a reference frame R whose angular velocity in E is taken to be

$$^E\boldsymbol{\omega}^R = -\omega \cos \phi \mathbf{u} \qquad \text{(n)}$$

In accordance with Equation (2.13), $\dot{\mathbf{r}}$ and $\ddot{\mathbf{r}}$ can then be expressed as

$$\dot{\mathbf{r}} = \frac{^R d\mathbf{r}}{dt} - \omega \cos \phi \mathbf{u} \times \mathbf{r}$$

$$\ddot{\mathbf{r}} = \frac{^R d^2\mathbf{r}}{dt^2} - 2\omega \cos \phi \mathbf{u} \times \frac{^R d\mathbf{r}}{dt} + \omega^2 \cos^2 \phi \mathbf{u} \times (\mathbf{u} \times \mathbf{r}) \qquad \text{(o)}$$

and substitution into Equation (m) gives

$$\frac{^R d^2\mathbf{r}}{dt^2} + p^2\mathbf{r} = 0 \qquad \text{(p)}$$

if, as before, terms involving ω^2 are dropped. The solution of Equation (p) will reveal how P' moves in reference frame R; and, since the angular velocity of R in E is given by Equation (n), and is thus known, a detailed description of the motion of the pendulum relative to the earth will then be at hand.

Equation (p) has the general solution

$$\mathbf{r} = \mathbf{A} \cos pt + \mathbf{B} \sin pt \tag{q}$$

where $\mathbf{A}$ and $\mathbf{B}$ are vectors fixed in reference frame R. By reference to initial conditions, $\mathbf{A}$ and $\mathbf{B}$ can be expressed in physically meaningful terms. Suppose, for example, that P is initially at rest in E, with $q_1 = \alpha$ and $q_2 = 0$, all of which can be realized physically by attaching one end of a thread to P and fastening the other end to a suitable support. If the thread is then burned, and time t is measured from the instant at which P becomes free, the initial values of $\mathbf{r}$ and $\dot{\mathbf{r}}$ are (see Figure 5.3)

$$\mathbf{r}\Big|_{t=0} = L\alpha\mathbf{s}\Big|_{t=0} \tag{r}$$

and

$$\dot{\mathbf{r}}\Big|_{t=0} = 0 \tag{s}$$

Now

$$\mathbf{r}\Big|_{t=0} \underset{(q)}{=} \mathbf{A} \tag{t}$$

so that

$$\mathbf{A} \underset{(r),(t)}{=} L\alpha\mathbf{s}\Big|_{t=0} \tag{u}$$

and

$$\dot{\mathbf{r}}\Big|_{t=0} \underset{(o)}{=} \frac{{}^R d\mathbf{r}}{dt}\Big|_{t=0} - \omega \cos \phi \mathbf{u}\Big|_{t=0} \times \mathbf{r}\Big|_{t=0} \tag{v}$$

which, together with Equations (s) and (r), gives

$$\frac{{}^R d\mathbf{r}}{dt}\Big|_{t=0} = L\alpha\omega \cos \phi \mathbf{e}\Big|_{t=0} \tag{w}$$

But

$$\frac{{}^R d\mathbf{r}}{dt}\Big|_{t=0} \underset{(q)}{=} \mathbf{B}p \tag{x}$$

Hence,

$$\mathbf{B} \underset{(w),(x)}{=} \frac{L\alpha\omega}{p} \cos \phi \mathbf{e}\Big|_{t=0} \tag{y}$$

The solution of Equation (p) corresponding to the initial conditions under consideration is thus

$$\mathbf{r} \underset{(q),(u),(y)}{=} L\alpha\left[\cos pt\mathbf{s}_0 + \frac{\omega}{p} \cos \phi \sin pt\mathbf{e}_0\right] \tag{z}$$

where $\mathbf{s}_0$ and $\mathbf{e}_0$ denote the values of $\mathbf{s}$ and $\mathbf{e}$ in reference frame R at time $t = 0$.

Equation (z) shows that P' moves on an elliptical path (see Figure 5.4)

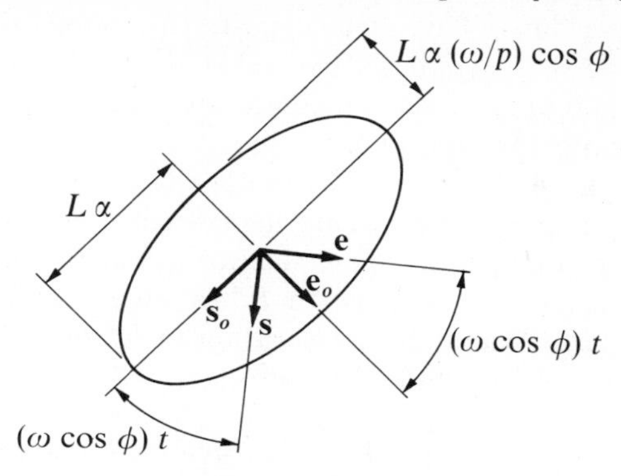

Figure 5.4

fixed in reference frame R. P' traverses this path once every $2\pi/p$ units of time; the major axes of the ellipse are parallel to $\mathbf{s}_0$ and $\mathbf{e}_0$; the lengths of the associated semidiameters are $L\alpha$ and $L\alpha(\omega/p)\cos\phi$; and [see Equation (n)] the ellipse rotates in E at a rate of $\omega\cos\phi$ radians per unit of time, the rotation being clockwise when viewed from point Q (see Figure 5.1), provided point O be situated in the Northern Hemisphere.

There is a pronounced qualitative difference between this description of the motion of P' and that corresponding to the assumption that E is an inertial reference frame, because ω affects both the shape of the ellipse and the rate at which the ellipse turns. Specifically, the previously mentioned deletion of ω-dependent terms leads to the prediction that P' simply moves on a straight line fixed in E. However, the predicted difference between the two motions is not necessarily physically apparent; for, as ω/p is a very small number, and $\cos\phi \leqq 1$, the ellipse may be expected to have the appearance of a straight line, and the angle through which the ellipse turns in a short time interval is small. An experiment intended to reveal this difference must, therefore, be performed with considerable care. Foucault's experiments, carried out at the Panthéon in Paris, and subsequently duplicated in numerous other places, revealed that the ellipse does, indeed, rotate at a rate that is in good agreement with the predictions based on the assumption that F is an inertial reference frame.

5.2 Reference Frames That Move Relative to an Inertial Reference Frame. Astronomical observations furnish a large body of data showing that the reference frame F of the example in Section 5.1 is not an inertial reference frame, because the use of Equation (5.1) with the assumption that F is such a reference frame results in descriptions of motions that disagree with the data. This fact suggests the following question: Why does the hypothesis that F is an inertial reference frame lead to a satisfactory description of motions of a pendulum, but to an incorrect description of, for example, the motion of the moon? This question and a number of related ones can be answered in the light of the following theorem.

If R' is a reference frame having a prescribed motion relative to an inertial reference frame R^*, then R' is an inertial reference frame if and only if the motion of R' relative to R^* is such that both the angular velocity $\boldsymbol{\omega}$ of R' in R^* and the acceleration in R^* of at least one point Q fixed in R' are at all times equal to zero.

Proof: If F_r is a generalized active force for a system S in R', then it is also a generalized active force for S in R^*, but generalized inertia forces F_r' for S in R' and F_r^* for S in R^* may differ from each other if the accelerations ${}^{R'}\mathbf{a}^P$ and ${}^{R^*}\mathbf{a}^P$ of particles P of S in R' and in R^* differ from each other. Now,

$$ {}^{R^*}\mathbf{a}^P \underset{(2.24)}{=} {}^{R'}\mathbf{a}^P + {}^{R^*}\mathbf{a}^{\bar{R}'} + 2\boldsymbol{\omega} \times {}^{R'}\mathbf{v}^P \qquad \text{(a)} $$

and, since Q is a point fixed in R',

$$ {}^{R^*}\mathbf{a}^{\bar{R}'}\mathbf{a} \underset{(2.23)}{=} {}^{R^*Q} + \boldsymbol{\alpha} \times \mathbf{r} + \boldsymbol{\omega} \times (\boldsymbol{\omega} \times \mathbf{r}) \qquad \text{(b)} $$

where $\mathbf{r}$ is the position vector of P (or $\bar{R}'$) relative to Q. Consequently, the hypothesis that

$$ \boldsymbol{\omega} = 0 \qquad \text{(c)} $$

and

$$ {}^{R^*}\mathbf{a}^Q = 0 \qquad \text{(d)} $$

which implies that

$$ \boldsymbol{\alpha} \underset{(c)}{=} 0 \qquad \text{(e)} $$

so that

$$ {}^{R^*}\mathbf{a}^{\bar{R}'} \underset{(b)}{=} \underset{(d)}{0} + \underset{(e)}{0} + \underset{(c)}{0} = 0 \qquad \text{(f)} $$

and

$$ {}^{R^*}\mathbf{a}^P \underset{(a)}{=} {}^{R'}\mathbf{a}^P + \underset{(f)}{0} + \underset{(c)}{0} \qquad \text{(g)} $$

leads to the conclusion that

$$ F_r^* = F_r' \qquad \text{(h)} $$

The further hypothesis that R^* is an inertial reference frame permits one to write

$$ F_r + F_r^* \underset{(5.1)}{=} 0 \qquad \text{(i)} $$

and it follows that

$$F_r + F_r' \underset{(h)\quad(i)}{=} 0$$

which, in accordance with Section 5.1, shows that R' is an inertial reference frame.

To prove the second part of the theorem, let S consists of a single particle P of mass m. Then, if $\boldsymbol{\omega} \neq 0$, it is always possible to make P move in such a way that

$$^{R^*}\mathbf{a}^{\bar{R}'} + 2\boldsymbol{\omega} \times {}^{R'}\mathbf{v}^P \neq 0 \tag{j}$$

so that

$$^{R^*}\mathbf{a}^P \underset{(a),(j)}{\neq} {}^{R'}\mathbf{a}^P \tag{k}$$

Or, if $^{R^*}\mathbf{a}^Q \neq 0$ for any point Q fixed in R', let P remain fixed in R'. Then

$$^{R'}\mathbf{a}^P = 0 \tag{l}$$

whereas

$$^{R^*}\mathbf{a}^P \neq 0 \tag{m}$$

so that, once again

$$^{R^*}\mathbf{a}^P \underset{(l),(m)}{\neq} {}^{R'}\mathbf{a}^P$$

In either case, therefore, there exist motions such that

$$F_r^* \neq F_r' \tag{n}$$

and, if R^* is an inertial reference frame, so that

$$F_r + F_r^* \underset{(5.1)}{=} 0 \tag{o}$$

then

$$F_r + F_r' \underset{(n),(o)}{\neq} 0$$

which means that R' is not an inertial reference frame.

Returning now to the question raised earlier, let A be a reference frame in which both the sun S and the orientation of the reference frame F of the example in Section 5.1 remain fixed (that is, $^A\boldsymbol{\omega}^F = 0$), but in which E^* moves on a plane curve. A, S, F and the earth E are depicted in Figure 5.5, which also shows the moon M.

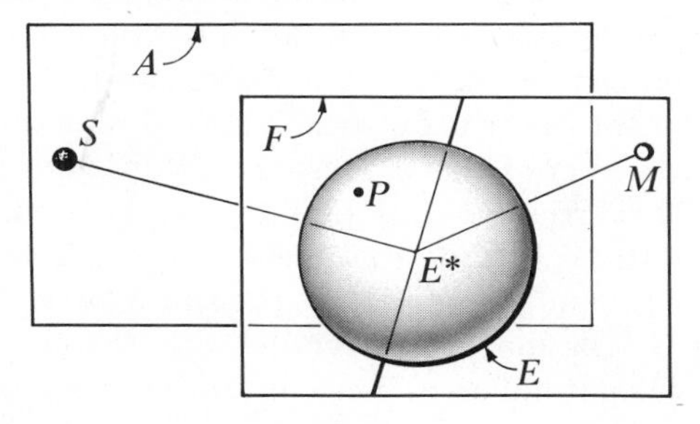

Figure 5.5

Suppose that A is an inertial reference frame. Then F cannot possibly be inertial, because there exists no point of F whose acceleration in A is equal to zero. As a matter of fact, the acceleration in A of every point that is fixed in F is equal to the acceleration ${}^A\mathbf{a}^{E^*}$ of the center E^* of E; and this acceleration has a magnitude that can be estimated with sufficient accuracy for present purposes by assuming that E^* moves in A on a circle of radius $R \approx 93 \times 10^6$ miles, described once per year. Now, how important is this acceleration? That depends on the acceleration in F of the particles of the body under consideration. Since the angular velocity and, hence, the angular acceleration of F in A are equal to zero, the acceleration ${}^A\mathbf{a}^P$ of a particle P in A differs from the acceleration ${}^F\mathbf{a}^P$ of P in F by precisely ${}^A\mathbf{a}^{E^*}$. Errors introduced by regarding F, rather than A, as inertial thus grow in importance as the ratio of $|{}^A\mathbf{a}^{E^*}|$ to $|{}^F\mathbf{a}^P|$ increases. Hence, in studying the moon, which moves in F, essentially, on a circle of radius $r \approx 240{,}000$ miles, completing approximately twelve such orbits per year, it must make a substantial difference whether A or F is assigned the role of inertial reference frame, because this ratio has the value

$$\frac{|{}^A\mathbf{a}^{E^*}|}{|{}^F\mathbf{a}^M|} = \left(\frac{R}{r}\right)\left(\frac{1}{12}\right)^2 \approx 2.7$$

which shows that ${}^A\mathbf{a}^{E^*}$ cannot be considered negligible in comparison with ${}^F\mathbf{a}^M$. For the pendulum considered in the example in Section 5.1, on the other hand, it may be inferred from the description of the motion of P that $g\alpha$ is a reasonable upper bound for $|{}^F\mathbf{a}^P|$, so that, with $g \approx 32.2$ ft/s² and a value of α even as small as 0.01 rad, one obtains

$$\frac{|{}^A\mathbf{a}^{E^*}|}{|{}^F\mathbf{a}^P|} \approx 0.06$$

Consequently, so far as numerical results are concerned, it matters far less in this case whether A or F is regarded as inertial.

It was shown in the example in Section 5.1 that the pendulum experiments of Foucault support the hypothesis that F is an inertial reference frame. Now it appears that these experiments support the same hypothesis for A, but that comparisons of actual and predicted motions of, for example, the moon can reveal which of these two reference frames is the stronger contender for the title of "true" inertial reference frame. It turns out that A is the winner of this contest. But this is not to say either that A is, in fact, truly inertial, or that Equation (5.1) should always be used in conjunction with A (rather than with F). Such phenomena as the nutation of the earth and the motion of the sun relative to the galaxy as a whole show that A falls short of perfection; and the use of A in place of F is desirable only when it makes a discernible difference, because, otherwise, it merely complicates matters. Similarly, a reference frame fixed relative to the earth is superior both to A and to F whenever its use leads to analytical simplifications unaccompanied by significant losses of accuracy, as is the case in a large number of situations encountered in engineering. Hence, unless explicitly exempted, every application of Equation (5.1) occurring in the remainder of this book will involve the assumption that any reference frame fixed relative to the earth is an inertial reference frame.

5.3 Use of Constraint Equations. If $m = 0$ (that is, if S is a holonomic system), the dynamical equations (5.1) furnish second-order differential equations in a number sufficient for the determination of the generalized coordinates of S. If $m \neq 0$, these equations must be supplemented with m constraint equations.

▪ EXAMPLE

Figure 5.6 shows the system previously considered in the examples in

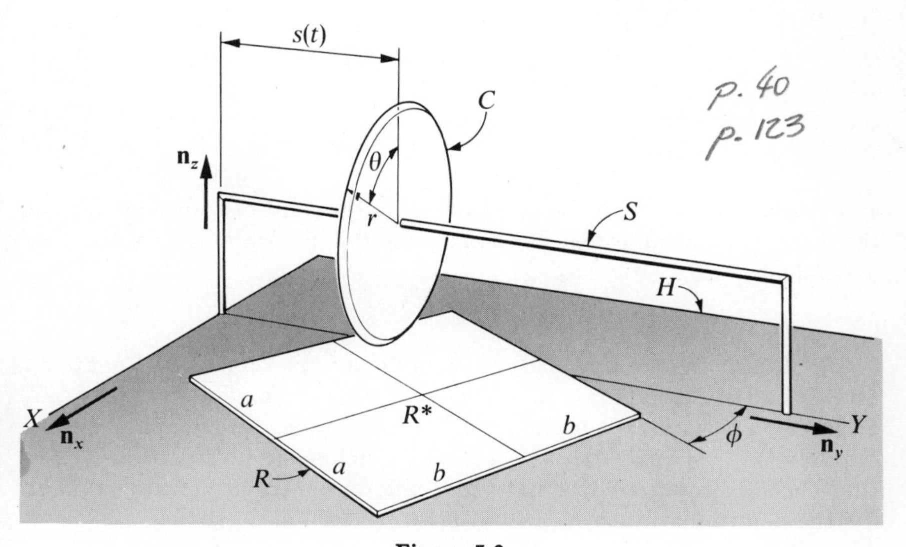

Figure 5.6

Sections 2.22, 2.23, and 3.24. The following facts were brought to light in the earlier discussions.

The system possesses two degrees of freedom (see the example in Section 2.22). If four generalized coordinates are defined as

$$q_1 = x \qquad q_2 = \phi \qquad q_3 = y \qquad q_4 = \theta$$

then these are related to each other by the constraint equations (see the example in Section 2.23)

$$\dot{q}_3 = \dot{s} - q_1\dot{q}_2 \tag{a}$$

$$\dot{q}_4 = \frac{1}{r}\left[(q_3 - s)\dot{q}_2 - \dot{q}_1\right] \tag{b}$$

and the generalized inertia forces corresponding to q_1 and q_2 are (see the example in Section 3.24)

$$F_1^* = -M\ddot{q}_1 + \frac{\ddot{q}_4 mr}{2}$$

$$F_2^* = Mq_1\ddot{q}_3 - \frac{\ddot{q}_2 M(a^2 + b^2)}{3} - \frac{\ddot{q}_4 mr(q_3 - s)}{2}$$

It may be verified by reference to Sections 3.3, 3.5, and 3.8 that the generalized active forces corresponding to q_1 and q_2 are equal to zero. Consequently, the dynamical equations (5.1) for this system are the two second-order differential equations

$$-M\ddot{q}_1 + \frac{\ddot{q}_4 mr}{2} = 0 \tag{c}$$

$$Mq_1\ddot{q}_3 - \frac{\ddot{q}_2 M(a^2 + b^2)}{3} - \frac{\ddot{q}_4 mr(q_3 - s)}{2} = 0 \tag{d}$$

and these, together with the two first-order differential equations (a) and (b), furnish four equations that suffice for the determination of the time dependence of $q_1, \cdots, q_4$.

5.4 Simultaneous Use of Dynamical Equations, Constraint Equations, and Equations Defining u's. When the m constraint equations (2.27) and $n - m$ dynamical equations (5.1) for a simple nonholonomic system S involve $n - m$ quantities $u_1, \cdots, u_{n-m}$ defined as in Equations (2.31), then $2n - m$ simultaneous first-order differential equations are available for the determination of $q_1, \cdots, q_n, u_1, \cdots, u_{n-m}$.

▪ EXAMPLE

In Figure 5.7, S represents a uniform sphere of radius r and mass m that rolls on the interior surface of a fixed tube T of radius R, the axis of

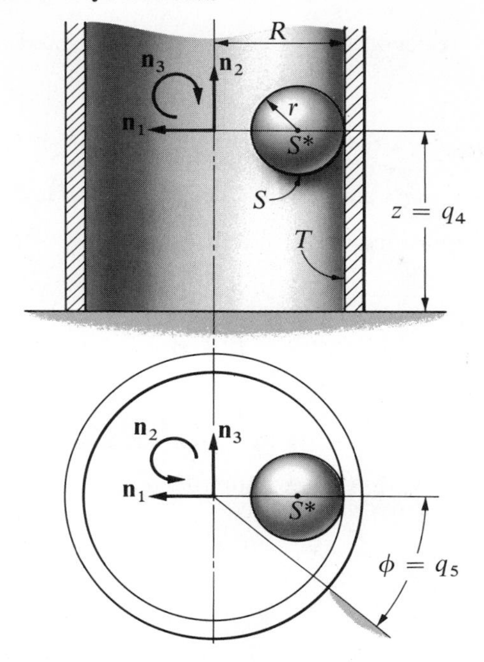

Figure 5.7

T being vertical. Five generalized coordinates may be introduced by letting $\mathbf{n}_1$, $\mathbf{n}_2$, and $\mathbf{n}_3$ be mutually perpendicular unit vectors directed at a typical instant as shown in the sketch and, further, taking $q_r = \theta_r$, for $r = 1, 2, 3$, $q_4 = z$, and $q_5 = \phi$, where z and ϕ measure a distance and an angle, respectively, as indicated in Figure 5.7, and θ_1, θ_2, and θ_3 are defined as follows: Let A_1, A_2, and A_3 be mutually perpendicular axes passing through S^* and parallel to $\mathbf{n}_1$, $\mathbf{n}_2$, and $\mathbf{n}_3$, respectively; let X_1, X_2, and X_3 be axes fixed in S and initially coincident with A_1, A_2, and A_3, respectively; and perform successive, right-handed rotations of amounts θ_1, θ_2, and θ_3 about X_1, X_2, and X_3, respectively.

If $\boldsymbol{\omega}$ and $\mathbf{v}$ denote the angular velocity and the velocity of the center S^* of S, the fact that S rolls on T can be expressed as

$$\mathbf{v} + \boldsymbol{\omega} \times (-r\mathbf{n}_1) \underset{(2.20)}{=} 0 \tag{a}$$

Now (see Figure 5.7)

$$\begin{aligned} \mathbf{v} &= \dot{z}\mathbf{n}_2 + (R - r)\dot{\phi}\mathbf{n}_3 \\ &= \dot{q}_4\mathbf{n}_2 + (R - r)\dot{q}_5\mathbf{n}_3 \end{aligned} \tag{b}$$

and

$$\boldsymbol{\omega} = \omega_1\mathbf{n}_1 + \omega_2\mathbf{n}_2 + \omega_3\mathbf{n}_3 \tag{c}$$

with

$$\left.\begin{aligned} \omega_1 &= \dot{q}_1 + s_2\dot{q}_3 \\ \omega_2 &= c_1\dot{q}_2 - s_1c_2\dot{q}_3 + \dot{q}_5 \\ \omega_3 &= s_1\dot{q}_2 + c_1c_2\dot{q}_3 \end{aligned}\right\} \qquad \text{(d)}$$

where

$$s_r = \sin q_r \quad c_r = \cos q_r \qquad r = 1, 2, 3$$

Hence, from Equations (a), (b), and (c),

and

$$\left.\begin{aligned} \dot{q}_4 - r\omega_3 &= 0 \\ (R - r)\dot{q}_5 + r\omega_2 &= 0 \end{aligned}\right\} \qquad \text{(e)}$$

If u_1, u_2, and u_3 are defined as

$$u_r = \omega_r \qquad r = 1, 2, 3 \qquad \text{(f)}$$

then the equations that here play the role of Equation (2.31) are [see Equations (d) and (f)]

$$\left.\begin{aligned} u_1 &= \dot{q}_1 + s_2\dot{q}_3 \\ u_2 &= c_1\dot{q}_2 - s_1c_2\dot{q}_3 + \dot{q}_5 \\ u_3 &= s_1\dot{q}_2 + c_1c_2\dot{q}_3 \end{aligned}\right\} \qquad \text{(g)}$$

Equations (e) and (f) permit the constraint equations to be expressed as

$$\left.\begin{aligned} \dot{q}_4 &= ru_3 \\ \dot{q}_5 &= -\frac{r}{R - r}\,u_2 \end{aligned}\right\} \qquad \text{(h)}$$

The three generalized inertia forces associated with u_1, u_2, and u_3 were found in the example in Section 3.25, and the corresponding generalized active forces are given by

$$F_r \underset{(3.7)}{=} -mg\mathbf{n}_2 \cdot \tilde{\mathbf{v}}_{u_r} \qquad r = 1, 2, 3$$

where (see the example in Section 3.25)

$$\tilde{\mathbf{v}}_{u_1} = 0 \qquad \tilde{\mathbf{v}}_{u_2} = -r\mathbf{n}_3 \qquad \tilde{\mathbf{v}}_{u_3} = r\mathbf{n}_2$$

Thus

$$F_1 = 0 \qquad F_2 = 0 \qquad F_3 = -mgr$$

and the dynamical equations (5.1) for the system are (see the example in Section 3.25 for F_1^*, F_2^*, and F_3^*)

$$\dot{u}_1 = \frac{r}{R - r}\,u_2u_3 \qquad \text{(i)}$$

$$\dot{u}_2 = 0 \qquad \text{(j)}$$

$$\dot{u}_3 = -\frac{1}{7}\left(\frac{2r}{R - r}\,u_1u_2 + 5\,\frac{g}{r}\right) \qquad \text{(k)}$$

The eight first-order differential equations (g) through (k) are, in principle, sufficient for the determination of the eight quantities $q_1, \cdots, q_5, u_1, \cdots, u_3$. In practice, however, not all of these can be found with equal ease. Specifically, a "closed form" solution of Equations (h) through (k) can be obtained readily; and this solution yields expressions for u_1, u_2, u_3, q_4, and q_5 involving only elementary functions of time; but the nonlinearities in Equations (g) are such as to prevent one from arriving at equally simple expressions for q_1, q_2, and q_3.

This sort of separation of equations of motion into two groups occurs frequently. Fortunately, it often turns out that the variables of primary importance are among those that can be found rather easily. For instance, in the present case, one might be interested only in the motion of the center S^* of S, rather than in the "attitude" of the sphere, in which case only q_4 and q_5 would have to be determined. To illustrate how one proceeds in such a situation, and because the results are rather interesting, a detailed description of the motion of S^* is given below.

Equation (j) has the general solution

$$u_2 = u_2^0 \tag{l}$$

where u_2^0 denotes the value of u_2 at time $t = 0$. From Equations (h) it then follows that

$$q_5 = -\frac{r}{R - r} u_2^0 t + q_5^0 \tag{m}$$

which means (see Figure 5.7) that S^* moves in a plane that rotates with a constant angular speed of magnitude $r|u_2^0|/(R - r)$ about the axis of the tube T. It remains to determine the (necessarily vertical) motion of S^* in this plane—that is, the time dependence of q_4.

Equation (k), solved for u_1, gives (for $u_2^0 \neq 0$)

$$u_1 \underset{(l)}{=} \frac{r - R}{2ru_2^0}\left(5\frac{g}{r} + 7\dot{u}_3\right)$$

and substitution into Equation (i) leads to

$$\ddot{u}_3 + p^2 u_3 \underset{(i),(m),(n)}{=} 0 \tag{n}$$

where p is defined as

$$p = \frac{1}{(R/r) - 1}\sqrt{\tfrac{2}{7}}\,|u_2^0| \tag{o}$$

The general solution of Equation (n) is

$$u_3 = u_3^0 \cos pt + (\dot{u}_3^0/p) \sin pt \tag{p}$$

where u_3^0 and $\dot{u}_3^0$ denote the initial values of u_3 and of $\dot{u}_3$, respectively;

and $\dot{u}_3{}^0$ can be expressed in terms of the initial values of u_1 and u_2 by reference to Equation (k):

$$\dot{u}_3{}^0 = -\frac{1}{7}\left(\frac{2r}{R - r}\, u_1{}^0 u_2{}^0 + 5\,\frac{g}{r}\right) \tag{q}$$

Only one further integration is required. If q_4 at $t = 0$ is taken equal to zero, then Equations (h) and (p) yield

$$q_4 = \frac{r}{p}\left[u_3{}^0 \sin pt + \frac{\dot{u}_3{}^0}{p}(1 - \cos pt)\right]$$

The predicted vertical motion of S^* is thus seen to have an oscillatory character. The circular frequency of the oscillations, p, depends [see Equation (o)] both on the initial value of ω_2 [see Equation (f)] and on the geometric parameter R/r. The amplitude of the oscillations can be adjusted to any desired value by appropriate choice [see Equation (q)] of the initial values of ω_1, ω_2, and ω_3. In particular, the amplitude can be made equal to zero—that is, the center of the sphere can be made to move on a horizontal circle—by taking ω_3 at $t = 0$ equal to zero and adjusting the initial values of ω_1 and ω_2 in such a way as to make $\dot{u}_3{}^0 = 0$.

These predictions may be in conflict with one's intuition; that is, one might expect the sphere to move downward, regardless of initial conditions. And, indeed, this is what happens in reality. The reason for the difference between the predicted and actual motions is that certain physically unavoidable dissipation effects—for example, frictional resistance to rotation—have not been taken into account in the analysis. [See Problem 11(d) for a more realistic approach.]

Finally, it should be noted that no explicit use was made of Equations (g) in the preceding analysis, so that the coordinates q_1, q_2, and q_3 need never have been mentioned. The use of u's in place of $\dot{q}$'s is particularly advantageous in situations of this kind.

5.5 Initial-Value Problems. The $2n - m$ first-order differential equations mentioned in Section 5.4 can always be expressed as

$$\sum_{s=1}^{n} V_{rs}\dot{q}_s = V_r \qquad r = 1, \cdots, n \tag{5.2}$$

$$\sum_{s=1}^{n-m} W_{rs}\dot{u}_s = W_r \qquad r = 1, \cdots, n - m \tag{5.3}$$

where V_{rs}, V_r, W_{rs}, and W_r are known functions of $q_1, \cdots, q_n, u_1, \cdots, u_{n-m}$, and t. When the values of $q_1, \cdots, q_n$ and $u_1, \cdots, u_{n-m}$ are known for one instant of time, this formulation is particularly convenient.

for a numerical integration. Hence, it is sometimes desirable to introduce $u_1, \cdots, u_{n-m}$ solely for the purpose of facilitating the solution of a so-called *initial-value problem.*

▪ EXAMPLE

It was shown in the example in Section 5.3 that the two first-order differential equations

$$\dot{q}_3 = \dot{s} - q_1\dot{q}_2 \tag{a}$$

$$\dot{q}_4 = \frac{1}{r}[(q_3 - s)\dot{q}_2 - \dot{q}_1] \tag{b}$$

together with the two second-order differential equations

$$-M\ddot{q}_1 + \frac{\ddot{q}_4 mr}{2} = 0 \tag{c}$$

$$Mq_1\ddot{q}_3 - \frac{\ddot{q}_2 M(a^2 + b^2)}{3} - \frac{\ddot{q}_4 mr(q_3 - s)}{2} = 0 \tag{d}$$

govern $q_1, \cdots, q_4$. If the state of the system at time $t = t_0$ is known, and one wishes to determine $q_1, \cdots, q_4$ for $t > t_0$, how can these equations be brought into a form suitable for numerical integration, and what are appropriate initial conditions?

Introduce u_1 and u_2 by letting

$$\dot{q}_1 = u_1 \tag{e}$$

and

$$\dot{q}_2 = u_2 \tag{f}$$

Then

$$\dot{q}_3 \underset{(a),(f)}{=} \dot{s} - q_1 u_2 \tag{g}$$

$$\dot{q}_4 \underset{(b),(e),(f)}{=} \frac{1}{r}[(q_3 - s)u_2 - u_1] \tag{h}$$

and

$$\ddot{q}_1 \underset{(e)}{=} \dot{u}_1 \tag{i}$$

$$\ddot{q}_2 \underset{(f)}{=} \dot{u}_2 \tag{j}$$

$$\ddot{q}_3 \underset{(g)}{=} \ddot{s} - \dot{q}_1 u_2 - q_1\dot{u}_2$$

$$\underset{(e)}{=} \ddot{s} - u_1 u_2 - q_1\dot{u}_2 \tag{k}$$

$$\ddot{q}_4 \underset{(b)}{=} \frac{1}{r}[(\dot{q}_3 - \dot{s})u_2 + (q_3 - s)\dot{u}_2 - \dot{u}_1]$$

$$\underset{(g)}{=} \frac{1}{r}[-q_1 u_2^2 + (q_3 - s)\dot{u}_2 - \dot{u}_1] \tag{l}$$

Consequently,

$$-\underset{(i)}{M\dot{u}_1} + \frac{m}{2}[-q_1u_2^2 + \underset{(l)}{(q_3 - s)}\dot{u}_2 - \dot{u}_1] \underset{(c)}{=} 0 \tag{m}$$

and

$$M q_1(\ddot{s} - \underset{(k)}{u_1u_2} - q_1\dot{u}_2) - \underset{(j)}{\frac{\dot{u}_2 M(a^2 + b^2)}{3}}$$

$$- \frac{m}{2}(q_3 - s)[-\underset{(l)}{q_1u_2^2} + (q_3 - s)\dot{u}_2 - \dot{u}_1] \underset{(d)}{=} 0 \tag{n}$$

or, after rearrangement,

$$\left(M + \frac{m}{2}\right)\dot{u}_1 - \frac{m}{2}(q_3 - s)\dot{u}_2 \underset{(m)}{=} -\frac{m}{2}q_1u_2^2 \tag{o}$$

and

$$\frac{m}{2}(q_3 - s)\dot{u}_1 - \left[\frac{M}{3}(a^2 + b^2) + Mq_1^2 + \frac{m}{2}(q_3 - s)^2\right]\dot{u}_2$$

$$\underset{(n)}{=} q_1\left[M(u_1u_2 - \ddot{s}) - \frac{m}{2}(q_3 - s)u_2^2\right] \tag{p}$$

Equations (e) through (h) have the form of Equation (5.2), and (o) and (p) are analogous to (5.3). A numerical integration can be performed as soon as the values of $q_1, \cdots, q_4, u_1$, and u_2 at t_0 have been determined. In the case of $q_1, \cdots, q_4$, this can be done directly by reference to a description of the state of the system at t_0; u_1 and u_2 at t_0 are found by substituting known initial values of two of the four quantities $\dot{q}_1, \cdots, \dot{q}_4$ into two of the four Equations (e)–(h). The only restriction is that $\dot{q}_2$ and $\dot{q}_3$ cannot be specified independently, because

$$\dot{q}_3 \underset{(f),(g)}{=} \dot{s} - q_1\dot{q}_2$$

5.6 The Principle of Virtual Work. The proposition that follows is called the *principle of virtual work*. It is equivalent to Lagrange's form of D'Alembert's principle (see Section 5.1), in the sense that either can be derived from the other.

There exist reference frames R, called inertial reference frames, such that, if S is any simple nonholonomic system possessing $n - m$ degrees of freedom in R, and δW and δW^* are, respectively, the virtual work of the active forces and the virtual work of the inertia forces for S in R (see Section 4.14) for *any* set of mutually compatible virtual displacements of the particles of S (see Section 2.28), then the equation

$$\delta W + \delta W^* = 0 \tag{5.4}$$

is satisfied during every motion of S.

Proofs: Assume the validity of Section 5.1. Then

$$\delta W + \delta W^* \underset{(4.30),(4.31)}{=} \sum_{r=1}^{n-m} (F_r + F_r^*)\delta q_r \underset{(5.1)}{=} 0$$

Conversely, assume the validity of the principle of virtual work, and form δW and δW^* for a set of mutually compatible virtual displacements of the particles of S, taking $\delta q_r = 0$ for all values of r except, say, $r = 1$. Then

$$\delta W \underset{(4.30)}{=} F_1 \delta q_1 \qquad \delta W^* \underset{(4.31)}{=} F_1^* \delta q_1$$

and

$$\delta W + \delta W^* = (F_1 + F_1^*)\delta q_1 \underset{(5.2)}{=} 0$$

from which it follows that

$$F_1 + F_1^* = 0$$

because, by hypothesis, $\delta q_1 \neq 0$. Similarly, for $r = 2$,

$$F_2 + F_2^* = 0$$

and so forth.

▪ EXAMPLE

In the example in Section 4.16 it was assumed that the angle θ can remain constant. Values of θ for which this is actually possible are found by noting that $\delta W + \delta W^*$ vanishes for all values of δx and $\delta\theta$ if the coefficients of these quantities are equal to zero—that is, if

$$g(2M + M' + m) \sin \beta - \ddot{x}\left(2M + M' + m + \frac{2I}{r^2}\right) = 0 \tag{a}$$

and

$$\ddot{x} \cos (\beta - \theta) - g \sin \theta = 0 \tag{b}$$

When $\ddot{x}$ is eliminated from these equations, it appears that θ must satisfy the equation

$$\mu \sin \beta \cos (\beta - \theta) - \sin \theta \underset{(a),(b)}{=} 0 \tag{c}$$

where

$$\mu = \frac{2M + M' + m}{2M + M' + m + 2I/r^2} \tag{d}$$

It follows that θ must be smaller than β, because, if θ exceeds β, then

$$\mu \cos(\beta - \theta) \underset{(c)}{>} 1 \tag{e}$$

so that

$$\cos(\beta - \theta) \underset{(e),(d)}{>} 1$$

which is impossible.

5.7 Lagrange's Equations of the First and of the Second Kind. If S is a holonomic system, Equations (5.1) can be replaced with

$$\frac{d}{dt}\frac{\partial K}{\partial \dot{q}_r} - \frac{\partial K}{\partial q_r} = F_r \qquad r = 1, \cdots, n \tag{5.5}$$

where K is the kinetic energy of S in an inertial reference frame; and if, furthermore, S is a conservative holonomic system (see Section 4.4) with a potential energy P in R, and a quantity L, defined as

$$L = K - P \tag{5.6}$$

is regarded as a function of $q_1, \cdots, q_n, \dot{q}_1, \cdots, \dot{q}_n$, and t, then Equation (5.5) can be replaced with

$$\frac{d}{dt}\frac{\partial L}{\partial \dot{q}_r} - \frac{\partial L}{\partial q_r} = 0 \qquad r = 1, \cdots, n \tag{5.7}$$

L is called the *Lagrangian* or the *kinetic potential* of S in R, and Equations (5.5) and (5.7) are known as *Lagrange's equations of motion* of the first kind and Lagrange's equations of motion of the second kind, respectively. The use of these equations in place of Equation (5.1) frequently reduces the amount of labor required for the formulation of dynamical equations, because it eliminates the necessity to determine accelerations.

Proofs: Equations (5.5) are an immediate consequence of (5.1) and (4.11); and if S possesses a potential energy P in R, then

$$F_r \underset{(4.6)}{=} -\frac{\partial P}{\partial q_r}$$

Consequently,

$$\frac{d}{dt}\frac{\partial K}{\partial \dot{q}_r} - \frac{\partial K}{\partial q_r} \underset{(5.5)}{=} -\frac{\partial P}{\partial q_r}$$

which can be replaced with

$$\frac{d}{dt}\frac{\partial}{\partial \dot{q}_r}(K - P) - \frac{\partial}{\partial q_r}(K - P) = 0$$

because P is independent of $\dot{q}_1, \cdots, \dot{q}_n$. Hence, to obtain Equations (5.7), it is only necessary to define L as in Equation (5.6).

5.8 The Activity and Activity-Energy Principles.

The following observations are intended to shed light on Equations (5.1) and (5.5) by reference to analogies between these equations and other, perhaps more familiar, relationships.

Consider a single particle P, and let $\mathbf{F}$ be the resultant of all contact and body forces acting on P, while $\mathbf{F}^*$ is the inertia force for P in an inertial reference frame R. Then, from D'Alembert's principle,

$$\mathbf{F} + \mathbf{F}^* = 0 \tag{a}$$

When this equation is dot-multiplied with the velocity $\mathbf{v}$ of P, there results

$$\mathbf{v} \cdot \mathbf{F} + \mathbf{v} \cdot \mathbf{F}^* = 0 \tag{b}$$

and, if two scalars, A and A^*, are defined as

$$A = \mathbf{v} \cdot \mathbf{F} \qquad A^* = \mathbf{v} \cdot \mathbf{F}^* \tag{c}$$

it then follows that

$$A + A^* = 0 \tag{d}$$

A and A^* are called the *activity* of the force $\mathbf{F}$ and the activity of the force $\mathbf{F}^*$, respectively, and Equation (d) is a statement of the *activity principle.*

If $\mathbf{a}$ denotes the acceleration of P in R, the activity A^* can be expressed as

$$\begin{aligned} A^* &\underset{(c)}{=} \mathbf{v} \cdot \mathbf{F}^* \\ &= \mathbf{v} \cdot (-m\mathbf{a}) \\ &= -m\mathbf{v} \cdot \frac{d\mathbf{v}}{dt} \\ &= -\frac{m}{2}\frac{d}{dt}(\mathbf{v}^2) \\ &= -\frac{d}{dt}\frac{m\mathbf{v}^2}{2} \\ &= -\frac{dK}{dt} \end{aligned} \tag{e}$$

where K is the kinetic energy of P in R. It then follows from Equations (d) and (e) that

$$\frac{dK}{dt} = A \tag{f}$$

This equation, which is simply another form of Equation (d), expresses the so-called *activity-energy principle.*

As both Equations (d) and (f) are scalar equations, neither can furnish sufficient information for the solution of any problem in which P has more than one degree of freedom. To this extent, both Equations (d) and (f) are weaker than (a), which is equivalent to three scalar equations. But (d) and (f) possess the following advantage over (a): If $\mathbf{F}$ contains any contributions from (unknown) constraint forces, these will certainly affect (a), but they frequently play no part in (d) and (f), this elimination being accomplished by the dot multiplication performed in deriving (b) from (a). To arrive at a formulation which, on the one hand, contains sufficient information for the solution of problems in which P has more than one degree of freedom, and, on the other hand, automatically eliminates unknown constraint forces, one may replace Equation (b) with

$$\mathbf{v}_{\dot{q}_r} \cdot \mathbf{F} + \mathbf{v}_{\dot{q}_r} \cdot \mathbf{F}^* = 0 \tag{b$'$}$$

where $\mathbf{v}_{\dot{q}_r}$ is the partial rate of change with respect to q_r of the position of P in R (see Section 2.12). If F_r and $F_r{}^*$ are defined as

$$F_r = \mathbf{v}_{\dot{q}_r} \cdot \mathbf{F} \qquad F_r{}^* = \mathbf{v}_{\dot{q}_r} \cdot \mathbf{F}^* \tag{c$'$}$$

it then follows that

$$F_r + F_r{}^* = 0 \tag{d$'$}$$

F_r and $F_r{}^*$ are, of course, the generalized active force (see Section 3.2) and the generalized inertia force (see Section 3.9) for P in R; and Equation (d′) is a special case of (5.1).

An analogy to the transition from Equation (d) to (f) is obtained by expressing $F_r{}^*$ as

$$\begin{aligned}
F_r{}^* &\underset{(c')}{=} \mathbf{v}_{\dot{q}_r} \cdot \mathbf{F}^* \\
&= \mathbf{v}_{\dot{q}_r} \cdot (-m\mathbf{a}) \\
&= -m\mathbf{v}_{\dot{q}_r} \cdot \frac{d\mathbf{v}}{dt} \\
&\underset{(2.25)}{=} -\frac{m}{2}\left(\frac{d}{dt}\frac{\partial \mathbf{v}^2}{\partial \dot{q}_r} - \frac{\partial \mathbf{v}^2}{\partial q_r}\right) \\
&= -\frac{d}{dt}\frac{\partial}{\partial \dot{q}_r}\left(\frac{m\mathbf{v}^2}{2}\right) + \frac{\partial}{\partial q_r}\left(\frac{m\mathbf{v}^2}{2}\right) \\
&= -\frac{d}{dt}\frac{\partial K}{\partial \dot{q}_r} + \frac{\partial K}{\partial q_r}
\end{aligned} \tag{e$'$}$$

where K is, as before, the kinetic energy of P in R. It then follows from

Equations (d′) and (e′) that

$$\frac{d}{dt}\frac{\partial K}{\partial \dot{q}_r} - \frac{\partial K}{\partial q_r} = F_r \tag{f'}$$

which shows that Equation (5.1) may be regarded as analogous to the activity-energy principle.

5.9 Systems at Rest in an Inertial Reference Frame. Lagrange's form of D'Alembert's principle (see Section 5.1), the principle of virtual work (see Section 5.6), and Lagrange's equations of motion, both of the first and of the second kind (see Section 5.7), apply not only to systems moving in an inertial reference frame, but also to systems at rest in such a reference frame, because rest is a special form of motion. For a system at rest, Equations (5.1), (5.4), and (5.5) through (5.7) reduce to

$$F_r = 0 \qquad r = 1, \cdots, n - m \tag{5.8}$$

$$\delta W = 0 \tag{5.9}$$

$$\frac{\partial P}{\partial q_r} = 0 \qquad r = 1, \cdots, n \tag{5.10}$$

Equations (5.8) and (5.9) apply both to holonomic and simple nonholonomic systems, whereas Equations (5.10) are valid only for holonomic systems.

▪ EXAMPLE

Figure 5.8 shows a frame F supported by wheels W_1, W_2, and W_3, the

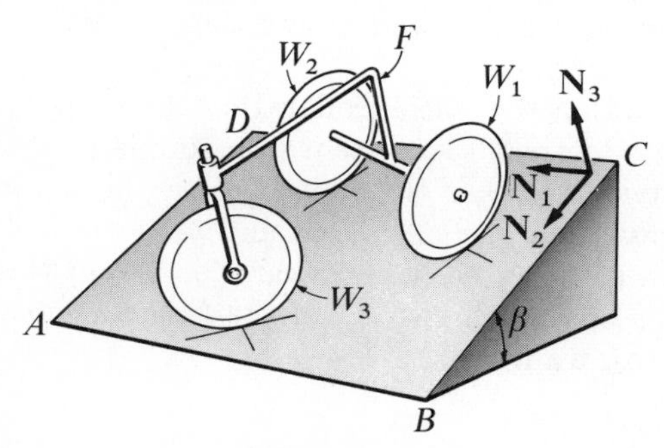

Figure 5.8

whole assembly resting on a plane $ABCD$ that is inclined at an angle β to the horizontal. W_1 can rotate freely about its axis; rotation of W_2 is resisted by a "braking" couple whose torque has a magnitude T; and

W_3 is mounted in a "fork," in such a way that it is free to rotate both about its own axis and that of the fork.

The configuration of this system can be described in terms of wheel rotation angles θ_1, θ_2, θ_3, a "steering" angle ψ, an "orientation" angle ϕ, and Cartesian coordinates x and y of the mass center S^*. These are indicated in a top view (Figure 5.9).

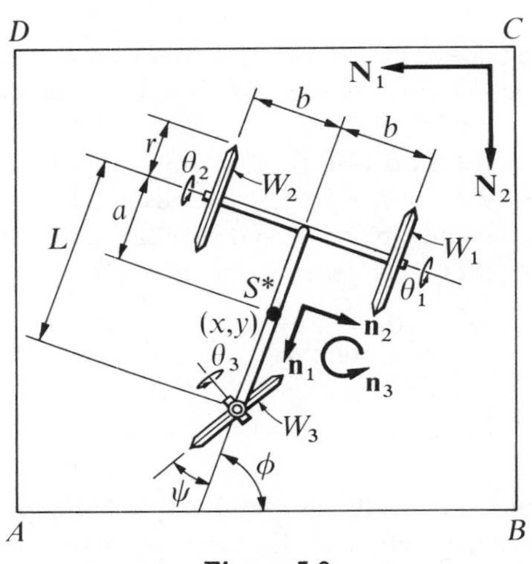

Figure 5.9

Assuming that there is sufficient friction to prevent the wheels from slipping on the supporting plane, the value of T required for equilibrium is to be expressed in terms of the inclination angle β, the weight W of the system, the dimensions r, a, b, and L and the quantities θ_1, θ_2, θ_3, ψ, ϕ, x, and y.

The assumption that the wheels roll, rather than slip, on the supporting plane permits one to treat the system as a simple nonholonomic system possessing only two degrees of freedom, because this assumption makes it possible to express each of $\dot{\theta}_1$, $\dot{\theta}_2$, $\dot{\phi}$, $\dot{x}$, and $\dot{y}$ (but not $\dot{\psi}$) in terms of $\dot{\theta}_3$ and one or more of θ_1, θ_2, θ_3, ψ, ϕ, x, and y. To do this, note that the angular velocities of the wheels and of the frame are (see Figure 5.9 for the unit vectors $\mathbf{n}_1$, $\mathbf{n}_2$, $\mathbf{n}_3$)

$$\boldsymbol{\omega}^{W_1} = \dot{\theta}_1\mathbf{n}_2 + \dot{\phi}\mathbf{n}_3 \tag{a}$$
$$\boldsymbol{\omega}^{W_2} = \dot{\theta}_2\mathbf{n}_2 + \dot{\phi}\mathbf{n}_3 \tag{b}$$
$$\boldsymbol{\omega}^{W_3} = \dot{\theta}_3(\sin\psi\mathbf{n}_1 + \cos\psi\mathbf{n}_2) + (\dot{\phi} - \dot{\psi})\mathbf{n}_3 \tag{c}$$
$$\boldsymbol{\omega}^{F} = \dot{\phi}\mathbf{n}_3 \tag{d}$$

If no slip occurs at the points of contact of the wheels with the plane, the velocities $\mathbf{v}_1$, $\mathbf{v}_2$, and $\mathbf{v}_3$ of the centers of W_1, W_2, and W_3 can be expressed as

$$\mathbf{v}_1 = \boldsymbol{\omega}^{W_1} \times (r\mathbf{n}_3) = r\dot{\theta}_1\mathbf{n}_1 \tag{e}$$
$$\mathbf{v}_2 = \boldsymbol{\omega}^{W_2} \times (r\mathbf{n}_3) = r\dot{\theta}_2\mathbf{n}_1 \tag{f}$$
$$\mathbf{v}_3 = \boldsymbol{\omega}^{W_3} \times (r\mathbf{n}_3) = r\dot{\theta}_3(\cos\psi\mathbf{n}_1 - \sin\psi\mathbf{n}_2) \tag{g}$$

But $\mathbf{v}_1$ and $\mathbf{v}_2$ are also given by

$$\mathbf{v}_1 \underset{(2.20)}{=} \mathbf{v}_3 + \boldsymbol{\omega}^F \times (-L\mathbf{n}_1 + b\mathbf{n}_2) \tag{h}$$

and

$$\mathbf{v}_2 \underset{(2.20)}{=} \mathbf{v}_3 + \boldsymbol{\omega}^F \times (-L\mathbf{n}_1 - b\mathbf{n}_2) \tag{i}$$

Substitution from Equations (d), (e), and (g) into (h) leads to the two scalar equations

$$r\dot{\theta}_1 = r\dot{\theta}_3 \cos\psi - b\dot{\phi}$$
$$0 = -r\dot{\theta}_3 \sin\psi - L\dot{\phi}$$

so that

$$\dot{\phi} = -\frac{r\dot{\theta}_3}{L}\sin\psi \tag{j}$$

and

$$\dot{\theta}_1 = \dot{\theta}_3\left[\cos\psi + \frac{b}{L}\sin\psi\right] \tag{k}$$

Similarly, from Equations (d), (f), and (g),

$$\dot{\theta}_2 = \dot{\theta}_3\left[\cos\psi - \frac{b}{L}\sin\psi\right] \tag{l}$$

Finally, similar expression can be found for $\dot{x}$ and $\dot{y}$ by using the fact that the velocity $\mathbf{v}^*$ of S^* is given both by (see Figure 5.9 for the unit vectors $\mathbf{N}_1$ and $\mathbf{N}_2$)

$$\mathbf{v}^* = \dot{x}\mathbf{N}_1 + \dot{y}\mathbf{N}_2$$

and by

$$\mathbf{v}^* \underset{(2.20)}{=} \mathbf{v}_3 + \boldsymbol{\omega}^F \times [-(L-a)\mathbf{n}_1]$$
$$\underset{(d),(g),(j)}{=} r\dot{\theta}_3\left(\cos\psi\mathbf{n}_1 - \frac{a}{L}\sin\psi\mathbf{n}_2\right) \tag{m}$$

Two generalized forces, F_1 and F_2, are formed by taking into account the weight force $\mathbf{W}$, applied at S^* and given by (see Figure 5.9 for the unit vectors $\mathbf{N}_2$ and $\mathbf{N}_3$)

$$\mathbf{W} = W(\sin\beta\mathbf{N}_2 - \cos\beta\mathbf{N}_3)$$
$$= W[\sin\beta\,(\sin\phi\mathbf{n}_1 + \cos\phi\mathbf{n}_2) - \cos\beta\mathbf{n}_3] \tag{n}$$

and torques $\mathbf{T}^F$ and $\mathbf{T}^{W_2}$, representing the action of the braking couple on the frame F and wheel W_2, respectively. These are given by

$$\mathbf{T}^F = -\mathbf{T}^{W_2} = T\mathbf{n}_2 \tag{o}$$

If θ_3 and ψ are now regarded as generalized coordinates q_1 and q_2, respectively, then F_1 and F_2 can be expressed as [see Equations (3.6) and (3.5)]

$$F_r = \tilde{\mathbf{v}}_{\dot{q}_r}{}^* \cdot \mathbf{W} + \boldsymbol{\omega}_{\dot{q}_r}{}^F \cdot \mathbf{T}^F + \tilde{\boldsymbol{\omega}}_{\dot{q}_r}{}^{W_2} \cdot \mathbf{T}^{W_2} \qquad r = 1, 2 \tag{p}$$

where

$$\tilde{\mathbf{v}}_{\dot{q}_1}{}^* \underset{(m)}{=} r\left(\cos\psi\mathbf{n}_1 - \frac{a}{L}\sin\psi\mathbf{n}_2\right) \tag{q}$$

$$\tilde{\boldsymbol{\omega}}_{\dot{q}_1}{}^F \underset{(d),(j)}{=} -\frac{r}{L}\sin\psi\mathbf{n}_3 \tag{r}$$

$$\tilde{\boldsymbol{\omega}}_{\dot{q}_1}{}^{W_2} \underset{(b),(j),(l)}{=} \left(\cos\psi - \frac{b}{L}\sin\psi\right)\mathbf{n}_2 - \frac{r}{L}\sin\psi\mathbf{n}_3 \tag{s}$$

and

$$\tilde{\mathbf{v}}_{\dot{q}_2}{}^* = \tilde{\boldsymbol{\omega}}_{\dot{q}_2}{}^F = \tilde{\boldsymbol{\omega}}_{\dot{q}_2}{}^{W_2} = 0 \tag{t}$$

Hence

$$\begin{aligned} F_1 \underset{(p)}{=} rW\sin\beta&\left(\cos\psi\sin\phi \underset{(n),(q)}{-} \frac{a}{L}\sin\psi\cos\phi\right) \\ &+ \underset{(o),(r)}{0} + (-T)\left(\cos\psi \underset{(o),(s)}{-} \frac{b}{L}\sin\psi\right) \end{aligned} \tag{u}$$

and

$$F_2 \underset{(p),(t)}{=} 0$$

One of the equilibrium conditions [Equation (5.8)] is thus seen to be satisfied identically, and the other yields the desired expression for T, namely

$$T = rW\sin\beta\,\frac{\cos\psi\sin\phi - (a/L)\sin\psi\cos\phi}{\cos\psi - (b/L)\sin\psi}$$

which, however, is valid only when the denominator is not equal to zero. If

$$\cos\psi - \frac{b}{L}\sin\psi = 0 \tag{v}$$

then it follows from Equation (u) that F_1 can vanish only if either $\beta = 0$ or

$$\cos\psi\sin\phi - \frac{a}{L}\sin\psi\cos\phi = 0$$

which, in view of Equation (v), is equivalent to

$$\tan\phi = \frac{a}{b} \tag{w}$$

These results may be stated in geometric terms, as follows: If $\beta \neq 0$ and [see Equation (v)] the axis of wheel W_3 passes through the center of wheel W_2, then equilibrium is possible only if [see Equation (w)] the line passing through S^* and the center of W_2 is parallel to line BC in Figure 5.9.

5.10 The Principle of Stationary Potential Energy. Equations (5.10) express the *principle of stationary potential energy*. This principle furnishes an especially convenient method for the derivation of equilibrium conditions because it requires nothing more than partial differentiations of a single scalar function of the generalized coordinates.

▪ EXAMPLE

Two uniform bars, B_1 and B_2 each of length L and weight W, are supported by pins, as indicated in Figure 5.10, and are attached to each

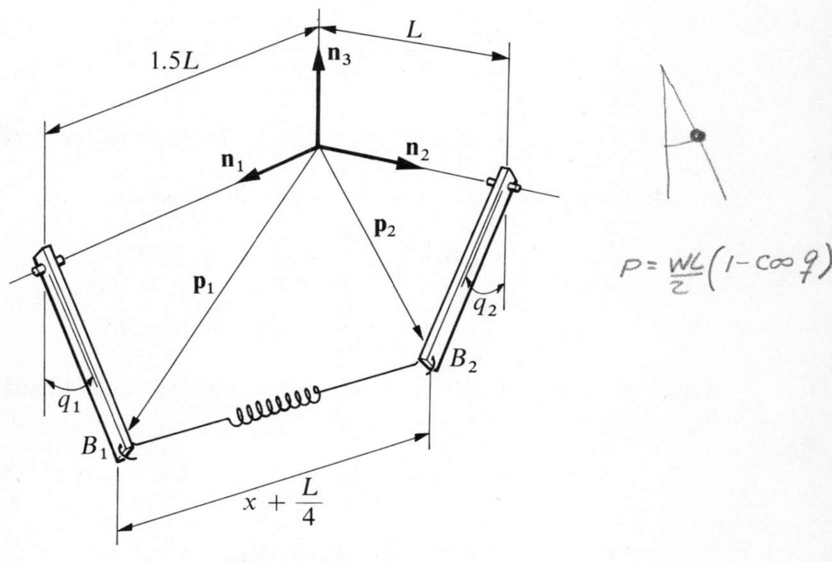

Figure 5.10

other by a linear spring of natural length $L/4$ and spring constant W/L. The elongation of the spring when the system is at rest is to be determined.

The potential energy P of the system can be expressed as [see Equations (4.2) and (4.5)]

$$P = -\frac{WL}{2}c_1 - \frac{WL}{2}c_2 + \frac{1}{2}\frac{W}{L}x^2$$

where c_1 and c_2 denote the cosines of the angles q_1 and q_2 shown in Figure 5.10 and x measures the elongation of the spring. With q_1 and q_2 used as generalized coordinates, Equations (5.10) lead to the equilibrium conditions

$$\frac{\partial P}{\partial q_1} = \frac{WL}{2} s_1 + \frac{W}{L} x \frac{\partial x}{\partial q_1} = 0 \tag{a}$$

$$\frac{\partial P}{\partial q_2} = \frac{WL}{2} s_2 + \frac{W}{L} x \frac{\partial x}{\partial q_2} = 0 \tag{b}$$

where $s_i = \sin q_i$, $i = 1, 2$.

A third equation is obtained by noting that

$$\left(x + \frac{L}{4}\right)^2 = (\mathbf{p}_1 - \mathbf{p}_2)^2 \tag{c}$$

where $\mathbf{p}_1$ and $\mathbf{p}_2$ are the position vectors shown in Figure 5.10. These are given by (see Figure 5.10 for $\mathbf{n}_1$, $\mathbf{n}_2$, $\mathbf{n}_3$)

$$\mathbf{p}_1 = L(1.5\mathbf{n}_1 + s_1\mathbf{n}_2 - c_1\mathbf{n}_3)$$

and

$$\mathbf{p}_2 = L(s_2\mathbf{n}_1 + \mathbf{n}_2 - c_2\mathbf{n}_3)$$

Consequently,

$$\left(x + \frac{L}{4}\right)^2 \underset{(c)}{=} L^2[(1.5 - s_2)^2 + (1 - s_1)^2 + (c_1 - c_2)^2] \tag{d}$$

and differentiation with respect to q_1 and q_2 leads to

$$\left(x + \frac{L}{4}\right) \frac{\partial x}{\partial q_1} = (s_1c_2 - c_1)L^2$$

$$\left(x + \frac{L}{4}\right) \frac{\partial x}{\partial q_2} = (c_1s_2 - 1.5c_2)L^2$$

Solution of these equations for $\partial x/\partial q_1$ and $\partial x/\partial q_2$, and substitution into Equations (a) and (b) then gives

$$\left(x + \frac{L}{4}\right) s_1 + 2x(s_1c_2 - c_1) = 0 \tag{e}$$

$$\left(x + \frac{L}{4}\right) s_2 + 2x(c_1s_2 - 1.5c_2) = 0 \tag{f}$$

Equations (d), (e), and (f) comprise a set of simultaneous, nonlinear equations in the unknowns q_1, q_2, and x. A method for solving such equations with a high-speed digital computer is discussed in the next section. Applied to the present problem, it yields the two sets of solutions

$$x = 0.65L \qquad q_1 = 35.55 \text{ degrees} \qquad q_2 = 44.91 \text{ degrees}$$

and

$$x = 2.91L \qquad q_1 = -114.69 \text{ degrees} \qquad q_2 = -94.77 \text{ degrees}$$

and it may be verified by substitution that Equations (d), (e), and (f) are satisfied by either set of values.

5.11 Real Solutions of Sets of Nonlinear Simultaneous Equations.

The necessity to find real solutions of a set of nonlinear simultaneous equations arises frequently in mechanics, as well as in other areas of applied mathematics. What follows is the description of a method that is well suited for use with a high-speed digital computer.

The most general set of n simultaneous equations in n unknowns $x_1, \cdots, x_n$ can be expressed as

$$\left.\begin{aligned} F_1(x_1, \cdots, x_n) &= 0 \\ F_2(x_1, \cdots, x_n) &= 0 \\ &\vdots \\ F_n(x_1, \cdots, x_n) &= 0 \end{aligned}\right\} \tag{5.11}$$

Let $y_1(\tau), \cdots, y_n(\tau)$ be a set of functions of a variable τ, $0 \leq \tau \leq 1$; take

$$y_i(0) = k_i \qquad i = 1, \cdots, n \tag{5.12}$$

where k_i is selected arbitrarily; and require that $y_1(\tau), \cdots, y_n(\tau)$ satisfy the equations

$$\left.\begin{aligned} F_1(y_1, \cdots, y_n) &= F_1(k_1, \cdots, k_n)(1-\tau) \\ F_2(y_1, \cdots, y_n) &= F_2(k_1, \cdots, k_n)(1-\tau) \\ &\vdots \\ F_n(y_1, \cdots, y_n) &= F_n(k_1, \cdots, k_n)(1-\tau) \end{aligned}\right\} \tag{5.13}$$

Then, as the right-hand sides of Equations (5.13) vanish at $\tau = 1$, the functions $y_1(\tau), \cdots, y_n(\tau)$ satisfy, at $\tau = 1$, the same equations as $x_1, \cdots, x_n$; and $y_1(1), \cdots, y_n(1)$ may be found as follows: Differentiate Equation (5.13) with respect to τ to obtain the set of first-order ordinary differential equations

$$\left.\begin{aligned} \frac{\partial F_1}{\partial y_1}\frac{dy_1}{d\tau} + \cdots + \frac{\partial F_1}{\partial y_n}\frac{dy_n}{d\tau} &= -F_1(k_1, \cdots, k_n) \\ &\vdots \\ \frac{\partial F_n}{\partial y_1}\frac{dy_1}{d\tau} + \cdots + \frac{\partial F_n}{\partial y_n}\frac{dy_n}{d\tau} &= -F_n(k_1, \cdots, k_n) \end{aligned}\right\} \tag{5.14}$$

and solve these with a high-order integration scheme[1] on a digital computer, using Equations (5.12) for initial conditions and terminating the integration at $\tau = 1$.

▪ EXAMPLE

If q_1, q_2, and x of the example in Section 5.10 are replaced with x_1, x_2, and Lx_3, respectively, and Equations (d), (e), and (f) are compared with Equation (5.11), the functions that play the parts of F_1, F_2, and F_3 are seen to be

$$\begin{aligned}
F_1(x_1, x_2, x_3) &= (1.5 - \sin x_2)^2 + (1 - \sin x_1)^2 \\
&\quad + (\cos x_1 - \cos x_2)^2 - (x_3 + 0.25)^2 \\
F_2(x_1, x_2, x_3) &= (x_3 + 0.25)\sin x_1 \\
&\quad + 2x_3(\sin x_1 \cos x_2 - \cos x_1) \\
F_3(x_1, x_2, x_3) &= (x_3 + 0.25)\sin x_2 \\
&\quad + 2x_3(\cos x_1 \sin x_2 - 1.5\cos x_2)
\end{aligned}$$

The equations corresponding to (5.14) are

$$\begin{aligned}
&2(\cos y_2 \sin y_1 - \cos y_1)\frac{dy_1}{d\tau} + 2(\cos y_1 \sin y_2 - 1.5\cos y_2)\frac{dy_2}{d\tau} \\
&\quad - 2(y_3 + 0.25)\frac{dy_3}{d\tau} = -[(1.5 - \sin k_2)^2 + (1 - \sin k_1)^2 \\
&\qquad + (\cos k_1 - \cos k_2)^2 - (k_3 + 0.25)^2]
\end{aligned}$$

$$\begin{aligned}
&[(y_3 + 0.25)\cos y_1 + 2y_3(\cos y_1 \cos y_2 + \sin y_1)]\frac{dy_1}{d\tau} \\
&\quad - 2y_3 \sin y_1 \sin y_2 \frac{dy_2}{d\tau} + [\sin y_1 + 2(\sin y_1 \cos y_2 - \cos y_1)]\frac{dy_3}{d\tau} \\
&\qquad = -[(k_3 + 0.25)\sin k_1 + 2k_3(\sin k_1 \cos k_2 - \cos k_1)]
\end{aligned}$$

$$\begin{aligned}
&-2y_3 \sin y_1 \sin y_2 \frac{dy_1}{d\tau} \\
&\quad + [(y_3 + 0.25)\cos y_2 + 2y_3(\cos y_1 \cos y_2 + 1.5\sin y_2)]\frac{dy_2}{d\tau} \\
&\quad + [\sin y_2 + 2(\cos y_1 \sin y_2 - 1.5\cos y_2)]\frac{dy_3}{d\tau} \\
&\qquad = -[(k_3 + 0.25)\sin k_2 + 2k_3(\cos k_1 \sin k_2 - 1.5\cos k_2)]
\end{aligned}$$

[1] Fox, L., *Numerical Solution of Ordinary and Partial Differential Equations.* New York: Pergamon Press, 1962.

A numerical integration of these equations in the interval $0 \leq \tau \leq 1$ was performed on a digital computer. In accordance with Equations (5.12), the initial conditions for the integration were taken to be

$$y_i(0) = k_i \qquad i = 1, 2, 3$$

and, with k_1, k_2, and k_3 set equal to zero, this led to

$$y_1(1) = 35.55 \text{ degrees} \qquad y_2(1) = 44.91 \text{ degrees} \qquad y_3(1) = 0.65$$

Identical results were obtained with $k_1 = k_2 = 10$ degrees and $k_3 = 0.1$, but a different solution was found by taking $k_1 = k_2 = -100$ degrees and $k_3 = 2$; that is,

$$y_1(1) = -114.69 \text{ degrees} \qquad y_2(1) = -94.77 \text{ degrees} \qquad y_3(1) = 2.91$$

This is not surprising, for a set of nonlinear equations can possess many sets of solutions, and the second one simply corresponds to an equilibrium configuration in which the bars lie above, rather than below, the horizontal plane determined by the axes of the supporting pins.

5.12 Connected Rigid Bodies. Use of Equations (5.8), (5.9), or (5.10) is especially advantageous when a system is composed of a number of connected rigid bodies, because it facilitates elimination of forces of interaction between these bodies (see Sections 3.6 and 4.16).

▪ EXAMPLE

Figure 5.11 shows a pin-connected linkage. Two of the links are pinned

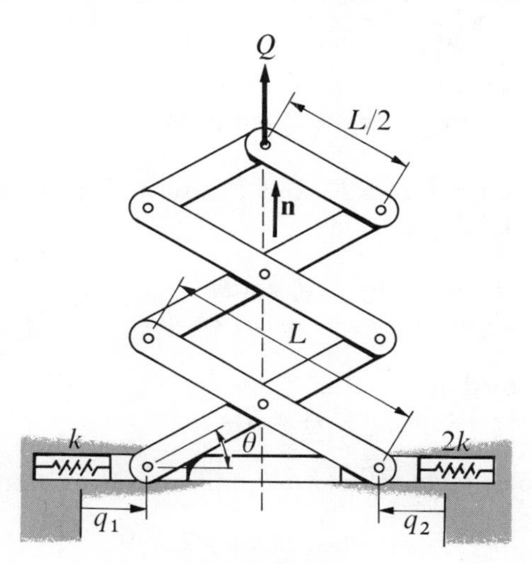

Figure 5.11

to sliders, which are, in turn, connected to fixed supports by means of linear springs; a force **Q** is applied to the linkage, as shown in the sketch. The springs have spring constants k and $2k$; the angle θ is equal to 30 degrees when both springs are undeformed; and the links have lengths L and $L/2$.

Assuming that weight forces are negligible in comparison with the force **Q** and forces exerted by the springs, **Q** is to be determined such that the system remains at rest when θ is equal to 45 degrees.

The system possesses two degrees of freedom, and the quantities q_1 and q_2 shown in Figure 5.11 may be used as generalized coordinates. If these are chosen such that both are equal to zero when θ is equal to thirty degrees, then q_1, q_2, and θ are related to each other as follows:

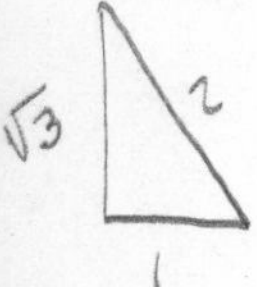

$$q_1 + q_2 + L \cos \theta = \frac{L\sqrt{3}}{2} \tag{a}$$

The velocity **v** of the point of application of the force **Q** is given by

$$\begin{aligned} \mathbf{v} &= \frac{d}{dt}\left(\frac{5L}{2} \sin \theta\right) \mathbf{n} + \cdots \\ &= \tfrac{5}{2} L \cos \theta \dot{\theta} \mathbf{n} + \cdots \end{aligned} \tag{b}$$

where **n** is a unit vector directed as shown in Figure 5.11 and the dots represent a vector perpendicular to **n**. To eliminate $\dot{\theta}$ from Equation (b), differentiate (a) with respect to t, solve for $\dot{\theta}$, and substitute. This gives

$$\mathbf{v} \underset{(a),(b)}{=} \tfrac{5}{2} \cot \theta (\dot{q}_1 + \dot{q}_2) \mathbf{n} + \cdots \tag{c}$$

and it follows that

$$\mathbf{v}_{\dot{q}_1} = \mathbf{v}_{\dot{q}_2} = \tfrac{5}{2} \cot \theta \mathbf{n} + \cdots \tag{d}$$

The generalized active forces F_1 and F_2 are now given by (see Sections 3.2, 3.6, 4.1, and 4.3)

$$\begin{aligned} F_1 &= -kq_1 + (\tfrac{5}{2} \cot \theta \mathbf{n} + \cdots) \cdot \mathbf{Q} \\ F_2 &= -2kq_2 + (\tfrac{5}{2} \cot \theta \mathbf{n} + \cdots) \cdot \mathbf{Q} \end{aligned}$$

and, if **Q** is expressed as

$$\mathbf{Q} = Q\mathbf{n} \tag{e}$$

then Equation (5.8) yields the two equilibrium conditions

$$\begin{aligned} -kq_1 + \tfrac{5}{2} \cot \theta Q &= 0 \\ -2kq_2 + \tfrac{5}{2} \cot \theta Q &= 0 \end{aligned}$$

These equations may be solved for q_1 and q_2, respectively, and substitution into Equation (a) then shows that Q and θ are related as follows whenever the system is at rest:

$$\frac{Q}{k}\frac{15}{4}\cot\theta + L\cos\theta = \frac{L\sqrt{3}}{2} \tag{f}$$

Hence, for $\theta = \pi/4$,

$$Q \underset{\text{(e),(f)}}{=} \frac{2kL(\sqrt{3}-\sqrt{2})}{15}\mathbf{n}$$

5.13 Constraint Forces and Forces of Interaction. When constraint forces or forces of interaction between parts of a system are, themselves, of interest, use of Equations (5.8), (5.9), and (5.10) would appear to be self-defeating (see Section 5.12). This difficulty can be overcome by replacing the given system temporarily with one that possesses additional degrees of freedom and introducing associated contact forces.

▪ EXAMPLE

A pin-connected frame is loaded by a force of magnitude P, as shown in Figure 5.12. The reaction of bar CD on bar AB is equivalent to a force $\mathbf{R}$

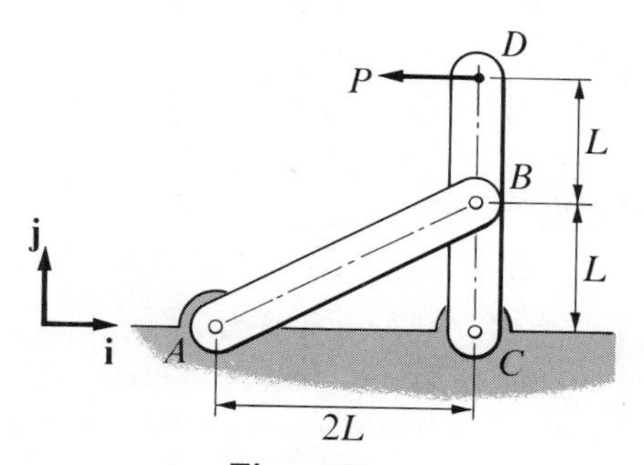

Figure 5.12

applied at B and given by

$$\mathbf{R} = H\mathbf{i} + V\mathbf{j}$$

where $\mathbf{i}$ and $\mathbf{j}$ are unit vectors directed as shown. To determine H and V, replace the given system temporarily with the two-degree-of-freedom system shown in Figure 5.13, using q_1 and q_2 as generalized coordinates, and

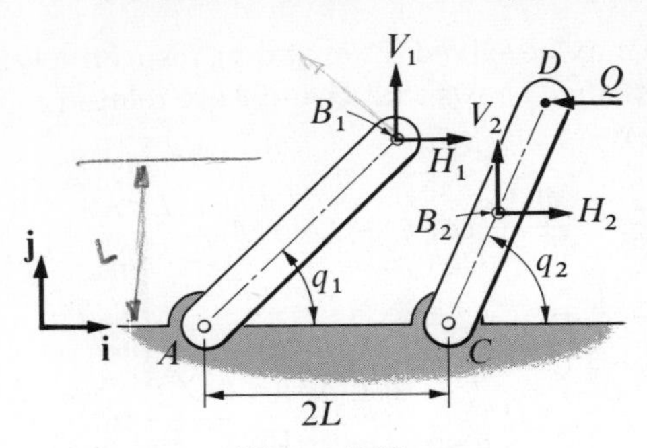

Figure 5.13

introduce forces as indicated. The generalized active forces F_1 and F_2 are then given by

$$F_r = \mathbf{v}_{\dot{q}_r}{}^{B_1} \cdot (H_1\mathbf{i} + V_1\mathbf{j}) + \mathbf{v}_{\dot{q}_r}{}^{B_2} \cdot (H_2\mathbf{i} + V_2\mathbf{j}) + \mathbf{v}_{\dot{q}_r}{}^{D} \cdot (-Q\mathbf{i}) \qquad r = 1, 2$$

and, when the system occupies the configuration of interest—that is, the one depicted in Figure 5.12—the velocities of points B_1, B_2, and D are

$$\mathbf{v}^{B_1} = L\dot{q}_1(-\mathbf{i} + 2\mathbf{j}) \qquad \mathbf{v}^{B_2} = -L\dot{q}_2\mathbf{i} \qquad \mathbf{v}^{D} = -2L\dot{q}_2\mathbf{i}$$

and (use the law of action and reaction)

$$H_1 = -H_2 = H \qquad V_1 = -V_2 = V \qquad Q = P$$

Hence

$$F_1 = L(-H + 2V) \underset{(5.8)}{=} 0$$

and

$$F_2 = L(H + 2P) \underset{(5.8)}{=} 0$$

H and V thus have the values

$$H = -2P \qquad V = -P$$

If only one component of the force **R** is of interest—for example, the component parallel to **j**—the given system can be replaced with a single-degree-of-freedom system in which AB is pinned to a slider that is attached to CD and in which forces of magnitude S_1, S_2, and Q are applied to AB and CD as shown in Figure 5.14 (S_1 acts on AB, S_2 on CD). The velocities $\mathbf{v}_1$, $\mathbf{v}_2$, and $\mathbf{v}$ of the points of application of these forces, when $q = 90$ degrees, are given by [see Equation (2.21)]

$$\mathbf{v}_1 = L\dot{q}(\mathbf{i} - 2\mathbf{j}) \qquad \mathbf{v}_2 = L\dot{q}\mathbf{i} \qquad \mathbf{v} = 2L\dot{q}\mathbf{i}$$

and S_1, S_2, and Q then have the values

$$S_1 = -S_2 = V \qquad Q = P$$

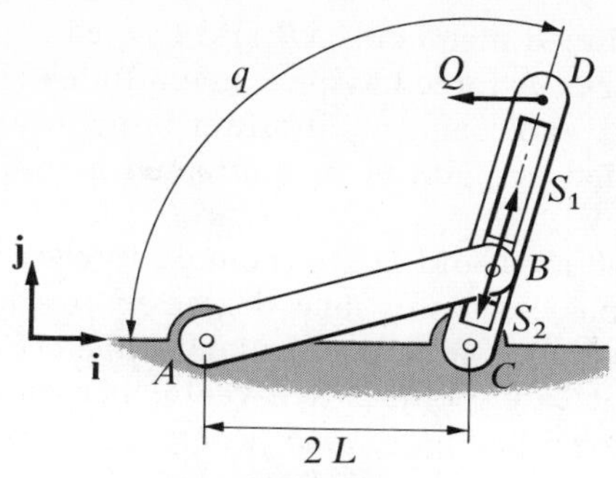

Figure 5.14

so that the generalized force F becomes

$$\begin{aligned} F &= L(\mathbf{i} - 2\mathbf{j}) \cdot (S_1\mathbf{j}) + L\mathbf{i} \cdot (-S_2\mathbf{j}) + 2L\mathbf{i} \cdot (-Q\mathbf{i}) \\ &= -2LV - 2LP = 0 \end{aligned} \tag{5.8}$$

and, as before,

$$V = -P$$

5.14 A Graphical Method. Equation (5.9) provides the underlying basis for a graphical method of solution of statics problems. This method consists of identifying the forces that contribute to δW (see Section 4.16); making a scale drawing of a set of mutually compatible virtual displacements (see Section 2.28) of the points of application of these forces; forming δW; and using Equation (5.9). The method is particularly helpful when only limited numerical accuracy is required and the geometry of the system under consideration is relatively complex.

▪ EXAMPLE

Figure 5.15 is a schematic representation of a "quick return" mechanism

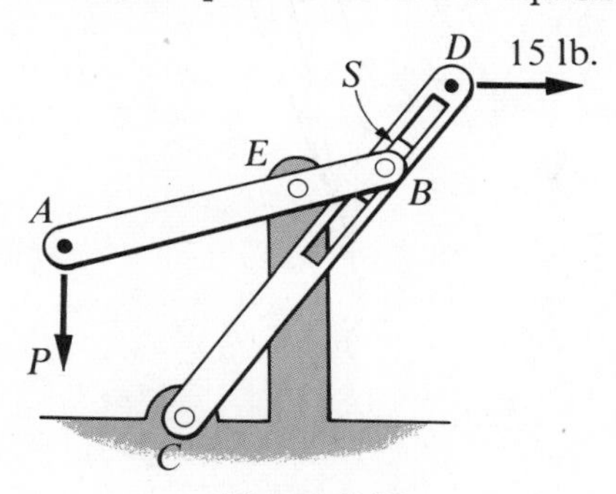

Figure 5.15

consisting of two pinned members, AB and CD, connected to each other by means of a slider S. A force having a magnitude of 15 lb and directed as shown is applied at D, and equilibrium is maintained by a force of magnitude P, applied at point A and directed as indicated. P is to be determined.

The forces applied at A and D are the only forces contributing to δW. Since the system possesses only one degree of freedom, and since the velocity of point A must be perpendicular to line AB, the most general virtual displacement $\delta\mathbf{a}$ of point A is a vector perpendicular to this line. Such a vector is shown in Figure 5.15, where numbers in circles indicate the order in which lines are drawn. Both the sense and the magnitude of $\delta\mathbf{a}$ have been selected arbitrarily. Next, lines ② and ③ are drawn through E and through the tip of $\delta\mathbf{a}$, and through B and perpendicular to line AB, respectively, to obtain a virtual displacement $\delta\mathbf{b}$ of B compatible with $\delta\mathbf{a}$. Now, $\delta\mathbf{b}$ can also be regarded as the sum of a virtual displacement $\delta\mathbf{b}_1$

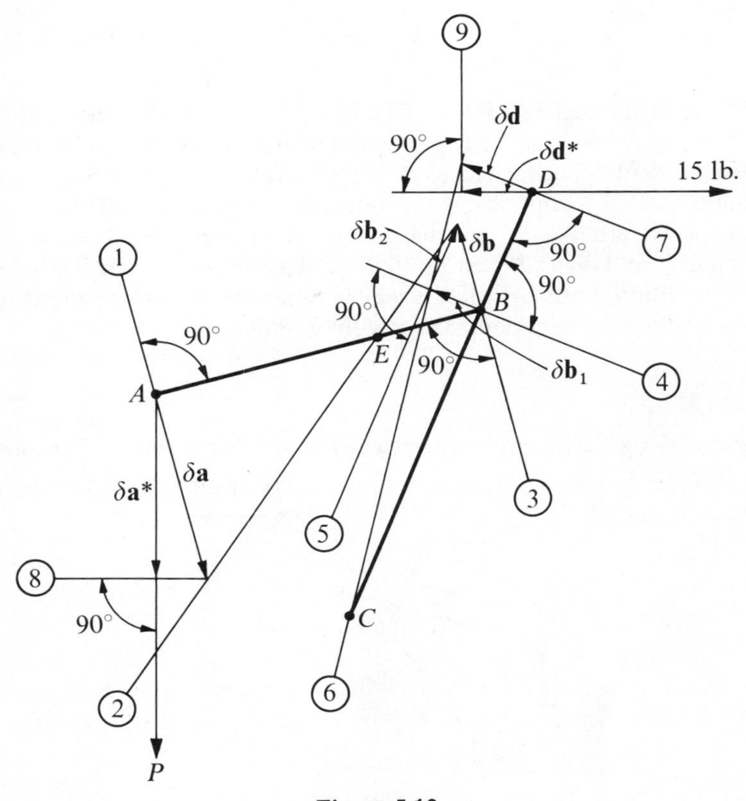

Figure 5.16

of a point of bar CD that coincides with B and a virtual displacement $\delta\mathbf{b}_2$ of B in a reference frame attached to bar CD. These virtual displacements must be perpendicular to CD and parallel to CD, respectively. $\delta\mathbf{b}_1$ and $\delta\mathbf{b}_2$ are thus found by drawing lines ④ and ⑤ as shown; and the virtual displacement $\delta\mathbf{d}$ of D compatible with $\delta\mathbf{b}_1$, and hence with $\delta\mathbf{a}$, is obtained by drawing line ⑥ through C and through the tip of $\delta\mathbf{b}_1$, and line ⑦ through D and perpendicular to CD. Finally, $\delta\mathbf{a}$ and $\delta\mathbf{d}$ are resolved into components parallel and perpendicular to the forces applied at the points A and D (lines ⑧ and ⑨), the components parallel to the forces being called $\delta\mathbf{a}^*$ and $\delta\mathbf{d}^*$, respectively; and δW can then be expressed as

$$\delta W = P|\delta\mathbf{a}^*| - 15|\delta\mathbf{d}^*|$$

Hence, for equilibrium, that is, when

$$\delta W = 0$$

P is given by

$$P = \frac{15|\delta\mathbf{a}^*|}{|\delta\mathbf{d}^*|} \text{ lb}$$

and the numerical value of the ratio $|\delta\mathbf{a}^*|/|\delta\mathbf{d}^*|$ can be found after expressing the lengths of the vectors $\delta\mathbf{a}^*$ and $\delta\mathbf{d}^*$ in Figure 5.16 in any convenient units.

5.15 Motions Resembling States of Rest. Certain systems can perform motions that bear a close resemblance to states of rest. Specifically, a holonomic system S possessing n generalized coordinates $q_1, \cdots, q_n$ in an inertial reference frame R may be able to move in such a way that, throughout some time interval,

$$\dot{q}_r = 0 \qquad r = 1, \cdots, n \tag{5.15}$$

This can occur if the kinetic energy K of S does not depend explicitly on the time t; and $q_1, \cdots, q_n$ must satisfy the equations

$$\bar{F}_r + \frac{\partial \bar{K}}{\partial q_r} = 0 \qquad r = 1, \cdots, n \tag{5.16}$$

where $\bar{F}_r$ and $\bar{K}$ are formed by setting $\dot{q}_1, \cdots, \dot{q}_n$ equal to zero in the generalized force F_r (see Section 3.2) and in K, respectively. If S possesses a potential energy P in R, Equations (5.16) can be replaced with

$$\frac{\partial \bar{L}}{\partial q_r} = 0 \qquad r = 1, \cdots, n \tag{5.17}$$

where $\bar{L}$ is defined as

$$\bar{L} = \bar{K} - P \tag{5.18}$$

Proof: If K does not depend explicitly on t, K can be regarded as a function of the $2n$ variables $q_1, \cdots, q_n$ and $\dot{q}_1, \cdots, \dot{q}_n$; and the partial derivatives $\partial K/\partial \dot{q}_r$, $r = 1, \cdots, n$, are then functions of precisely the same $2n$ variables; that is,

$$\frac{\partial K}{\partial \dot{q}_r} = f_r(q_1, \cdots, q_n, \dot{q}_1, \cdots, \dot{q}_n)$$

The total time derivative of $\partial K/\partial \dot{q}_r$ is thus given by

$$\frac{d}{dt}\frac{\partial K}{\partial \dot{q}_r} = \sum_{s=1}^{n} \left(\frac{\partial f_r}{\partial q_s} \dot{q}_s + \frac{\partial f_r}{\partial \dot{q}_s} \ddot{q}_s \right) \underset{(5.15)}{=} 0$$

and the generalized inertial force $F_r{}^*$ becomes

$$F_r{}^* \underset{(4.11)}{=} \frac{\partial K}{\partial q_r} \qquad r = 1, \cdots, n$$

Furthermore, it does not matter whether $\dot{q}_1, \cdots, \dot{q}_n$ are set equal to zero before or after differentiation with respect to q_r; that is,

$$\left. \frac{\partial K}{\partial q_r} \right|_{\dot{q}_1 = \cdots = \dot{q}_n = 0} = \frac{\partial \bar{K}}{\partial q_r} \qquad r = 1, \cdots, n$$

Hence,

$$F_r{}^* = \frac{\partial \bar{K}}{\partial q_r} \qquad r = 1, \cdots, n$$

and, from Equations (5.1),

$$\bar{F}_r + \frac{\partial \bar{K}}{\partial q_r} = 0 \qquad r = 1, \cdots, n$$

Finally, if S possesses a potential energy P, then

$$\bar{F}_r \underset{(4.6)}{=} -\frac{\partial P}{\partial q_r}$$

so that

$$-\frac{\partial P}{\partial q_r} + \frac{\partial \bar{K}}{\partial q_r} = 0 \qquad r = 1, \cdots, n$$

or, after using Equation (5.18),

$$\frac{\partial \bar{L}}{\partial q_r} = 0 \qquad r = 1, \cdots, n$$

▪ EXAMPLE

Two thin uniform bars, B_1 and B_2, each of mass m and length L, are connected by a pin, and B_1 is pinned to a vertical shaft S that is made to rotate with constant angular speed Ω, as indicated in Figure 5.17. (The

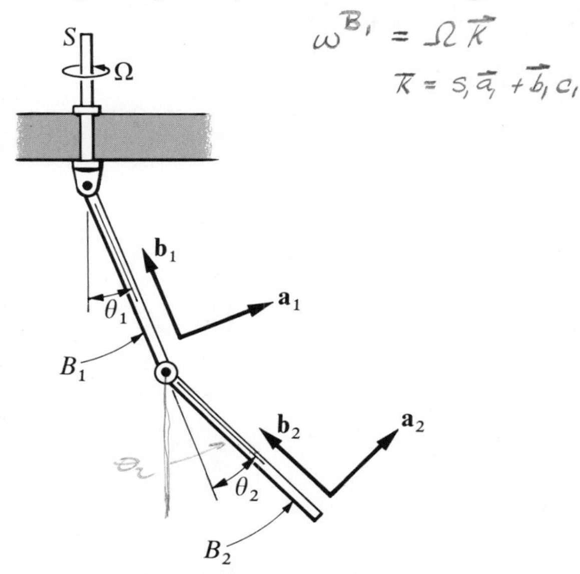

Figure 5.17

two pins are parallel to each other and horizontal.) This system can move in such a way that θ_1 and θ_2 remain constant. Equations governing θ_1 and θ_2 during such motions are to be derived.

When $\dot{\theta}_1 = \dot{\theta}_2 = 0$, the angular velocities of B_1 and B_2 are

$$\begin{aligned} \boldsymbol{\omega}^{B_1} &= \Omega(s_1\mathbf{a}_1 + c_1\mathbf{b}_1) \\ \boldsymbol{\omega}^{B_2} &= \Omega(s_2\mathbf{a}_2 + c_2\mathbf{b}_2) \end{aligned}$$

where $\mathbf{a}_i$ and $\mathbf{b}_i$, with $i = 1, 2$, are unit vectors directed as shown in Figure 5.17, and c_i and s_i denote $\cos \theta_i$ and $\sin \theta_i$, respectively; and the velocity $\mathbf{v}_2$ of the mass center of B_2 can be expressed as

$$\mathbf{v}_2 = L\Omega\left(s_1 + \frac{s_2}{2}\right)\mathbf{b}_1 \times \mathbf{a}_1$$

For the motion in question, the kinetic energy $\bar{K}$ is thus given by

$$\bar{K} = \underset{(4.20)}{\frac{1}{2}\frac{mL^2}{3}}\,\Omega^2 s_1{}^2 + \tfrac{1}{2}mL^2\Omega^2 \underset{(4.12)}{\left(s_1 + \frac{s_2}{2}\right)^2}$$

$$\underset{(4.16)}{+}\ \tfrac{1}{2}\,\frac{mL^2}{12}\,\Omega^2 s_2{}^2$$

$$= \frac{mL^2\Omega^2}{2}\,(\tfrac{4}{3}s_1{}^2 + s_1 s_2 + \tfrac{1}{3}s_2{}^2)$$

while the potential energy P becomes

$$P = -mgL(\tfrac{1}{2}c_1 + c_1 + \tfrac{1}{2}c_2) = -\frac{mgL}{2}\,(3c_1 + c_2)$$

Consequently,

$$\bar{L} \underset{(5.18)}{=} \frac{mL^2\Omega^2}{2}\,(\tfrac{4}{3}s_1{}^2 + s_1 s_2 + \tfrac{1}{3}s_2{}^2) + \frac{mgL}{2}\,(3c_1 + c_2)$$

and

$$\frac{\partial \bar{L}}{\partial \theta_1} = \frac{mL^2\Omega^2}{2}\,(\tfrac{8}{3}s_1 c_1 + c_1 s_2) - \frac{3mgL}{2}\,s_1$$

while

$$\frac{\partial \bar{L}}{\partial \theta_2} = \frac{mL^2\Omega^2}{2}\,(s_1 c_2 + \tfrac{2}{3}s_2 c_2) - \frac{mgL}{2}\,s_2$$

The desired equations are thus

$$\frac{L\Omega^2 c_1}{g}\,(8s_1 + 3s_2) - 9s_1 \underset{(5.17)}{=} 0$$

and

$$\frac{L\Omega^2 c_2}{g}\,(3s_1 + 2s_2) - 3s_2 \underset{(5.17)}{=} 0$$

5.16 Steady Motion. A simple nonholonomic system S possessing $n - m$ degrees of freedom in an inertial reference frame R (see Section 2.22) is said to be in a state of *steady motion* in R when $\dot{q}_1, \cdots, \dot{q}_{n-m}$, (but not necessarily $\dot{q}_{n-m+1}, \cdots, \dot{q}_n$) have constant values. (The motions discussed in Section 5.15 are thus steady motions.) The conditions under which this is possible can be found by using Equations (5.1). When doing so, one may replace $\dot{q}_1, \cdots, \dot{q}_{n-m}$ with constants to form expressions for generalized inertial forces, but partial rates of change of position and partial rates of change of orientation must first be constructed without reference to the fact that $\dot{q}_1, \cdots, \dot{q}_{n-m}$ are to remain constant.

▪ EXAMPLE

A sharp-edged circular disk can roll with constant speed on a horizontal plane in such a way that the center of the disk moves on a circular path while the normal to the middle plane of the disk makes a constant angle with the vertical. But this motion can occur only if the speed of the center, the radii of the disk and of the circular path, the inclination angle, and the inertial properties of the disk are related to each other in a certain way. To discover this relationship, refer to the example in Section 2.24, noting that $n - m = 3$; that the motion in question is a steady motion in which $\dot{\psi}$, $\dot{\phi}$, and $\dot{\theta}$ (but not $\dot{x}$ and $\dot{y}$) remain constant; and that partial rates of change of orientation of D and partial rates of change of position of D^* are

$$\tilde{\boldsymbol{\omega}}_{u_1} = \mathbf{n}_1 \qquad \tilde{\boldsymbol{\omega}}_{u_2} = \mathbf{n}_2 \qquad \tilde{\boldsymbol{\omega}}_{u_3} = \mathbf{n}_3 \tag{a}$$

$$\tilde{\mathbf{v}}_{u_1}{}^{D^*} = r\mathbf{n}_3 \qquad \tilde{\mathbf{v}}_{u_2}{}^{D^*} = 0 \qquad \tilde{\mathbf{v}}_{u_3}{}^{D^*} = -r\mathbf{n}_1 \tag{b}$$

where u_1, u_2, u_3, defined as

$$u_1 = -\dot{\theta} \qquad u_2 = \dot{\phi}\cos\theta \qquad u_3 = \dot{\psi} + \dot{\phi}\sin\theta \tag{c}$$

can be used to express the angular velocity of D as

$$\boldsymbol{\omega} = u_1\mathbf{n}_1 + u_2\mathbf{n}_2 + u_3\mathbf{n}_3 \tag{d}$$

When θ remains constant and D^* moves with a velocity $V\mathbf{n}_1$ on a circle of radius R, then

$$\dot{\theta} = 0 \tag{e}$$

and $V\mathbf{n}_1$ can be expressed both as

$$V\mathbf{n}_1 = R\dot{\phi}\mathbf{n}_1 \tag{f}$$

and as

$$V\mathbf{n}_1 = \boldsymbol{\omega} \times (r\mathbf{n}_2) \tag{g}$$

Consequently,

$$\dot{\phi} \underset{\text{(f)}}{=} \frac{V}{R} \tag{h}$$

and

$$V \underset{\text{(g),(d)}}{=} -ru_3 \tag{i}$$

During this motion, u_1, u_2, and u_3 are thus given by

$$u_1 \underset{\text{(c) (e)}}{=} 0 \qquad u_2 \underset{\text{(c),(h)}}{=} \frac{V}{R}\cos\theta \qquad u_3 \underset{\text{(i)}}{=} \frac{-V}{r} \tag{j}$$

and $\boldsymbol{\omega}$ becomes

$$\boldsymbol{\omega} \underset{(d),(j)}{=} \frac{V}{R}\left(\cos\theta\,\mathbf{n}_2 - \frac{R}{r}\,\mathbf{n}_3\right) \tag{k}$$

where $\mathbf{n}_2$, $\mathbf{n}_3$, and θ appear as shown in Figure 5.18.

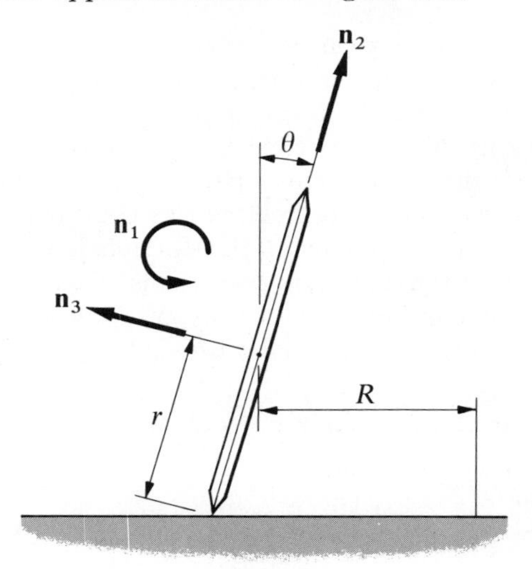

Figure 5.18

The acceleration of D^* is given by

$$\mathbf{a}^{D^*} = \frac{V^2}{R}\,(\sin\theta\,\mathbf{n}_2 - \cos\theta\,\mathbf{n}_3)$$

and the angular acceleration of D is

$$\boldsymbol{\alpha} = \frac{d\boldsymbol{\omega}}{dt} \underset{(k)}{=} -\frac{V^2}{R^2}\left(\frac{R}{r} + \sin\theta\right)\cos\theta\,\mathbf{n}_1$$

Let m be the mass of D, and let I and J denote the moments of inertia of D with respect to lines that pass through D^* and are, respectively, perpendicular and parallel to $\mathbf{n}_3$. The inertia force F^* and the inertia torque $\mathbf{T}^*$ can then be expressed as

$$\begin{aligned}\mathbf{F}^* &= -m\mathbf{a}^{D^*} \\ &= -\frac{mV^2}{R}\,(\sin\theta\,\mathbf{n}_2 - \cos\theta\,\mathbf{n}_3)\end{aligned}$$

and

$$\mathbf{T}^* \underset{(3.39)}{=} \frac{V^2}{R^2}\left(I\sin\theta + \frac{R}{r}J\right)\cos\theta\,\mathbf{n}_1$$

The generalized forces F_1 and F_1^* are given by

$$F_1 \underset{(3.7)}{=} -mg\,(\cos\theta\,\mathbf{n}_2 + \sin\theta\,\mathbf{n}_3)\cdot\tilde{\mathbf{v}}_{u_1}{}^{D^*}$$
$$\underset{(b)}{=} -mgr\sin\theta$$

and

$$F_1^* \underset{(3.48)}{=} \tilde{\mathbf{v}}_{u_1}{}^{D^*}\cdot\mathbf{F}^* + \tilde{\boldsymbol{\omega}}_{u_1}\cdot\mathbf{T}^*$$
$$= m\frac{V^2}{R}r\cos\theta + \frac{V^2}{R^2}\left(I\sin\theta + \frac{R}{r}J\right)\cos\theta$$
$$= \frac{V^2}{R}\left(mr + \frac{I}{R}\sin\theta + \frac{J}{r}\right)\cos\theta$$

while the generalized forces F_2, F_3, F_2^*, and F_3^* vanish. Consequently, the last two of the equations

$$F_r + F_r^* = 0 \qquad r = 1, 2, 3$$

are satisfied identically, and the first yields the relationship

$$-mgr\sin\theta + \frac{V^2}{R}\left(mr + \frac{I}{R}\sin\theta + \frac{J}{r}\right)\cos\theta = 0$$

5.17 Constraint Forces and Forces of Interaction. Constraint forces and forces of interaction between parts of a moving system can be found by proceeding as in the case of systems at rest (see Section 5.13); that is, the system under consideration may be replaced temporarily with one that possesses additional degrees of freedom and is subjected to the action of additional forces.

▪ EXAMPLE

Figure 5.19 shows a rod of mass $2m$ whose ends A and B move in a vertical and in a horizontal slot, respectively. At any instant during this motion, the system of forces exerted by the portion BC of the rod on the portion AC can be replaced (see Section 3.1) with a couple of torque $M_1\mathbf{n}_1 + M_2\mathbf{n}_2 + M_3\mathbf{n}_3$, together with a force applied at C; and if the surfaces of the slots are smooth, the system of forces exerted on the rod by the vertical slot is equivalent to a couple whose torque is perpendicular to $\mathbf{n}_3$, together with a force $P_1\mathbf{n}_1 + P_3\mathbf{n}_3$ applied at A. On grounds of symmetry, M_1, M_2, and P_3 can be assumed to be equal to zero. P_1 and M_3 can be found as follows.

So long as contact is maintained between the rod and lines X and Y,

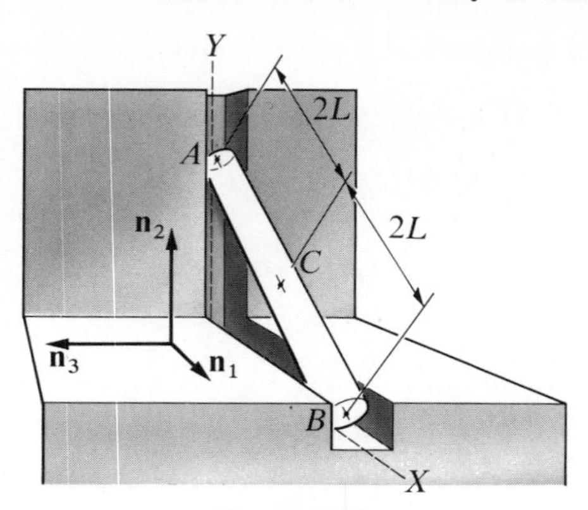

Figure 5.19

the rod possesses only a single degree of freedom. For the determination of P_1 and M_3, the rod is temporarily replaced with a three-degree-of-freedom system consisting of two pin-connected rods, R_1 and R_2, as shown in Figure 5.20. To recover the original system, one then chooses P_1 and M_3 in such a way that $q_1 = 0$ and $q_2 = q_3$.

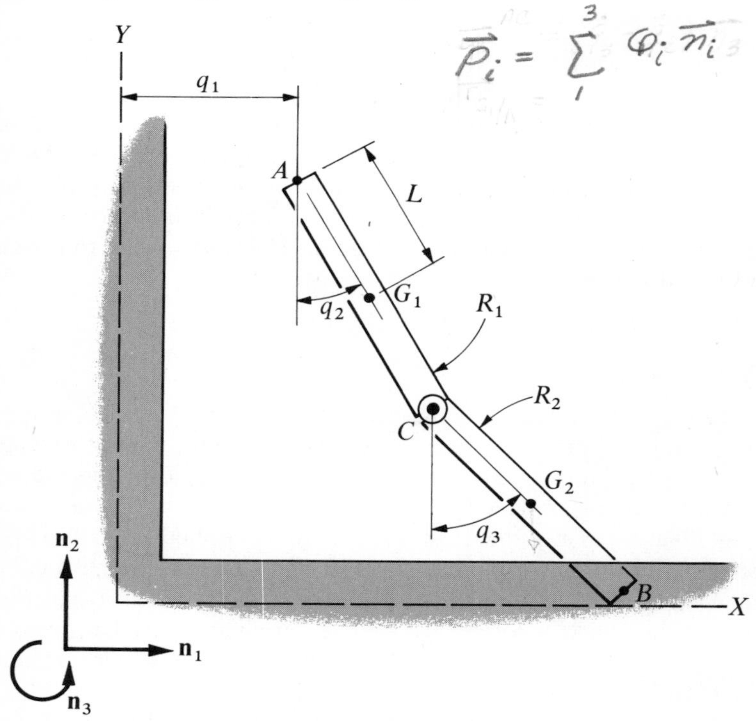

Figure 5.20

The velocities of points A, G_1, and G_2 can be expressed as

$$\begin{aligned} \mathbf{v}^A &= \dot{q}_1\mathbf{n}_1 + \cdots \\ \mathbf{v}^{G_1} &= (\dot{q}_1 + Lc_2\dot{q}_2)\mathbf{n}_1 - (Ls_2\dot{q}_2 + 2Ls_3\dot{q}_3)\mathbf{n}_2 \\ \mathbf{v}^{G_2} &= (\dot{q}_1 + 2Lc_2\dot{q}_2 + Lc_3\dot{q}_3)\mathbf{n}_1 - Ls_3\dot{q}_3\mathbf{n}_2 \end{aligned}$$

where s_i and c_i denote sin q_i and cos q_i, respectively. The angular velocities of R_1 and R_2 are

$$\boldsymbol{\omega}^{R_1} = \dot{q}_2\mathbf{n}_3 \qquad \boldsymbol{\omega}^{R_2} = \dot{q}_3\mathbf{n}_3$$

and the accelerations of G_1 and G_2, evaluated only for motions during which $q_1 = 0$ and $q_2 = q_3$, are given by

$$\begin{aligned} \mathbf{a}^{G_1} &= L[(-s_3\dot{q}_3^2 + c_3\ddot{q}_3)\mathbf{n}_1 - 3(c_3\dot{q}_3^2 + s_3\ddot{q}_3)\mathbf{n}_2] \\ \mathbf{a}^{G_2} &= L[3(-s_3\dot{q}_3^2 + c_3\ddot{q}_3)\mathbf{n}_1 - (c_3\dot{q}_3^2 + s_3\ddot{q}_3)\mathbf{n}_2] \end{aligned}$$

The function P, defined as

$$P = mgL(c_2 + 3c_3)$$

is a potential function for the gravitational forces acting on R_1 and R_2, and the generalized active forces F_1, F_2, and F_3 can now be expressed as

$$F_r = \mathbf{v}_{\dot{q}_r}{}^A \cdot (P_1\mathbf{n}_1) + \boldsymbol{\omega}_{\dot{q}_r}{}^{R_1} \cdot (M_3\mathbf{n}_3) + \boldsymbol{\omega}_{\dot{q}_r}{}^{R_2} \cdot (-M_3\mathbf{n}_3) - \frac{\partial P}{\partial q_r} \qquad r = 1, 2, 3$$

Consequently, for motions during which $q_1 = 0$ and $q_2 = q_3$,

$$\begin{aligned} F_1 &= P_1 \\ F_2 &= mgLs_3 + M_3 \\ F_3 &= 3mgLs_3 - M_3 \end{aligned}$$

If $\mathbf{T}_1{}^*$ and $\mathbf{T}_2{}^*$ are the inertia torques for R_1 and R_2, then the generalized inertia forces $F_1{}^*$, $F_2{}^*$, and $F_3{}^*$ assume the form

$$\begin{aligned} F_r{}^* = &-m\mathbf{v}_{\dot{q}_r}{}^{G_1} \cdot \mathbf{a}^{G_1} - m\mathbf{v}_{\dot{q}_r}{}^{G_2} \cdot \mathbf{a}^{G_2} \\ &+ \boldsymbol{\omega}_{\dot{q}_r}{}^{R_1} \cdot \mathbf{T}_1{}^* + \boldsymbol{\omega}_{\dot{q}_r}{}^{R_2} \cdot \mathbf{T}_2{}^* \qquad r = 1, 2, 3 \end{aligned}$$

Hence

$$\begin{aligned} F_1{}^* &= -m\mathbf{n}_1 \cdot \mathbf{a}^{G_1} - m\mathbf{n}_1 \cdot \mathbf{a}^{G_2} \\ F_2{}^* &= -mL[(c_3\mathbf{n}_1 - s_3\mathbf{n}_2) \cdot \mathbf{a}^{G_1} + 2c_3\mathbf{n}_1 \cdot \mathbf{a}^{G_2}] + \mathbf{n}_3 \cdot \mathbf{T}_1{}^* \\ F_3{}^* &= -mL[-2s_3\mathbf{n}_2 \cdot \mathbf{a}^{G_1} + (c_3\mathbf{n}_1 - s_3\mathbf{n}_2) \cdot \mathbf{a}^{G_2}] + \mathbf{n}_3 \cdot \mathbf{T}_2{}^* \end{aligned}$$

The moments of inertia of R_1 and R_2 about lines parallel to $\mathbf{n}_3$ and passing through G_1 and G_2, respectively, are both equal to $mL^2/3$. Thus,

$$n_3 \cdot \mathbf{T}_1{}^* = \mathbf{n}_3 \cdot \mathbf{T}_2{}^* \underset{(3.46)}{=} -\frac{\ddot{q}_3 mL^2}{3}$$

and the equations

$$F_r + F_r{}^* \underset{(5.1)}{=} 0 \qquad r = 1, 2, 3$$

lead to

$$
\begin{aligned}
P_1 &= 4mL(c_3\ddot{q}_3 - s_3\dot{q}_3^2) \\
M_3 &= mL^2[(7c_3^2 + 3s_3^2 + \tfrac{1}{3})\ddot{q}_3 - 4s_3c_3\dot{q}_3^2] - mgLs_3 \\
M_3 &= -mL^2[(7s_3^2 + 3c_3^2 + \tfrac{1}{3})\ddot{q}_3 + 4s_3c_3\dot{q}_3^2] + 3mgLs_3
\end{aligned}
$$

When $\dot{q}_3$ and $\ddot{q}_3$ are known, the first of these equations gives P_1, and either the second or third can be used to find M_3. Moreover, information about q_3 may be obtained by eliminating M_3 from the last two equations, which gives

$$\ddot{q}_3 = \frac{3gs_3}{8L}$$

or, after integration,

$$\dot{q}_3^2 = \frac{3g}{4L}(\bar{c}_3 - c_3)$$

where $\bar{c}_3$ is the value of c_3 when $\dot{q}_3 = 0$. P_1 can thus be expressed as

$$P_1 = 3mg(\tfrac{3}{2}c_3 - \bar{c}_3)s_3$$

Since P_1 cannot take on negative values (that is, the slot cannot pull on the rod), it now appears that the motion under consideration can take place only until q_3 reaches a value such that

$$c_3 = \tfrac{2}{3}\bar{c}_3$$

When q_3 reaches this value, point A loses contact with line Y.

5.18 Generalized Impulse and Generalized Momentum. When a system is subjected to the action of "impulsive" forces—that is, forces that become very large during a very small time interval—the velocities of various particles of the system may change substantially during this time interval while the configuration of the system remains essentially unaltered. This happens, for example, when two bodies collide. Although such phenomena may appear to be more complex than motions that proceed smoothly, they can frequently be treated analytically with comparatively simple methods, because the presumption that the configuration of the system remains unchanged makes it possible to integrate the dynamical equations (5.1) in general terms, and a theory involving algebraic rather than differential equations can thus be constructed. Specifically, if S is a simple nonholonomic system possessing $n - m$ degrees of freedom (see Section 2.22) in an inertial reference frame R, the *generalized impulse* I_r is defined as

Impulse simple nonhol

$$I_r = \sum_{i=1}^{N} \tilde{\mathbf{v}}_{u_r}{}^{P_i}(t_1) \cdot \int_{t_1}^{t_2} \mathbf{F}_i \, dt \qquad r = 1, \ldots, n - m \tag{5.19}$$

where N is the number of particles comprising S, t_1 and t_2 are the initial and final instants of a time interval during which the configuration remains essentially unchanged, $\tilde{\mathbf{v}}_{u_r}{}^{P_i}(t)$ is a nonholonomic partial rate of change of position of a typical particle P_i of S (see Section 2.24) at time t, and $\mathbf{F}_i$ is the resultant of all contact and body forces acting on P_i. The *generalized momentum* p_r is defined as

$$p_r(t) = \sum_{i=1}^{N} m_i \tilde{\mathbf{v}}_{u_r}{}^{P_i}(t) \cdot \mathbf{v}^{P_i}(t) \qquad r = 1, \cdots, n - m \tag{5.20}$$

where $\mathbf{v}^{P_i}(t)$ is the velocity of P_i in R at time t; and I_r and p_r are then related as follows:

$$I_r \approx p_r(t_2) - p_r(t_1) \qquad r = 1, \cdots, n - m \tag{5.21}$$

[In Equations (5.19) and (5.20), u_r may be replaced with $\dot{q}_r$.]

Proof: In general, $\tilde{\mathbf{v}}_{u_r}{}^{P_i}$ is a function of $q_1, \cdots, q_n$, and t in R. If $t_2 \approx t_1$, and $q_r(t_2) \approx q_r(t_1)$, then $\tilde{\mathbf{v}}_{\dot{q}_r}{}^{P_i}$ remains nearly constant (and nearly equal to its value at time t_1) throughout the time interval $t_2 - t_1$, and

$$\int_{t_1}^{t_2} F_r\, dt \underset{(3.3)}{\approx} \sum_{i=1}^{N} \tilde{\mathbf{v}}_{u_r}{}^{P_i}(t_1) \cdot \int_{t_1}^{t_2} \mathbf{F}_i\, dt \underset{(5.19)}{=} I_r \qquad r = 1, \cdots, n - m$$

Integration of F_r^* gives

$$\int_{t_1}^{t_2} F_r^*\, dt \underset{(3.8)}{=} \sum_{i=1}^{N} \int_{t_1}^{t_2} \tilde{\mathbf{v}}_{u_r}{}^{P_i}(t) \cdot \mathbf{F}_i^*$$

$$\underset{(3.9)}{=} - \sum_{i=1}^{N} m_i \int_{t_1}^{t_2} \tilde{\mathbf{v}}_{u_r}{}^{P_i}(t) \cdot \frac{d\mathbf{v}^{P_i}}{dt}\, dt$$

$$\approx - \sum_{i=1}^{N} m_i \tilde{\mathbf{v}}_{u_r}{}^{P_i}(t_1) \cdot \int_{t_1}^{t_2} \frac{d\mathbf{v}^{P_i}}{dt}\, dt$$

$$= - \sum_{i=1}^{N} m_i \tilde{\mathbf{v}}_{u_r}{}^{P_i}(t_1) \cdot [\mathbf{v}^{P_i}(t_2) - \mathbf{v}^{P_i}(t_1)]$$

$$\approx - \sum_{i=1}^{N} m_i \tilde{\mathbf{v}}_{u_r}{}^{P_i}(t_2) \cdot \mathbf{v}^{P_i}(t_2)$$

$$+ \sum_{i=1}^{N} m_i \tilde{\mathbf{v}}_{u_r}{}^{P_i}(t_1) \cdot \mathbf{v}^{P_i}(t_1) \underset{(5.20)}{=} -p_r(t_2) + p_r(t_1)$$

Now

$$\int_{t_1}^{t_2} F_r\, dt + \int_{t_1}^{t_2} F_r^*\, dt \underset{(5.1)}{=} 0$$

Hence

$$I_r - p_r(t_2) + p_r(t_1) \approx 0$$

▪ EXAMPLE

When a particle P_1 of mass m_1 collides with a particle P_2 of mass m_2, the contact force **R** exerted on P_1 by P_2 may be regarded as an impulsive force, whereas the gravitational forces $\mathbf{G}_1$ and $\mathbf{G}_2$ acting on the particles are not impulsive forces. The latter may, therefore, be neglected in comparison with **R** when $\mathbf{F}_1$ and $\mathbf{F}_2$, the resultants of the body and contact forces acting on P_1 and P_2, respectively, are integrated with respect to time; that is,

$$\int_{t_1}^{t_2} \mathbf{F}_1\, dt = \int_{t_1}^{t_2} (\mathbf{R} + \mathbf{G}_1)\, dt \approx \int_{t_1}^{t_2} \mathbf{R}\, dt = \mathbf{S} \tag{a}$$

and

$$\int_{t_1}^{t_2} \mathbf{F}_2\, dt = \int_{t_1}^{t_2} (-\mathbf{R} + G_2)\, dt \approx -\int_{t_1}^{t_2} \mathbf{R}\, dt = -\mathbf{S} \tag{b}$$

where **S** has been introduced as an abbreviation.

The system has six degrees of freedom; and the velocities $\mathbf{v}^{P_1}$ and $\mathbf{v}^{P_2}$ of P_1 and P_2 can always be expressed as

$$\left.\begin{aligned} \mathbf{v}^{P_1} &= u_1\mathbf{n}_1 + u_2\mathbf{n}_2 + u_3\mathbf{n}_3 \\ \mathbf{v}^{P_2} &= u_4\mathbf{n}_1 + u_5\mathbf{n}_2 + u_6\mathbf{n}_3 \end{aligned}\right\} \tag{c}$$

where $\mathbf{n}_1$, $\mathbf{n}_2$, and $\mathbf{n}_3$ are mutually perpendicular unit vectors fixed in an inertial reference frame and $u_1, \cdots, u_6$ are functions of the six generalized coordinates of the system and their time derivatives. It follows immediately that

$$\begin{aligned} I_1 &\underset{(5.19)}{=} \mathbf{v}_{u_1}{}^{P_1} \cdot \int_{t_1}^{t_2} \mathbf{F}_1\, dt + \mathbf{v}_{u_1}{}^{P_2} \cdot \int_{t_1}^{t_2} \mathbf{F}_2\, dt \\ &\underset{(a),(c)}{=} \mathbf{n}_1 \cdot \mathbf{S} + \underset{(c)}{0} \\ I_2 &= \mathbf{n}_2 \cdot \mathbf{S} \qquad I_3 = \mathbf{n}_3 \cdot \mathbf{S} \\ I_4 &= \mathbf{v}_{u_4}{}^{P_1} \cdot \int_{t_1}^{t_2} \mathbf{F}_1\, dt + \mathbf{v}_{u_4}{}^{P_2} \cdot \int_{t_1}^{t_2} \mathbf{F}_2\, dt \\ &\underset{(a),(b)}{=} 0 - \mathbf{n}_1 \cdot \mathbf{S} \\ I_5 &= -\mathbf{n}_2 \cdot \mathbf{S} \quad I_6 = -\mathbf{n}_3 \cdot \mathbf{S} \end{aligned}$$

Furthermore,

$$\begin{aligned} p_1 &\underset{(5.20)}{=} m_1\mathbf{v}_{u_1}{}^{P_1} \cdot \mathbf{v}^{P_1} + m_2\mathbf{v}_{u_1}{}^{P_2} \cdot \mathbf{v}^{P_2} = m_1u_1 + 0 \\ p_2 &= m_1u_2 \qquad p_3 = m_1u_3 \end{aligned}$$

and

$$p_4 = m_2u_4 \qquad p_5 = m_2u_5 \qquad p_6 = m_2u_6$$

For $r = 1, 2, 3$, Equations (5.21) thus lead to

$$\mathbf{n}_r \cdot \mathbf{S} \approx m_1[u_r(t_2) - u_r(t_1)] \tag{d}$$

and for $r = 4, 5, 6$ to

$$-\mathbf{n}_{r-3} \cdot \mathbf{S} \approx m_2[u_r(t_2) - u_r(t_1)] \tag{e}$$

The six equations (d) and (e) are not sufficient for the determination of the velocities of P_1 and P_2 at time t_2, even if the velocities of P_1 and P_2 at time t_1 are known, since (d) and (e) contain nine unknown quantities, namely $\mathbf{n}_1 \cdot \mathbf{S}$, $\mathbf{n}_2 \cdot \mathbf{S}$, $\mathbf{n}_3 \cdot \mathbf{S}$, and $u_r(t_2)$, for $r = 1, \cdot \cdot \cdot, 6$. If additional information is available, for example, if it is known that P_1 and P_2 become attached to each other during the collision, so that

$$\mathbf{v}^{P_1}(t_2) = \mathbf{v}^{P_2}(t_2) \tag{f}$$

then $\mathbf{v}^{P_1}(t_2)$ can be found by noting that

$$u_r(t_2) \underset{\text{(c),(f)}}{=} u_{r+3}(t_2) \qquad r = 1, 2, 3$$

and, after elimination of $\mathbf{n}_r \cdot \mathbf{S}$, with $r = 1, 2, 3$, from Equations (d) and (e),

$$u_r(t_2) = u_{r+3}(t_2) \approx \frac{m_1 u_r(t_1) + m_2 u_{r+3}(t_1)}{m_1} \qquad r = 1, 2, 3 \tag{g}$$

The velocities of P_1 and P_2 at time t_2 are then given by

$$\mathbf{v}^{P_1}(t_2) \underset{\text{(f)}}{=} \mathbf{v}^{P_2}(t_2) \underset{\text{(c),(g)}}{\approx} \frac{m_1 \mathbf{v}^{P_1}(t_1) + m_2 \mathbf{v}^{P_2}(t_1)}{m_1 + m_2}$$

5.19 Generalized Momentum and Kinetic Energy. The evaluation of generalized impulses [see Equations (5.19)] is facilitated by the fact that the total contribution to I_r of all forces that do not have an impulsive character or that do not contribute to F_r (see Sections 3.3, 3.4, 3.6, and 3.8) is equal to zero; and the derivation of expressions for generalized momenta [see Equations (5.20)] is frequently simplified by making use of the following relationship between generalized momenta and kinetic energy: If the kinetic energy K of a simple nonholonomic system S (see Section 2.22) is regarded as a function of the $2n - m + 1$ independent variables $q_1, \cdot \cdot \cdot, q_n, u_1, \cdot \cdot \cdot, u_{n-m}$, and t, then the generalized momentum p_r can be expressed as

$$p_r = \frac{\partial K}{\partial u_r} \qquad r = 1, \cdot \cdot \cdot, n - m \tag{5.22}$$

[Here, as in Equations (5.20), u_r may be replaced with $\dot{q}_r$.]

Proof:

$$\frac{\partial K}{\partial u_r} \underset{(4.10)}{=} \frac{\partial}{\partial u_r}\left[\frac{1}{2}\sum_{i=1}^{N} m_i(\mathbf{v}^{P_i})^2\right]$$

$$= \sum_{i=1}^{N} m_i \frac{\partial \mathbf{v}^{P_i}}{\partial u_r} \cdot \mathbf{v}^{P_i}$$

$$\underset{(2.32)}{=} \sum_{i=1}^{N} m_i \tilde{\mathbf{v}}_{u_r}{}^{P_i} \cdot \mathbf{v}^{P_i} \underset{(5.20)}{=} p_r$$

▪ EXAMPLE

A gear train consisting of three identical gears G_1, G_2, and G_3, each having a radius r and moment of inertia J about its axis, is set into motion when G_1 is meshed suddenly with a gear G' of radius r' and moment of inertia J', G' having an angular speed Ω' at the instant of contact. The angular speed Ω acquired by G_1 is to be determined.

Before G_1 and G' are brought into contact, the system possesses two degrees of freedom, and the two angles θ and θ' shown in Figure 5.21

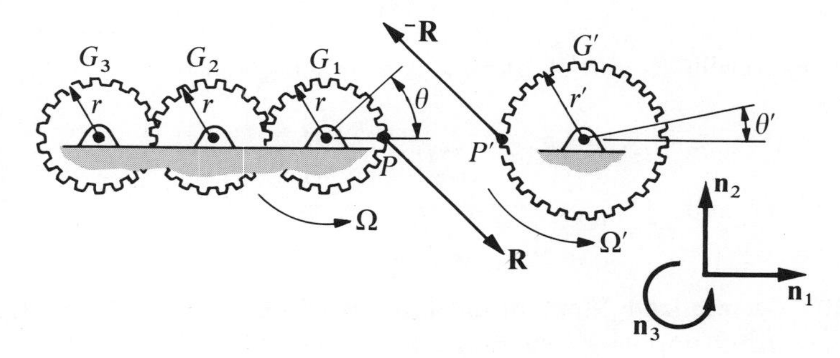

Figure 5.21

can be used as generalized coordinates. If u_1 and u_2 are defined as $u_1 = \dot{\theta}$ and $u_2 = \dot{\theta}'$, the kinetic energy K of the system is given by

$$K = \tfrac{3}{2}Ju_1{}^2 + \tfrac{1}{2}J'u_2{}^2$$

and the generalized momenta p_1 and p_2 are thus

$$p_1 \underset{(5.22)}{=} \frac{\partial K}{\partial u_1} = 3Ju_1 \qquad p_2 \underset{(5.22)}{=} J'u_2$$

The only forces that contribute to the generalized impulses I_1 and I_2 are the forces $\mathbf{R}$ and $-\mathbf{R}$ exerted on G_1 at P and on G' at P' (see Figure 5.21) when G' is brought into contact with G_1, because forces exerted on

G_1, G_2, and G_3 by their supports, as well as forces exerted by these gears on each other, do not contribute to generalized active forces (see Sections 3.3 and 3.8). The velocities of P and P' are

$$\mathbf{v}^P = ru_1\mathbf{n}_2 \quad \mathbf{v}^{P'} = -r'u_2\mathbf{n}_2 \tag{a}$$

Hence,

$$I_1 \underset{(5.19)}{=} \mathbf{v}_{u_1}{}^P \cdot \int_{t_1}^{t_2} \mathbf{R}\,dt - \mathbf{v}_{u_1}{}^{P'} \cdot \int_{t_1}^{t_2} \mathbf{R}\,dt$$
$$= r\mathbf{n}_2 \cdot \int_{t_1}^{t_2} \mathbf{R}\,dt + 0$$

and

$$I_2 = \mathbf{v}_{u_2}{}^P \cdot \int_{t_1}^{t_2} \mathbf{R}\,dt - \mathbf{v}_{u_2}{}^{P'} \cdot \int_{t_1}^{t_2} \mathbf{R}\,dt$$
$$= 0 + r'\mathbf{n}_2 \cdot \int_{t_1}^{t_2} \mathbf{R}\,dt$$

From Equations (5.21) one thus obtains

$$r\mathbf{n}_2 \cdot \int_{t_1}^{t_2} \mathbf{R}\,dt \approx 3J[u_1(t_2) - u_1(t_1)]$$

and

$$r'\mathbf{n}_2 \cdot \int_{t_1}^{t_2} \mathbf{R}\,dt \approx J'[u_2(t_2) - u_2(t_1)]$$

or, after eliminating $\mathbf{n}_2 \cdot \int_{t_1}^{t_2} R\,dt$,

$$\frac{3J}{r}[u_1(t_2) - u_1(t_1)] \approx \frac{J'}{r'}[u_2(t_2) - u_2(t_1)] \tag{b}$$

where $u_1(t_1)$ and $u_2(t_1)$ are known to have the values

$$u_1(t_1) = 0 \qquad u_2(t_1) = \Omega'$$

Furthermore, the velocities of P and P' are equal to each other at time t_2. Hence,

$$ru_1(t_2) \underset{(a)}{=} -r'u_2(t_2) \tag{c}$$

and, if $u_1(t_2)$ is called Ω, it follows that

$$u_2(t_2) = -\frac{r\Omega}{r'}$$

Consequently,

$$\frac{3J\Omega}{r} \underset{(b)}{\approx} \frac{J'}{r'}\left(-\frac{r\Omega}{r'} - \Omega'\right) \tag{d}$$

and

$$\Omega \underset{(d)}{\approx} -\frac{\Omega' J'}{r'}\left[\frac{3J}{r} + \left(\frac{r}{r'}\right)\frac{J'}{r'}\right]^{-1}$$

5.20 Collisions. When a system S is involved in a collision beginning at an instant t_1 and terminating at an instant t_2, the motion of S at time t_2 frequently cannot be determined solely by use of Equations (5.21) together with a complete description of the motion at time t_1. Generally, some additional information about the velocities of one or more particles of S at time t_2 must be obtained from an independent source, such as direct observation, and this information must be expressed in a suitable mathematical form [see, for example, Equation (f) of the example in Section 5.18 and Equation (c) of the of example in Section 5.19]. The propositions that follow represent an attempt to come to grips with this problem by formulating two assumptions that, as indicated by experiments, are valid in a number of situations of practical interest.

In Figure 5.22, P and P' designate points that come into contact with

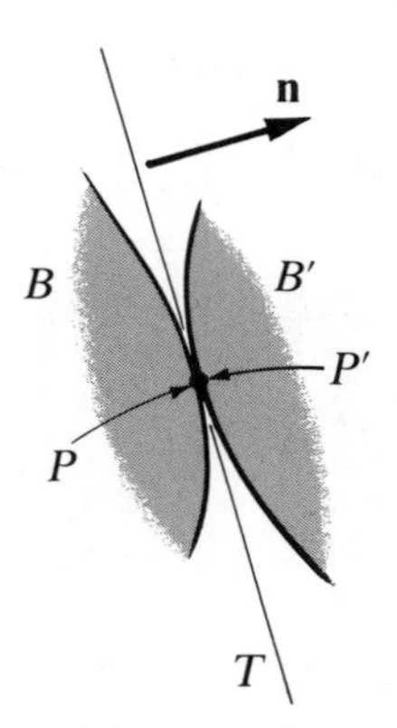

Figure 5.22

each other during a collision of two bodies B and B'; T represents the (common) tangent plane to the surfaces of B and B' at their point of contact; and $\mathbf{n}$ is a unit vector perpendicular to T. If $\mathbf{v}^P(t)$ and $\mathbf{v}^{P'}(t)$ denote the velocities of P and P' at time t, then $\mathbf{v}^{P/P'}(t_1)$ and $\mathbf{v}^{P/P'}(t_2)$, defined as

$$\left.\begin{aligned} \mathbf{v}^{P/P'}(t_1) &= \mathbf{v}^P(t_1) - \mathbf{v}^{P'}(t_1) \\ \mathbf{v}^{P/P'}(t_2) &= \mathbf{v}^P(t_2) - \mathbf{v}^{P'}(t_2) \end{aligned}\right\} \tag{5.23}$$

are called the *velocity of approach* and the *velocity of separation*, respectively. Each of these can be resolved into two components, one parallel to $\mathbf{n}$, called the *normal component*, the other perpendicular to $\mathbf{n}$, called the *tangential component*.

The first assumption is that the normal component of the velocity of separation has a magnitude proportional to the magnitude of the normal component of the velocity of approach, the constant of proportionality being a quantity e whose value depends on the physical constitution, but not on the motions, of B and B'. This can be stated analytically as

$$\mathbf{n} \cdot \mathbf{v}^{P/P'}(t_2) = -e\mathbf{n} \cdot \mathbf{v}^{P/P'}(t_1) \tag{5.24}$$

The constant e is called a *coefficient of restitution*. In practice, it is found to have values in the range from zero to one. When $e = 0$, the collision is said to be *inelastic;* and $e = 1$ characterizes an idealized case called a *perfectly elastic* collision. The negative sign in Equation (5.24) is necessary to ensure that the normal components of the velocity of approach and of the velocity of separation have opposite directions.

The second assumption involves both the velocity of separation and the impulsive force $\mathbf{R}$ exerted on B by B' during the collision. If $\mathbf{R}$ is integrated with respect to t in the interval $t_2 - t_1$, and the resulting vector is resolved into two components, one parallel to $\mathbf{n}$, called the *normal impulse* $\boldsymbol{\nu}$, the other perpendicular to $\mathbf{n}$, called the *tangential impulse* $\boldsymbol{\tau}$, then it is assumed that these impulses are related to $\boldsymbol{\sigma}$, the *tangential component of the velocity of separation*, as follows: There is no slipping at time t_2, which means that

$$\boldsymbol{\sigma} = 0 \tag{5.25}$$

if and only if

$$|\boldsymbol{\tau}| < \mu|\boldsymbol{\nu}| \tag{5.26}$$

where μ is a constant, called the *coefficient of friction*. Otherwise, there is slip at time t_2, Equation (5.26) is replaced by

$$|\boldsymbol{\tau}| = \mu|\boldsymbol{\nu}| \tag{5.27}$$

and $\boldsymbol{\sigma}$ and $\boldsymbol{\tau}$ have opposite directions, so that

$$\frac{\boldsymbol{\sigma}}{|\boldsymbol{\sigma}|} = \frac{-\boldsymbol{\tau}}{|\boldsymbol{\tau}|} \tag{5.28}$$

▪ EXAMPLE

While moving in a vertical plane, a circular hoop H of radius b and mass m collides with a horizontal plane. Assuming that the motion can be considered planar, the system possesses three degrees of freedom. The velocity $\mathbf{v}$ of the center of H and the angular velocity of H can be expressed as

$$\mathbf{v} = u_1\mathbf{n}_1 + u_2\mathbf{n}_2$$

and

$$\boldsymbol{\omega} = u_3\mathbf{n}_3$$

where $\mathbf{n}_1$, $\mathbf{n}_2$, and $\mathbf{n}_3$ are unit vectors directed as shown in Figure 5.23.

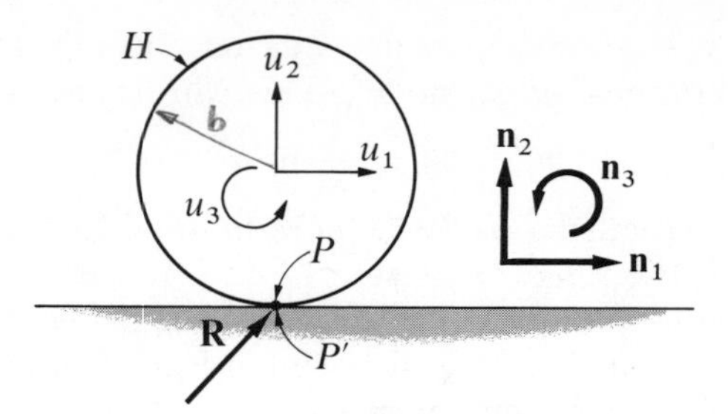

Figure 5.23

The kinetic energy K is given by [see Equations (4.14) and (4.15)]

$$K = \tfrac{1}{2}m\mathbf{v}^2 + \tfrac{1}{2}mb^2\boldsymbol{\omega}^2$$
$$= \tfrac{1}{2}m(u_1{}^2 + u_2{}^2 + b^2u_3{}^2)$$

and, from Equations (5.22), the generalized momenta are

$$p_1 = mu_1 \qquad p_2 = mu_2 \qquad p_3 = mb^2u_3$$

If S_1, S_2, and S_3 are defined such that

$$S_1\mathbf{n}_1 + S_2\mathbf{n}_2 + S_3\mathbf{n}_3 = \int_{t_1}^{t_2} \mathbf{R}\,dt$$

where $\mathbf{R}$ is the impulsive force acting on the point P of H that comes into contact with the horizontal plane during the collision (see Figure 5.23), then the generalized impulses are given by

$$I_r \underset{(5.19)}{=} \mathbf{v}_{u_r}{}^P \cdot (S_1\mathbf{n}_1 + S_2\mathbf{n}_2 + S_3\mathbf{n}_3) \qquad r = 1, 2, 3$$

Now

$$\mathbf{v}^P \underset{(2.20)}{=} \mathbf{v} + \boldsymbol{\omega} \times (-b\mathbf{n}_2) = (u_1 + bu_3)\mathbf{n}_1 + u_2\mathbf{n}_2$$

Consequently,

$$I_1 = S_1 \qquad I_2 = S_2 \qquad I_3 = bS_1$$

and, from Equations (5.21),

$$S_1 \approx m[u_1(t_2) - u_1(t_1)] \qquad \text{(a)}$$
$$S_2 \approx m[u_2(t_2) - u_2(t_1)] \qquad \text{(b)}$$
$$S_1 \approx mb[u_3(t_2) - u_3(t_1)] \qquad \text{(c)}$$

The velocity of the point P' with which P comes into contact during the collision (see Figure 5.23) is equal to zero at all times. Hence the velocities of approach and of separation are

$$\mathbf{v}^{P/P'}(t_1) \underset{(5.23)}{=} \mathbf{v}^P(t_1) = [u_1(t_1) + bu_3(t_1)]\mathbf{n}_1 + u_2(t_1)\mathbf{n}_2$$

and

$$\mathbf{v}^{P/P'}(t_2) = \mathbf{v}^P(t_2) = [u_1(t_2) + bu_3(t_2)]\mathbf{n}_1 + u_2(t_2)\mathbf{n}_2 \tag{d}$$

The condition

$$\mathbf{n}_2 \cdot \mathbf{v}^{P/P'}(t_2) \underset{(5.24)}{=} -e\mathbf{n}_2 \cdot \mathbf{v}^{P/P'}(t_1)$$

thus leads to

$$u_2(t_2) = -eu_2(t_1) \tag{e}$$

The normal impulse $\boldsymbol{\nu}$ is given by

$$\boldsymbol{\nu} = S_2\mathbf{n}_2 \tag{f}$$

and, if it is assumed that the impulsive force $\mathbf{R}$ is always perpendicular to $\mathbf{n}_3$, so that $S_3 = 0$, then the tangential impulse $\boldsymbol{\tau}$ becomes

$$\boldsymbol{\tau} = S_1\mathbf{n}_1 \tag{g}$$

Furthermore, $\boldsymbol{\sigma}$, the tangential component of the velocity of separation, is given by

$$\boldsymbol{\sigma} \underset{(d)}{=} [u_1(t_2) + bu_3(t_2)]\mathbf{n}_1 \tag{h}$$

There are now two possibilities: Either there is no slip at t_2, in which case [see Equation (5.25)]

$$u_1(t_2) + bu_3(t_2) \underset{(h)}{=} 0 \tag{i}$$

and [see Equation (5.26)]

$$|S_1| \underset{(f),(g)}{<} \mu|S_2| \tag{j}$$

or there is slip at t_2, so that [see Equation (5.27)]

$$|S_1| \underset{(f),(g)}{=} \mu|S_2| \tag{k}$$

and [see Equation (5.28)]

$$\frac{u_1(t_2) + bu_3(t_2)}{|u_1(t_2) + bu_3(t_2)|} \underset{(g),(h)}{=} -\frac{S_1}{|S_1|}$$

which implies (multiply both sides with S_1) that

$$S_1[u_1(t_2) + bu_3(t_2)] < 0 \tag{l}$$

These two possibilities will be examined separately. First, however, it is worth noting that Equations (a), (b), (c), and (e) apply in both cases and that (b) and (c) can be replaced with

$$S_2 \underset{(e)}{\approx} -m(1 + e)u_2(t_1) \tag{m}$$

and

$$bu_3(t_2) \underset{(a)}{\approx} bu_3(t_1) + u_1(t_2) - u_1(t_1) \tag{n}$$

respectively. A "no slip" collision is now characterized by

$$|u_1(t_2) \underset{(a)}{-} u_1(t_1)| \underset{(j)}{<} \mu(1 + e)\underset{(m)}{|}u_2(t_1)| \tag{o}$$

and

$$u_1(t_2) \underset{(i),(n)}{\approx} \tfrac{1}{2}[u_1(t_1) - bu_3(t_1)] \tag{p}$$

Consequently,

$$u_3(t_2) \underset{(n),(p)}{\approx} \tfrac{1}{2}\left[u_3(t_1) - \frac{u_1(t_1)}{b}\right] \tag{q}$$

and the motion at time t_2 is described by Equations (e), (p), and (q), provided that

$$|u_1(t_1) + bu_3(t_1)| \underset{(o),(p)}{<} -2\mu(1 + e)u_2(t_1) \tag{r}$$

[Here $|u_2(t_1)|$ has been replaced with $-u_2(t_1)$ because $u_2(t_1)$ must be negative; otherwise, no collision will occur.]

A "sliding" collision is said to occur when

$$|u_1(t_2) \underset{(a)}{-} u_1(t_1)| \underset{(k)}{=} -\mu(1 + e)\underset{(m)}{u_2}(t_1) \tag{s}$$

and

$$[u_1(t_2) \underset{(a)}{-} u_1(t_1)][2u_1(t_2) - \underset{(n)}{u_1(t_1)} + bu_3(t_1)] \underset{(l)}{<} 0 \tag{t}$$

One of two situations must now arise: If

$$u_1(t_2) > u_1(t_1)$$

then

$$u_1(t_2) \underset{(s)}{=} u_1(t_1) - \mu(1 + e)u_2(t_1) \tag{u}$$

and

$$u_3(t_2) \underset{(n),(u)}{\approx} u_3(t_1) - \frac{\mu(1 + e)u_2(t_1)}{b} \tag{v}$$

provided that

$$u_1(t_1) + bu_3(t_1) \underset{(t),(u)}{<} 2\mu(1 + e)u_2(t_1) \tag{w}$$

If, on the other hand,

$$u_1(t_2) < u_1(t_1)$$

then

$$u_1(t_2) \underset{(s)}{=} u_1(t_1) + \mu(1 + e)u_2(t_1) \tag{x}$$

$$u_3(t_2) \underset{(n),(x)}{\approx} u_3(t_1) + \frac{\mu(1 + e)u_2(t_1)}{b} \tag{y}$$

and the initial values of u_1, u_2, and u_3 must be such that

$$u_1(t_1) + bu_3(t_1) \underset{(t),(x)}{>} -2\mu(1 + e)u_2(t_1) \tag{z}$$

Some numerical examples may help one to understand the physical significance of these results.

Suppose that $e = 0.8$, $\mu = 0.25$, and the hoop has a "top spin" at the instant of contact with the support, which is the case if, for example,

$$u_1(t_1) = -u_2(t_1) = V \quad u_3(t_1) = \frac{-2V}{b}$$

Then

$$u_1(t_1) + bu_3(t_1) = -V$$

and

$$2\mu(1 + e)u_2(t_1) = -0.9V$$

and of the three conditions [Equations (r), (w), and (z)], only one, namely (w), is satisfied. Consequently,

$$u_1(t_2) \underset{(u)}{=} 1.45V$$

$$u_2(t_2) \underset{(e)}{=} 0.8V$$

$$u_3(t_2) \underset{(v)}{\approx} \frac{-1.55V}{b}$$

The first two of these can be used to compare the angles, θ_1 and θ_2, that the velocity vector of the center of the hoop makes with the vertical before and after the impact. θ_1 is equal to 45 degrees, whereas

$$\theta_2 = \text{arc tan}\,\frac{1.45}{0.8} = 61.1 \text{ degrees}$$

The top spin initially imparted to the hoop is thus seen to produce a "drop," a fact that will not surprise tennis players.

If the hoop initially has a sufficiently large "backspin," the velocity of the center can be altered even more drastically. For example, if e and μ have the same values as before, and

$$u_1(t_1) = V \qquad u_2(t_1) = -4V \qquad u_3(t_1) = \frac{4V}{b}$$

then Equation (z), but neither (r) nor (w), is satisfied, and

$$u_1(t_2) \underset{(x)}{=} -0.8V$$

$$u_2(t_2) \underset{(e)}{=} 3.2V$$

$$u_3(t_2) \underset{(y)}{\approx} \frac{2.2V}{b}$$

Hence, if the center were moving downward and to the right at time t_1, it would be moving upward and to the left at time t_2.

Both of the examples considered so far involve "sliding" collisions. To see that a "no-slip" collision can be produced easily, it is only necessary to study the collision that takes place when the hoop considered in the preceding example next strikes the support subsequent to the instant t_2. If t_3 is the time at which this occurs, then

$$u_1(t_3) = -0.8V$$
$$u_2(t_3) = -3.2V$$
$$u_3(t_3) = \frac{2.2V}{b}$$

and Equation (r) is the only one of the three conditions (r), (w), and (z) that is satisfied. Furthermore, it may be verified that all succeeding collisions are also of the same type.

Finally, one can estimate how long the hoop will continue to bounce. The maximum height reached by the center of the hoop during any bounce is, of course, smaller than that attained during the preceding one, so that this height approaches zero (and bouncing ceases) as the number n of bounces approaches infinity. The time T_n required for n bounces to occur is found by noting that the time that elapses between two successive impacts depends only on the value of u_2 at the end of the first of these; that is, if τ_n denotes the time required for the nth bounce, τ_1 is given by

$$\tau_1 = \frac{2u_2(t_2)}{g} \underset{(e)}{=} \frac{-2eu_2(t_1)}{g}$$

Next,

$$\tau_2 = \frac{2u_2(t_4)}{g} = \frac{-2eu_2(t_3)}{g}$$

But

$$u_2(t_3) = -u_2(t_2) \underset{(e)}{=} eu_2(t_1)$$

Hence,

$$\tau_2 = \frac{-2e^2u_2(t_1)}{g}$$

Similarly,

$$\tau_3 = \frac{-2e^3u_2(t_1)}{g}$$

and

$$\tau_n = \frac{-2e^nu_2(t_1)}{g}$$

Consequently,

$$T_n = \tau_1 + \cdots + \tau_n = \frac{-2(e + e^2 + \cdots + e^n)u_2(t_1)}{g}$$

Now,

$$e + e^2 + \cdots + e^n = \frac{e(1 - e^n)}{1 - e}$$

Thus,

$$T_n = \frac{-2e(1 - e^n)u_2(t_1)}{(1 - e)g}$$

and the total time T required for infinitely many bounces is given by

$$T = \lim_{n\to\infty} T_n = \frac{-2u_2(t_1)e}{(1 - e)g}$$

Since this value does not reflect the time consumed by the associated collisions (infinitely many), it should be regarded as a lower bound on the time required for bouncing to cease subsequent to time t_1. Applied to the last example, the expression just obtained gives

$$T = \frac{-2(-4V)(0.8)}{(1 - 0.8)g} = \frac{32V}{g}$$

Hence, if $V = 10$ ft/sec, the hoop may be expected to bounce for a little less than 10 seconds.

5.21 Nonlinear Equations. As pointed out previously, the impact theory set forth in Sections 5.18–5.20 gives rise to algebraic equations. Some of these equations may be nonlinear. When this is the case, the method described in Section 5.11 can be used to obtain a complete solution.

▪ EXAMPLE

In Figure 5.24, A represents a portion of a moving body, and B is a uniform sphere colliding with A. Assuming that slip occurs and that the motion of A is not affected by the collision (which is essentially the case if the mass of A is large in comparison with that of B), the change brought about in the motion of B by the collision is to be determined.

The velocity $\mathbf{v}^C$ of the center C of B and the angular velocity $\boldsymbol{\omega}$ of B can be expressed as

$$\mathbf{v}^C = u_1\mathbf{n}_1 + u_2\mathbf{n}_2 + u_3\mathbf{n}_3 \qquad \text{(a)}$$

and

$$\boldsymbol{\omega} = u_4\mathbf{n}_1 + u_5\mathbf{n}_2 + u_6\mathbf{n}_3 \qquad \text{(b)}$$

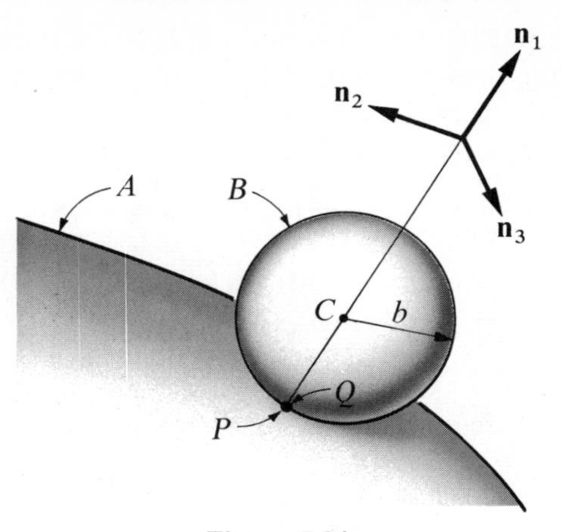

Figure 5.24

where $\mathbf{n}_1$, $\mathbf{n}_2$, and $\mathbf{n}_3$ form a right-handed set of mutually perpendicular unit vectors, with $\mathbf{n}_1$ normal to the surfaces of A and B at the point of contact of these surfaces, as shown in Figure 5.24. If m and I denote the mass of B and the moment of inertia of B about any line passing through C, the kinetic energy K of B is then given by

$$K \underset{(4.14)}{=} \tfrac{1}{2}[m(u_1^2 + u_2^2 + u_3^2) + I(u_4^2 + u_5^2 + u_6^2)] \tag{c}$$

and the generalized momenta $p_1, \cdots, p_6$ are

$$p_r \underset{(5.22)}{=} \begin{cases} mu_r & i = 1, 2, 3 \\ Iu_r & i = 4, 5, 6 \end{cases} \tag{d}$$

The velocity $\mathbf{v}^P$ of the point P of A that comes into contact with B is presumed known and may be expressed as

$$\mathbf{v}^P = v_1{}^P\,\mathbf{n}_1 + v_2{}^P\,\mathbf{n}_2 + v_3{}^P\,\mathbf{n}_3 \tag{e}$$

while the velocity $\mathbf{v}^Q$ of the point Q of B that comes into contact with A is given by

$$\mathbf{v}^Q = \mathbf{v}^C + \boldsymbol{\omega} \times (-b\mathbf{n}_1) \tag{f}$$

where b is the radius of B. Consequently,

$$\mathbf{v}^Q \underset{(a),(b),(f)}{=} u_1\mathbf{n}_1 + (u_2 - bu_6)\mathbf{n}_2 + (u_3 + bu_5)\mathbf{n}_3 \tag{g}$$

and the partial rates of change of position $\mathbf{v}_{u_r}{}^Q$, for $r = 1, \cdots, 6$, are

$$\mathbf{v}_{u_1}{}^Q = \mathbf{n}_1 \quad \mathbf{v}_{u_2}{}^Q = \mathbf{n}_2 \quad \mathbf{v}_{u_3}{}^Q = \mathbf{n}_3 \tag{h}$$

$$\mathbf{v}_{u_4}{}^Q = 0 \quad \mathbf{v}_{u_5}{}^Q = b\mathbf{n}_3 \quad \mathbf{v}_{u_6}{}^Q = -b\mathbf{n}_2 \tag{i}$$

If S_1, S_2, and S_3 are defined such that

$$S_1\mathbf{n}_1 + S_2\mathbf{n}_2 + S_3\mathbf{n}_3 = \int_{t_1}^{t_2} \mathbf{R}\,dt \tag{j}$$

where $\mathbf{R}$ is the impulsive force exerted on B by A during the collision, the generalized impulses I_r, for $r = 1, \cdots, 6$, become [see Equations (5.19)]

$$I_1 \underset{(h)}{=} S_1 \qquad I_2 \underset{(h)}{=} S_2 \qquad I_3 \underset{(h)}{=} S_3 \tag{k}$$

$$I_4 \underset{(i)}{=} 0 \qquad I_5 \underset{(i)}{=} bS_3 \qquad I_6 \underset{(i)}{=} -bS_2 \tag{l}$$

and substitution from Equations (d), (k), and (l) into Equations (5.21) gives

$$\left.\begin{aligned} S_1 &\approx m[u_1(t_2) - u_1(t_1)] \\ S_2 &\approx m[u_2(t_2) - u_2(t_1)] \\ S_3 &\approx m[u_3(t_2) - u_3(t_1)] \\ 0 &\approx I[u_4(t_2) - u_4(t_1)] \\ bS_3 &\approx I[u_5(t_2) - u_5(t_1)] \\ -bS_2 &\approx I[u_6(t_2) - u_6(t_1)] \end{aligned}\right\} \tag{m}$$

$\mathbf{v}^{Q/P}(t_1)$, the velocity of approach, and $\mathbf{v}^{Q/P}(t_2)$, the velocity of separation, are given by [see Equations (5.23) and (5.24)]

$$\begin{aligned} \mathbf{v}^{Q/P}(t_i) \underset{(g),(e)}{=} & [u_1(t_i) - v_1{}^P]\mathbf{n}_1 \\ & + [u_2(t_i) - bu_6(t_i) - v_2{}^P]\mathbf{n}_2 \\ & \qquad + [u_3(t_i) + bu_5(t_i) - v_3{}^P]\mathbf{n}_3 \end{aligned} \tag{n}$$

(In the case of $v_1{}^P$, $v_2{}^P$, and $v_3{}^P$, it is not necessary to distinguish values at t_1 from those at t_2 because it is assumed that the motion of A is not affected by the collision.)

If e denotes the coefficient of restitution, Equation (5.24) gives

$$\mathbf{n}_1 \cdot \mathbf{v}^{Q/P}(t_2) = -e\mathbf{n}_1 \cdot \mathbf{v}^{Q/P}(t_1) \tag{o}$$

so that

$$u_1(t_2) - v_1{}^P \underset{(n)}{=} -e[u_1(t_1) - v_1{}^P] \tag{p}$$

The normal impulse $\boldsymbol{\nu}$ and the tangential impulse $\boldsymbol{\tau}$ are

$$\boldsymbol{\nu} \underset{(j)}{=} S_1\mathbf{n}_1 \tag{q}$$

and

$$\boldsymbol{\tau} \underset{(j)}{=} S_2\mathbf{n}_2 + S_3\mathbf{n}_3 \tag{r}$$

while $\boldsymbol{\sigma}$, the tangential component of the velocity of separation is given by

$$\boldsymbol{\sigma} = [u_2(t_2) - bu_6(t_2) - v_2{}^P]\mathbf{n}_2 + [u_3(t_2) + bu_5(t_2) - v_3{}^P] \tag{s}$$

The assumption that slip occurs thus leads to [see Equation (5.27)]

$$(S_2{}^2 + S_3{}^2)^{1/2} \underset{(q),(r)}{=} \mu|S_1| \tag{t}$$

and to the two relations [see Equation (5.28)]

$$\frac{u_2(t_2) - bu_6(t_2) - v_2{}^P}{|\boldsymbol{\delta}|} \underset{(r),(s)}{=} \frac{-S_2}{(S_2{}^2 + S_3{}^2)^{1/2}} \tag{u}$$

$$\frac{u_3(t_2) + bu_5(t_2) - v_3{}^P}{|\boldsymbol{\delta}|} \underset{(r),(s)}{=} \frac{-S_3}{(S_2{}^2 + S_3{}^2)^{1/2}} \tag{v}$$

where

$$|\boldsymbol{\delta}| \underset{(s)}{=} \{[u_2(t_2) - bu_6(t_2) - v_2{}^P]^2 + [u_3(t_2) + bu_5(t_2) - v_3{}^P]^2\}^{1/2} \tag{w}$$

A complete description of the effect of the collision on the motion of B can now be obtained as follows.

From Equation (p), the difference $u_1(t_2) - u_1(t_1)$ is given by

$$u_1(t_2) - u_1(t_1) = (1 + e)[v_1{}^P - u_1(t_1)]$$

Hence,

$$S_1 \underset{(m)}{\approx} (1 + e)m[v_1{}^P - u_1(t_1)] \tag{x}$$

Next, from Equation (m), the quantities required for substitution into Equation (w) are found to be

$$u_2(t_2) - bu_6(t_2) - v_2{}^P = \frac{kS_2 + b_2}{m} \tag{y}$$

$$u_3(t_2) + bu_5(t_2) - v_3{}^P = \frac{kS_3 + b_3}{m} \tag{z}$$

where b_2, b_3, and k are defined as

$$b_2 = m[u_2(t_1) - bu_6(t_1) - v_2{}^P]$$
$$b_3 = m[u_3(t_1) + bu_5(t_1) - v_3{}^P]$$
$$k = 1 + \frac{mb^2}{I}$$

Consequently,

$$|\boldsymbol{\delta}| \underset{(w)}{=} \frac{1}{m}[(kS_2 + b_2)^2 + (kS_3 + b_3)^2]^{1/2}$$

and substitution from Equations (x), (y), (z), and (t) into (u) and (v) gives

$$kS_2 + b_2 = \frac{-S_2[(kS_2 + b_2)^2 + (kS_3 + b_3)^2]^{1/2}}{\mu(1 + e)m|v_1{}^P - u_1(t_1)|}$$

$$kS_3 + b_3 = \frac{-S_3[(kS_2 + b_2)^2 + (kS_3 + b_3)^2]^{1/2}}{\mu(1 + e)m|v_1{}^P - u_1(t_1)|}$$

S_1 is given in Equation (x), and S_2 and S_3 are the only unknowns appearing in the last two equations. Hence, once these (nonlinear) equations have been solved for S_2 and S_3, the changes in $u_1, \cdots, u_6$ resulting from the collision can be determined by reference to Equations (m).

Chapter **6**

Integration of Equations of Motion

6.1 Replacement of n Second-Order Equations with $2n$ First-Order Equations. Many of the differential equations governing motions of holonomic and simple nonholonomic systems in an inertial reference frame are nonlinear, and there exists no general, analytical method for their complete solution. However, complete or partial solutions can sometimes be effected by taking advantage of special properties of a system (see the example in Section 5.4); and, when such properties are shared by all members of a class of systems, comprehensive methods of solution can be devised. This is the case for conservative holonomic systems (see Section 4.4).

196

If S is a conservative holonomic system possessing n degrees of freedom in an inertial reference frame R, then n second-order differential equations governing the generalized coordinates $q_1, \cdots, q_n$ in R are obtained from Equations (5.7). The discussion of the solution of these equations is facilitated by introducing n additional dependent variables and replacing the n second-order equations with a set of $2n$ first-order equations. One way to do this is to take for the new variables the generalized momenta $p_1, \cdots, p_n$ as given by Equations (5.22) with $u_r = \dot{q}_r$ —that is, to let

$$p_r = \frac{\partial K}{\partial \dot{q}_r} \qquad r = 1, \cdots, n \tag{6.1}$$

where K is the kinetic energy of S in R. Since K is a function of the second degree in $\dot{q}_1, \cdots, \dot{q}_n$ (see Section 4.12), these equations are linear (but not necessarily homogeneous) in $\dot{q}_1, \cdots, \dot{q}_n$; and, solving them for these quantities, one obtains n equations of the form

$$\dot{q}_r = \sum_{s=1}^{n} \mu_{rs}(q_1, \cdots, q_n, t)p_s + \nu_r(q_1, \cdots, q_n, t) \qquad r = 1, \cdots, n \tag{6.2}$$

After using Equations (6.2) to eliminate $\ddot{q}_1, \cdots, \ddot{q}_n$ from Equations (5.7), one then has $2n$ first-order differential equations in the dependent variables $p_1, \cdots, p_n$ and $q_1, \cdots, q_n$.

▪ EXAMPLE

The potential energy P and kinetic energy K for the system described in the example in Section 3.2, found in the example in Sections 4.4 and 4.8, respectively, are

$$P = \tfrac{1}{2}k_1q_1^2 - k_2(q_2q_1 - \tfrac{1}{2}q_1^2) + \tfrac{1}{2}k_2q_2^2 + w(t) - g\cos\theta\,(m_1q_1 + m_2q_2)$$

and

$$K = \tfrac{1}{2}m_1[\dot{q}_1^2 + (L_1 + q_1)^2\dot{\theta}^2] + \tfrac{1}{2}m_2[\dot{q}_2^2 + (L_1 + L_2 + q_2)^2\dot{\theta}^2]$$

Use of Equations (5.7) leads to the two second-order differential equations of motion

$$m_1[\ddot{q}_1 - (L_1 + q_1)\dot{\theta}^2] + k_1q_1 - k_2(q_2 - q_1) - m_1g\cos\theta = 0$$
$$m_2[\ddot{q}_2 - (L_1 + L_2 + q_2)\dot{\theta}^2] + k_2(q_2 - q_1) - m_2g\cos\theta = 0$$

The generalized momenta p_1 and p_2, obtained by means of Equations (6.1), are

$$p_1 = \frac{\partial K}{\partial \dot{q}_1} = m_1\dot{q}_1 \qquad p_2 = \frac{\partial K}{\partial \dot{q}_2} = m_2\dot{q}_2$$

Consequently,

$$\dot{q}_1 = \frac{p_1}{m_1} \qquad \dot{q}_2 = \frac{p_2}{m_2} \tag{a}$$

and substitution into the equations of motion yields

$$\dot{p}_1 - m_1(L_1 + q_1)\dot{\theta}^2 + k_1q_1 - k_2(q_2 - q_1) - m_1g\cos\theta = 0 \tag{b}$$
$$\dot{p}_2 - m_2(L_1 + L_2 + q_2)\dot{\theta}^2 + k_2(q_2 - q_1) - m_2g\cos\theta = 0 \tag{c}$$

Equations (a) through (c) are the desired four first-order equations.

6.2 Matrix Notation for Differentiation. Equations (6.1) and (6.2), as well as many relationships in the sequel, can be expressed conveniently in matrix form. To this end, the following notations are introduced for dealing with derivatives.

If z is a column (row) matrix whose elements $z_1, \cdots, z_n$ are functions of t, then $\dot{z}$ or dz/dt denotes the column (row) matrix whose elements are the time derivatives $\dot{z}_1, \cdots, \dot{z}_n$ of $z_1, \cdots, z_n$; that is,

$$z = \begin{bmatrix} z_1 \\ \cdot \\ \cdot \\ \cdot \\ z_n \end{bmatrix} \Rightarrow \dot{z} = \frac{dz}{dt} = \begin{bmatrix} \dot{z}_1 \\ \cdot \\ \cdot \\ \cdot \\ \dot{z}_n \end{bmatrix} \tag{6.3}$$

and

$$z = [z_1 \quad \cdots \quad z_n] \Rightarrow \dot{z} = \frac{dz}{dt} = [\dot{z}_1 \quad \cdots \quad \dot{z}_n] \tag{6.4}$$

If f is a scalar function of the $2n + 1$ variables $x_1, \cdots, x_n, y_1, \cdots, y_n$, and t, then f_x and f_y denote the row matrices whose elements are $\partial f/\partial x_1, \cdots, \partial f/\partial x_n$ and $\partial f/\partial y_1, \cdots, \partial f/\partial y_n$, respectively, and f_t denotes the partial derivative of f with respect to t; that is,

$$f_x = \begin{bmatrix} \dfrac{\partial f}{\partial x_1} & \cdots & \dfrac{\partial f}{\partial x_n} \end{bmatrix} \tag{6.5}$$

$$f_y = \begin{bmatrix} \dfrac{\partial f}{\partial y_1} & \cdots & \dfrac{\partial f}{\partial y_n} \end{bmatrix} \tag{6.6}$$

$$f_t = \frac{\partial f}{\partial t} \tag{6.7}$$

If u is a column (row) matrix whose elements $u_1, \cdots, u_n$ are functions of $x_1, \cdots, x_n, y_1, \cdots, y_n$, and t, then u_x and u_y denote the square matrices whose rows (columns) are composed of the elements of the matrices $(u_1)_x, \cdots, (u_n)_x$ and $(u_1)_y, \cdots, (u_n)_y$, respectively, and u_t denotes the column (row) matrix whose elements are $(u_1)_t, \cdots, (u_n)_t$; that is,

$$u = \begin{bmatrix} u_1 \\ \cdot \\ \cdot \\ \cdot \\ u_n \end{bmatrix} \Rightarrow u_x = \begin{bmatrix} \dfrac{\partial u_1}{\partial x_1} & \cdots & \dfrac{\partial u_1}{\partial x_n} \\ \cdot & & \cdot \\ \cdot & & \cdot \\ \cdot & & \cdot \\ \dfrac{\partial u_n}{\partial x_n} & \cdots & \dfrac{\partial u_n}{\partial x_n} \end{bmatrix} \tag{6.8}$$

$$u = [u_1 \quad \cdots \quad u_n] \Rightarrow u_x = \begin{bmatrix} \dfrac{\partial u_1}{\partial x_1} & \cdots & \dfrac{\partial u_n}{\partial x_1} \\ \cdot & & \cdot \\ \cdot & & \cdot \\ \cdot & & \cdot \\ \dfrac{\partial u_1}{\partial x_n} & \cdots & \dfrac{\partial u_n}{\partial x_n} \end{bmatrix} \tag{6.9}$$

and similarly for u_y; and

$$u = \begin{bmatrix} u_1 \\ \cdot \\ \cdot \\ \cdot \\ u_n \end{bmatrix} \Rightarrow u_t = \begin{bmatrix} \dfrac{\partial u_1}{\partial t} \\ \cdot \\ \cdot \\ \cdot \\ \dfrac{\partial u_n}{\partial t} \end{bmatrix} \tag{6.10}$$

$$u = [u_1, \cdot \cdot \cdot , u_n] \Rightarrow u_t = \left[\frac{\partial u_1}{\partial t}, \cdot \cdot \cdot , \frac{\partial u_n}{\partial t}\right] \tag{6.11}$$

The following consequences of these definitions are of particular interest.

If u and x are n-dimensional column (row) matrices and $u = x$, then u_x is equal to the $n \times n$ *unit* or *identity matrix* U:

$$u_x = x_x \underset{(6.8),(6.9)}{=} \begin{bmatrix} 1 & 0 & \cdot \cdot \cdot & 0 \\ 0 & 1 & & 0 \\ \cdot & & & \cdot \\ \cdot & & & \cdot \\ \cdot & & & \cdot \\ 0 & 0 & \cdot \cdot \cdot & 1 \end{bmatrix} = U \tag{6.12}$$

If u and v are an n-dimensional row matrix and an n-dimensional column matrix, respectively, and the elements of u and v are functions of $x_1, \cdot \cdot \cdot , x_n, y_1, \cdot \cdot \cdot , y_n$, and t, then the partial derivatives of the product[1] uv with respect to $x_1, \cdot \cdot \cdot , x_n, y_1, \cdot \cdot \cdot , y_n$, and t are given by

$$(uv)_x = uv_x + v^T u_x{}^T \tag{6.13}$$
$$(uv)_y = uv_y + v^T u_y{}^T \tag{6.14}$$
$$(uv)_t = uv_t + v^T u_t{}^T \tag{6.15}$$

[1] If u_r and v_r are typical elements of u and v, then uv denotes the scalar quantity

$$\sum_{r=1}^{n} u_r v_r$$

and if, furthermore, w_{rs} is the element in the rth row and sth column of an $n \times n$ matrix w, then uw denotes the row matrix whose sth element is

$$\sum_{r=1}^{n} u_r w_{rs}$$

and wv denotes the column matrix whose rth element is

$$\sum_{s=1}^{n} w_{rs} v_s$$

where the superscript T denotes transposition—that is, the interchanging of rows and columns.

When a scalar quantity Q depends on $u_1, \cdots, u_n, v_1, \cdots, v_n$, and t, and u_r and v_r, with $r = 1, \cdots, n$, depend on $x_1, \cdots, x_n, y_1, \cdots, y_n$, and t, then Q can be represented either by a function of F of $u_1, \cdots, u_n, v_1, \cdots, v_n$, and t or by a function G of $x_1, \cdots, x_n, y_1, \cdots, y_n$, and t, and this can be indicated by writing

$$Q = F(u, v, t) = G(x, y, t) \tag{6.16}$$

where u, v, x, and y denote the column matrices whose elements are, respectively, u_r, v_r, x_r, and y_r, with $r = 1, \cdots, n$. Partial derivatives of F and G are then related as follows:

$$G_t = F_t \tag{6.17}$$
$$G_x = F_u u_x + F_v v_x \tag{6.18}$$
$$G_y = F_u u_y + F_v v_y \tag{6.19}$$

Furthermore, if $x_1, \cdots, x_n$ and $y_1, \cdots, y_n$ are, themselves, functions of t, then the total derivative of G with respect to t is given by

$$\frac{dG}{dt} = G_x \dot{x} + G_y \dot{y} + G_t \tag{6.20}$$

Equations (6.17) through (6.20) are valid also when F and G denote n-dimensional column matrices; but if F and G are row matrices, only Equation (6.17) remains applicable, whereas (6.18) through (6.20) must be replaced with

$$G_x = u_x{}^T F_u + v_x{}^T F_v \tag{6.21}$$
$$G_y = u_y{}^T F_u + v_y{}^T F_v \tag{6.22}$$
$$\frac{dG}{dt} = \dot{x}^T G_x + \dot{y}^T G_y + G_t \tag{6.23}$$

▪ EXAMPLE

In Equations (5.7), L represents the kinetic potential regarded as a function of $q_1, \cdots, q_n, \dot{q}_1, \cdots, \dot{q}_n$, and t. If q denotes the column matrix whose elements are $q_1, \cdots, q_n$, then, by Equation (6.3), $\dot{q}$ also denotes a column matrix; L_q and $L_{\dot{q}}$, by Equations (6.5) and (6.6), denote the row matrices whose elements are the partial derivatives in (5.7); and Equation (5.7) can, therefore, be expressed as

$$\frac{d}{dt} L_{\dot{q}} = L_q \tag{a}$$

Furthermore, the potential energy P is independent of $\dot{q}_1, \cdots, \dot{q}_n$, and can, therefore, be regarded as a function of $q_1, \cdots, q_n, \dot{q}_1, \cdots, \dot{q}_n$, and t such that

$$P_{\dot{q}} = 0 \tag{b}$$

It then follows from Equation (5.6) that

$$K_{\dot{q}} = L_{\dot{q}} \tag{c}$$

and, if p denotes the column matrix whose elements are $p_1, \cdots, p_n$, then Equations (6.1) are equivalent to

$$p^T = L_{\dot{q}} \tag{d}$$

Substitution into Equation (a) now yields the relationship

$$\dot{p}^T = L_q \tag{e}$$

which will be found useful later on.

The matrix form of Equations (6.2) is

$$\dot{q} = \mu p + \nu \tag{f}$$

where μ and ν denote a square matrix and a column matrix, respectively, both having elements which are functions of $q_1, \cdots, q_n$, and t. Consequently, the elements of $\dot{q}$ may be regarded as functions of $p_1, \cdots, p_n$, $q_1, \cdots, q_n$, and t. If the elements of p are also regarded as functions of these independent variables, then

$$p_p \underset{(6.12)}{=} U \tag{g}$$

while

$$p_q = 0 \tag{h}$$

and the evaluation of $(p^T\,\dot{q})_p$ and $(p^T\,\dot{q})_q$ provides examples in the use of Equations (6.13) and (6.14). Specifically, letting p^T, $\dot{q}$, p, and q play the parts of u, v, x, and y, respectively, one obtains

$$(p^T\,\dot{q})_p \underset{(6.13)}{=} p^T\,\dot{q}_p + \dot{q}^T\,p_p \underset{(g)}{=} p^T\,\dot{q}_p + \dot{q}^T \tag{i}$$

and

$$(p^T\,\dot{q})_q \underset{(6.14)}{=} p^T\,\dot{q}_q + \dot{q}^T\,p_q \underset{(h)}{=} p^T\,\dot{q}_q \tag{j}$$

Finally, as an illustration in the use of Equations (6.16) through (6.19), suppose that a column matrix z is defined as

$$z = q \tag{k}$$

so that the elements of z may be regarded as functions of $q_1, \cdots, q_n,$ $p_1, \cdots, p_n,$ and t, with

$$z_p = 0 \tag{l}$$

$$z_t = 0 \tag{m}$$

and

$$z_q \underset{(6.12)}{=} U \tag{n}$$

Then the kinetic potential L of S can be represented either by a function M of $z_1, \cdots, z_n, \dot{q}_1, \cdots, \dot{q}_n,$ and t or, in view of Equation (f), by a function N of $p_1, \cdots, p_n, q_1, \cdots, q_n,$ and t; that is,

$$L = M(z, \dot{q}, t) = N(p, q, t) \tag{o}$$

Identifying the symbols in Equation (o) with appropriate symbols in Equation (6.16), one thus obtains

$$N_p \underset{(6.18)}{=} M_z z_p + M_{\dot{q}} \dot{q}_p \underset{(l)}{=} M_{\dot{q}} \dot{q}_p$$

But $M_{\dot{q}}$ has the same meaning as $L_{\dot{q}}$ in Equation (d). Hence

$$N_p = p^T \dot{q}_p \tag{p}$$

Similarly,

$$N_q \underset{(6.19)}{=} M_z z_q + M_{\dot{q}} \dot{q}_q \underset{(n),(d)}{=} M_z + p^T \dot{q}_q$$

and M_z has the same meaning as L_q in Equation (e). Hence,

$$N_q = \dot{p}^T + p^T \dot{q}_q \tag{q}$$

6.3 Hamilton's Canonic Equations. The procedure described in Section 6.1 for obtaining $2n$ first-order differential equations in the dependent variables $p_1, \cdots, p_n$ and $q_1, \cdots, q_n$ can be formalized as follows: Let

$$\boxed{p_r = \frac{\partial K}{\partial \dot{q}_r}} \qquad r = 1, \cdots, n \tag{6.24}$$

and define H, called the *Hamiltonian* of S in R, as

$$\boxed{H = \sum_{r=1}^{n} p_r \dot{q}_r - L} \tag{6.25}$$

where L is the kinetic potential of S in R (see Section 5.7). Solve Equations (6.24) for $\dot{q}_1, \cdots, \dot{q}_n,$ and use the results to eliminate these quantities from Equation (6.25). H can then be regarded as a function of $p_1, \cdots,$

$p_n, q_1, \cdots, q_n$, and t, and the desired equations can be expressed as

$$\dot{p}_r = -\frac{\partial H}{\partial q_r} \qquad r = 1, \cdots, n \tag{6.26}$$

and

$$\dot{q}_r = \frac{\partial H}{\partial p_r} \qquad r = 1, \cdots, n \tag{6.27}$$

Equations (6.26) and (6.27) are known as *Hamilton's canonic equations.* In matrix notation, Equations (6.25) through (6.27) become

$$p = K_{\dot{q}}{}^T \tag{6.28}$$
$$H = p^T \dot{q} - L \tag{6.29}$$
$$\dot{p}^T = -H_q \tag{6.30}$$
$$\dot{q}^T = H_p \tag{6.31}$$

Proof: It was pointed out in the example in Section 6.2 that $p^T \dot{q}$ can be regarded as a function of $p_1, \cdots, p_n, q_1, \cdots, q_n$, and t such that [see Equations (i) and (j) of the example in Section 6.2]

$$(p^T \dot{q})_p = \dot{q}^T + p^T \dot{q}_p \tag{a}$$

and

$$(p^T \dot{q})_q = p^T \dot{q}_q \tag{b}$$

It was shown also that, when the kinetic potential is represented by a function N of $p_1, \cdots, p_n, q_1, \cdots, q_n$, and t, that is,

$$L = N(p, q, t) \tag{c}$$

then [see Equation (p)]

$$N_p = p^T \dot{q}_p \tag{d}$$

and [see Equation (q)]

$$N_q = \dot{p}^T + p^T \dot{q}_q \tag{e}$$

Hence,

$$\begin{aligned} H_p &\underset{(6.29)}{=} (p^T \dot{q})_p - \underset{(c)}{N_p} \\ &\underset{(a),(d)}{=} \dot{q}^T + p^T \dot{q}_p - p^T \dot{q}_p = \dot{q}^T \end{aligned}$$

and

$$\begin{aligned} H_q &\underset{(6.29)}{=} (p^T \dot{q})_q - \underset{(c)}{N_q} \\ &\underset{(b),(e)}{=} p^T \dot{q}_q - \dot{p}^T - p^T \dot{q}_q = -\dot{p}^T \end{aligned}$$

▪ EXAMPLE

The generalized momenta p_1 and p_2 for the system described in the example in Section 3.2 were shown, in the example in Section 6.1, to be given by

$$\dot{p}_1 = m_1\dot{q}_1 \qquad p_2 = m_2\dot{q}_2$$

Hence,

$$\dot{q}_1 = \frac{p_1}{m_1} \qquad \dot{q}_2 = \frac{p_2}{m_2} \tag{a}$$

and, using the expressions for potential energy and kinetic energy given in the example in Section 6.1, one can express the Lagrangian L as

$$\begin{aligned} L = K - P = {} & \tfrac{1}{2}m_1\left[\left(\frac{p_1}{m_1}\right)^2 + (L_1 + q_1)^2\dot{\theta}^2\right] \\ & + \tfrac{1}{2}m_2\left[\left(\frac{p_2}{m_2}\right)^2 + (L_1 + L_2 + q_2)^2\dot{\theta}^2\right] \\ & - \tfrac{1}{2}k_1q_1^2 + k_2(q_2q_1 - \tfrac{1}{2}q_1^2) - \tfrac{1}{2}k_2q_2^2 \\ & - w(t) + g\cos\theta(m_1q_1 + m_2q_2) \end{aligned} \tag{b}$$

The Hamiltonian is then given by

$$\begin{aligned} H \underset{(6.25)}{=} {} & p_1\dot{q}_1 + p_2\dot{q}_2 - L \\ \underset{(a),(b)}{=} {} & \tfrac{1}{2}m_1\left[\left(\frac{p_1}{m_1}\right)^2 - (L_1 + q_1)^2\dot{\theta}^2\right] \\ & + \tfrac{1}{2}m_2\left[\left(\frac{p_2}{m_2}\right)^2 - (L_1 + L_2 + q_2)^2\dot{\theta}^2\right] \\ & + \tfrac{1}{2}k_1q_1^2 - k_2(q_2q_1 - \tfrac{1}{2}q_1^2) + \tfrac{1}{2}k_2q_2^2 \\ & + w(t) - g\cos\theta(m_1q_1 + m_2q_2) \end{aligned} \tag{c}$$

and the canonic equations are [see Equations (6.26)]

$$\dot{p}_1 = -\frac{\partial H}{\partial q_1} \underset{(c)}{=} m_1(L_1 + q_1)\dot{\theta}^2 - k_1q_1 + k_2(q_2 - q_1) + m_1g\cos\theta \tag{d}$$

$$\dot{p}_2 = -\frac{\partial H}{\partial q_2} \underset{(c)}{=} m_2(L_1 + L_2 + q_2)^2\dot{\theta}^2 + k_2(q_1 - q_2) + m_2g\cos\theta \tag{e}$$

and [see Equations (6.27)]

$$\dot{q}_1 = \frac{\partial H}{\partial p_1} = \frac{p_1}{m_1} \tag{f}$$

$$\dot{q}_2 = \frac{\partial H}{\partial p_2} = \frac{p_2}{m_2} \tag{g}$$

Equations (d) and (e) agree with (b) and (c) of the example in Section 6.1, respectively; and (f) and (g) are the same as Equation (a) of that example. This is no coincidence: When the partial differentiations indicated in Equations (6.27) are carried out, one always obtains expressions for $\dot{q}_1$,

$\cdots, \dot{q}_n$ which are precisely those resulting from the solution of Equations 6.25.

6.4 Integrals of the Canonic Equations. An equation of the form

$$f(p_1, \cdots, p_n, q_1, \cdots, q_n, t) = \alpha \tag{6.32}$$

where α is an arbitrary constant, is called an *integral* of the canonic equations (see Section 6.3) if the total time derivative of f vanishes whenever $p_1, \cdots, p_n$ and $q_1, \cdots, q_n$ satisfy the canonic equations (6.27) and (6.28). The complete solution of the canonic equations consists of $2n$ independent integrals.

A function $f(p_1, \cdots, p_n, q_1, \cdots, q_n, t)$ furnishes an integral if and only if

$$\sum_{r=1}^{n} \left(\frac{\partial f}{\partial q_r} \frac{\partial H}{\partial p_r} - \frac{\partial f}{\partial p_r} \frac{\partial H}{\partial q_r} \right) + \frac{\partial f}{\partial t} = 0 \tag{6.33}$$

or, in matrix notation [see Equations (6.5) through (6.7)],

$$f_q H_p^T - f_p H_q^T + f_t = 0 \tag{6.34}$$

Proof: Let p and q be the column matrices whose elements are $p_1, \cdots, p_n$ and $q_1, \cdots, q_n$, respectively, and suppose that f is a function of $p_1, \cdots, p_n, q_1, \cdots, q_n$, and t, such that

$$\frac{df}{dt} = 0 \tag{a}$$

whenever p and q satisfy the equations

$$\dot{p}^T \underset{(6.30)}{=} -H_q \qquad \dot{q}^T \underset{(6.31)}{=} H_p \tag{b}$$

Then

$$0 \underset{(a)}{=} \frac{df}{dt} \underset{(6.20)}{=} f_p \dot{p} + f_q \dot{q} + f_t$$
$$\underset{(b)}{=} -f_p H_q^T + f_q H_p^T + f_t$$

and Equation (6.34) is seen to be satisfied. Conversely, if f satisfies the equation

$$f_q H_p^T - f_p H_q^T + f_t \underset{(6.34)}{=} 0 \tag{c}$$

and p and q satisfy Equations (b), then

$$0 \underset{(c),(b)}{=} f_q \dot{q} + f_p \dot{p} + f_t \underset{(6.20)}{=} \frac{df}{dt}$$

and Equation (a) is satisfied.

▪ EXAMPLE

Consider a particle P of mass m moving in a vertical plane under the action of gravity, and let q_1 and q_2 be Cartesian coordinates of P in this plane, the associated axes being horizontal and vertical, respectively. Then $\dot{q}_1$ and $\dot{q}_2$, expressed in terms of the generalized momenta p_1 and p_2 of P, are [see Equation (6.24)]

$$\dot{q}_1 = \frac{p_1}{m} \qquad \dot{q}_2 = \frac{p_2}{m}$$

and the kinetic potential of P is given by

$$L \underset{(5.6)}{=} \frac{1}{2m}(p_1^2 + p_2^2) - mgq_2$$

so that the Hamiltonian H becomes

$$H \underset{(6.25)}{=} \frac{1}{2m}(p_1^2 + p_2^2) + mgq_2 \tag{a}$$

with the partial derivatives

$$\frac{\partial H}{\partial p_1} = \frac{p_1}{m} \qquad \frac{\partial H}{\partial p_2} = \frac{p_2}{m} \qquad \frac{\partial H}{\partial q_1} = 0 \qquad \frac{\partial H}{\partial q_2} = mg \qquad \frac{\partial H}{\partial t} = 0 \tag{b}$$

Substitution from Equations (b) into (6.33) shows that a function f of p_1, p_2, q_1, q_2, and t furnishes an integral of the canonic equations whenever it satisfies

$$-mg\frac{\partial f}{\partial p_2} + \frac{p_1}{m}\frac{\partial f}{\partial q_1} + \frac{p_2}{m}\frac{\partial f}{\partial q_2} + \frac{\partial f}{\partial t} = 0 \tag{c}$$

Suppose f is taken to be the function $q_1 - p_1t/m$. Then

$$\frac{\partial f}{\partial p_2} = 0 \qquad \frac{\partial f}{\partial q_1} = 1 \qquad \frac{\partial f}{\partial q_2} = 0 \qquad \frac{\partial f}{\partial t} = -\frac{p_1}{m}$$

and substitution into the left-hand member of Equation (c) leads to the value zero. Hence the equation

$$q_1 - \frac{p_1t}{m} = c_1 \tag{d}$$

where c_1 is an arbitrary constant, is an integral of the motion of P.

Proceeding in the same way, one can verify that

$$p_1 = c_2 \tag{e}$$

$$q_2 - \frac{p_2t}{m} - \frac{gt^2}{2} = mc_3 \tag{f}$$

and

$$p_2 + mgt = mc_4 \tag{g}$$

where c_2, c_3, and c_4 are arbitrary constants, are three further integrals.

The four integrals (d) through (g) provide a complete description of all motions of P; for, when p_1 and p_2 are eliminated from Equations (d) and (f) by using (e) and (g), one obtains explicit expressions for q_1 and q_2 as functions of t:

$$q_1 \underset{(\mathrm{d}),(\mathrm{e})}{=} c_1 + c_2 t \tag{h}$$

$$q_2 \underset{(\mathrm{f}),(\mathrm{g})}{=} c_3 + c_4 t - \frac{gt^2}{2} \tag{i}$$

(The constants $c_1, \cdot \cdot \cdot , c_4$ can be interpreted as the values of q_1, $\dot{q}_1$, q_2, and $\dot{q}_2$ at time $t = 0$.)

It is worth noting that the function H as given in Equation (a) also furnishes an integral in the present problem; that is, Equation (c) is satisfied when f is replaced with H, as may be verified by making use of (b). Consequently,

$$\frac{1}{2m}(p_1^2 + p_2^2) + mgq_2 = C \tag{j}$$

where C is an arbitrary constant. But this integral is not independent of the four discussed previously; that is, it can be deduced from Equations (e) through (g) by purely algebraic operations.

So far, nothing has been said about the source of the four integrals (d) through (g)—that is, about the method used to find them. Indeed, the method is irrelevant to the purpose at hand, which is, simply, to provide illustrations. General statements about ways to find integrals are made in the sections that follow. In the present example, the integrals (d) through (g) were obtained by first determining the general solution of the canonic equations, which involved four arbitrary constants, and then solving the resulting four equations for these constants. Once the general solution of the canonic equations has been found, this method can always be used to generate $2n$ integrals, but it is, of course, pointless to do so if one's sole objective is to solve the canonic equations.

6.5 The Energy Integral. When the Hamiltonian of a system S (see Section 6.3) is expressed as a function H of $p_1, \cdot \cdot \cdot , p_n, q_1, \cdot \cdot \cdot , q_n$, and t, it may occur that t does not appear, so that

$$\frac{\partial H}{\partial t} = 0 \tag{6.35}$$

The canonic equations then possess the integral (see Section 6.4)

$$H = E \tag{6.36}$$

where E is an arbitrary constant. Equation (6.36) is known as the *energy integral.* This terminology is particularly suggestive when the kinetic energy K of S is homogeneous and of second degree in $\dot{q}_1, \cdots, \dot{q}_n$, for H is then equal to the total mechanical energy of S; that is,

$$H = K + P \tag{6.37}$$

where K and P are the kinetic and the potential energy of S. The energy integral [Equation (6.36)], assuming it exists, then becomes a statement of the *law of conservation of energy,*

$$K + P = E \tag{6.38}$$

Proof: Equation (6.34) is satisfied when f is replaced with H, provided $H_t = 0$, which is guaranteed by Equation (6.35). Hence, f may be replaced with H in Equation (6.32), and (6.36) follows when α is replaced with E.

Next,

$$\begin{aligned} H &\underset{(6.25)}{=} \sum_{r=1}^{n} p_r \dot{q}_r - L \\ &\underset{(6.24)}{=} \sum_{r=1}^{n} \frac{\partial K}{\partial \dot{q}_r} \dot{q}_r - L \\ &\underset{(4.27),(5.6)}{=} 2K - (K - P) = K + P \end{aligned}$$

which establishes the validity of Equation (6.37). Finally, Equation (6.38) follows directly whenever both (6.36) and (6.37) are satisfied.

▪ EXAMPLE

Two particles of mass m_1 and m_2 are free to slide in a circular tube of radius r and are connected to each other by a linear spring having a natural length r and modulus k (not necessarily time-independent). The tube is made to rotate with angular speed Ω (not necessarily time-independent) about a vertical axis, as indicated in Figure 6.1.

The kinetic energy K and the potential energy P of the system can be expressed as

$$K = \tfrac{1}{2}[m_1 r^2 \dot{q}_1^2 + m_2 r^2 \dot{q}_2^2 + r^2 \Omega^2 (m_1 s_1^2 + m_2 s_2^2)] \tag{a}$$

and

$$P = \tfrac{1}{2} k r^2 (q_2 - q_1 - 1)^2 - g r (m_1 c_1 + m_2 c_2) \tag{b}$$

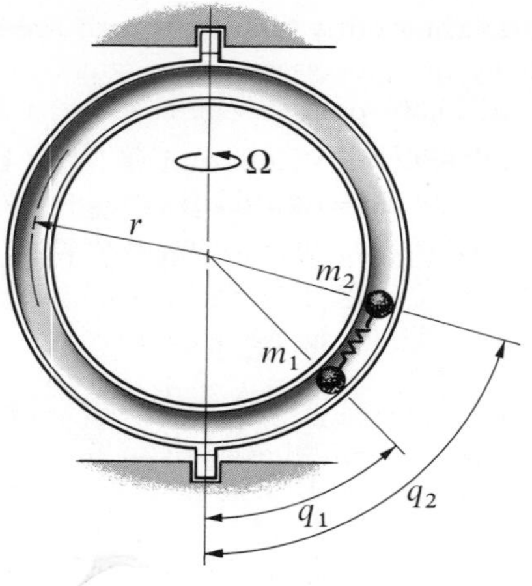

Figure 6.1

where $s_i = \sin q_i$, $c_i = \cos q_i$, and q_1 and q_2 are the angles shown in Figure 6.1. The generalized momenta, p_1 and p_2, are [see Equation (6.24)]

$$p_1 = m_1 r^2 \dot{q}_1 \qquad p_2 = m_2 r^2 \dot{q}_2 \tag{c}$$

and the Hamiltonian H is thus given by

$$H = \tfrac{1}{2}\left[\frac{p_1^{\,2}}{m_1 r^2} + \frac{p_2^{\,2}}{m_2 r^2} - r^2\Omega^2(m_1 s_1^{\,2} + m_2 s_2^{\,2})\right] + \tfrac{1}{2}kr^2(q_2 - q_1 - 1)^2 - gr(m_1 c_1 + m_2 c_2) \tag{d}$$

Suppose now that either Ω or k is time-dependent. Then $\partial H/\partial t$ does not vanish and the system does not possess an energy integral; that is, H does not remain constant. Conversely, if both Ω and k remain constant, then there exists the integral

$$\tfrac{1}{2}\left[\frac{p_1^{\,2}}{m_1 r^2} + \frac{p_2^{\,2}}{m_2 r^2} - r^2\Omega^2(m_1 s_1^{\,2} + m_2 s_2^{\,2})\right] + \tfrac{1}{2}kr^2(q_2 - q_1 - 1)^2 - gr(m_1 c_1 + m_2 c_2) = E$$

where E is a constant. However, the total mechanical energy—that is, $K + P$—does not remain constant unless $\Omega = 0$, for only then is K [see Equation (a)] a homogeneous quadratic function of $\dot{q}_1$ and $\dot{q}_2$. Finally, it appears from Equations (a) through (d) that $H = K + P$ whenever $\Omega = 0$, but that this condition is not sufficient to guarantee the existence of an energy integral: If k varies with t, then $\partial H/\partial t$ can differ from zero while $H = K + P$.

6.6 Systems Possessing Only One Degree of Freedom. When a conservative holonomic system possesses only a single degree of freedom, use of the energy integral (6.36) permits one to express the time derivative $\dot{q}$ of the generalized coordinate q as

$$\dot{q} = \pm v(q, q_0, \dot{q}_0) \tag{6.39}$$

where q_0 and $\dot{q}_0$ denote the values of q and $\dot{q}$, respectively, at an arbitrarily selected instant t_0, and v is the non-negative square root of a function of q, q_0, and $\dot{q}_0$. The time t is then given by

$$t = t_0 \pm \int_{q_0}^{q} \frac{d\zeta}{v(\zeta, q_0, \dot{q}_0)} \tag{6.40}$$

This relationship is particularly useful when the indicated quadrature can be executed in closed form—that is, when the result can be expressed in terms of tabulated functions. If, on the other hand, numerical methods must be employed, it may be preferable to apply them directly to the solution of the differential equations of motion.

▪ EXAMPLE

A horizontal shaft AB carries a uniform, rectangular plate of mass m and is made to rotate with constant angular speed Ω about a fixed vertical axis, as indicated in Figure 6.2. The plate is free to rotate, but not to slide, on the shaft.

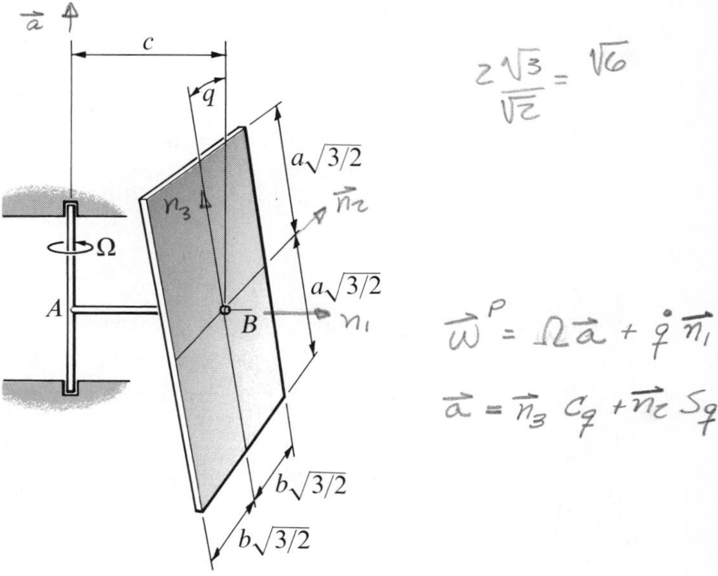

Figure 6.2

The potential energy P, kinetic energy K, and generalized momentum p are given by

$$P = 0$$

$$K = \tfrac{1}{2}m[(a^2 + b^2)\dot{q}^2 + (a^2 \sin^2 q + b^2 \cos^2 q + c^2)\Omega^2]$$

and

$$p \underset{(6.1)}{=} m(a^2 + b^2)\dot{q}$$

The Hamiltonian H can thus be expressed as

$$H \underset{(6.25),(5.6)}{=} p\dot{q} - K + P$$
$$= \tfrac{1}{2}m[(a^2 + b^2)\dot{q}^2 - (a^2 \sin^2 q + b^2 \cos^2 q + c^2)\Omega^2]$$

and the energy integral [Equation (6.36)] is

$$\tfrac{1}{2}m[(a^2 + b^2)\dot{q}^2 - (a^2 \sin^2 q + b^2 \cos^2 q + c^2)\Omega^2] = E$$

If q and $\dot{q}$ have the values q_0 and $\dot{q}_0$, respectively, at some instant t_0, then

$$E = \tfrac{1}{2}m[(a^2 + b^2)\dot{q}_0^2 - (a^2 \sin^2 q_0 + b^2 \cos^2 q_0 + c^2)\Omega^2]$$

Consequently,

$$(a^2 + b^2)\dot{q}^2 - (a^2 \sin^2 q + b^2 \cos^2 q + c^2)\Omega^2$$
$$= (a^2 + b^2)\dot{q}_0^2 - (a^2 \sin^2 q_0 + b^2 \cos^2 q_0 + c^2)\Omega^2$$

and

$$\dot{q} = \pm v(q, q_0, \dot{q}_0) \tag{a}$$

where

$$v(q, q_0, \dot{q}_0) = \left\{\dot{q}_0^2 + \frac{[a^2(\sin^2 q - \sin^2 q_0) + b^2(\cos^2 q - \cos^2 q_0)]\Omega^2}{a^2 + b^2}\right\}^{1/2} \tag{b}$$

Suppose now that $b > a$ and that q_0 and $\dot{q}_0$ have the values

$$q_0 = 0 \qquad \dot{q}_0 = \frac{\Omega}{k}\sqrt{\frac{b^2 - a^2}{b^2 + a^2}} \tag{c}$$

where k is a constant whose value lies between zero and one. Then

$$v \underset{(b),(c)}{=} \frac{\Omega}{k}\sqrt{\frac{b^2 - a^2}{b^2 + a^2}}\,(1 - k^2 \sin^2 q)^{1/2}$$

and v is seen to be intrinsically positive. Moreover, $\dot{q}$ must be positive at t_0 [see Equation (c)]. Hence the positive sign in (a) applies for all t; that is, the plate rotates on the shaft, and

$$t \underset{(6.40)}{=} t_0 + \frac{k}{\Omega}\sqrt{\frac{b^2 + a^2}{b^2 - a^2}} \int_0^q (1 - k^2 \sin^2 \zeta)^{-1/2}\, d\zeta$$

or, if $F(k, q)$ is defined as

$$F(k, q) = \int_0^q (1 - k^2 \sin^2 \zeta)^{-1/2} \, d\zeta$$

then

$$t = t_0 + \frac{k}{\Omega} \sqrt{\frac{b^2 + a^2}{b^2 - a^2}} F(k, q) \tag{d}$$

The function $F(k, q)$ is called the *incomplete elliptic integral* of the first kind, and has been tabulated extensively (see, for example, *Handbook of Mathematical Functions*, National Bureau of Standards, Applied Mathematics Series, 55). When $q = \pi/2$, the value of $F(k, q)$ is denoted by $K(k)$; that is,

$$K(k) = F(k, \pi/2) \tag{e}$$

and this function of k is called the *complete elliptic integral* of the first kind of modulus k. In the present problem, $K(k)$ has the following significance: When $q = \pi/2$, the plate has completed one fourth of a revolution. Hence, if T is the time required for one revolution, then t has the value $t_0 + T/4$ at the end of the first quarter revolution subsequent to t_0 and

$$\frac{T}{4} \underset{(d),(e)}{=} \frac{k}{\Omega} \sqrt{\frac{b^2 + a^2}{b^2 - a^2}} K(k)$$

6.7 Use of the Energy Integral To Check Numerical Integrations. When the differential equations of motion of a system are integrated numerically, the energy integral can be used to check the accuracy of the calculations. This is especially easy when the differential equations are the canonic equations (6.26) and (6.27), but it can also be done when variables other than the generalized coordinates and generalized momenta, such as, for example, the u's introduced in Section 2.24, occur in the equations of motion.

▪ EXAMPLE

Figure 6.3 represents a system that is similar to the one considered in the example in Section 6.6. Once again, AB is a horizontal shaft that is made to rotate with constant angular speed Ω about a fixed vertical axis, but this shaft now supports a rigid body R at a single point, the mass center of R, which coincides with the end B of the shaft.

Dynamical equations can be formulated most conveniently in terms of u_1, u_2, u_3, defined as follows: Let $\mathbf{n}_1$, $\mathbf{n}_2$, $\mathbf{n}_3$ be a right-handed set of mutually perpendicular unit vectors respectively parallel to principal

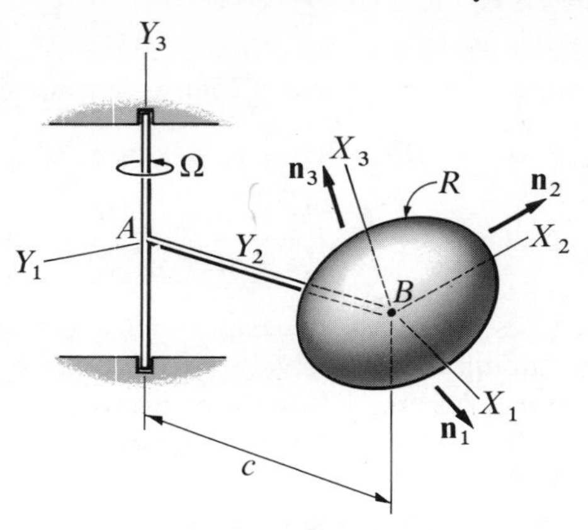

Figure 6.3

axes of inertia X_1, X_2, X_3 which are fixed in R, and take

$$u_r = \boldsymbol{\omega} \cdot \mathbf{n}_r \qquad r = 1, 2, 3 \tag{a}$$

where $\boldsymbol{\omega}$ is the angular velocity of R in an inertial reference frame. Then the generalized active forces F_1, F_2, F_3, are all equal to zero, and the generalized inertial forces F_1^*, F_2^*, F_3^*, are given by

$$F_r^* \underset{(3.48)}{=} \mathbf{n}_r \cdot \mathbf{T}^* \qquad r = 1, 2, 3$$

where

$$\begin{aligned} \mathbf{T}^* \underset{(3.44)}{=} & [u_2u_3(I_2 - I_3) - \dot{u}_1I_1]\mathbf{n}_1 \\ & + [u_3u_1(I_3 - I_1) - \dot{u}_2I_2]\mathbf{n}_2 \\ & \qquad + [u_1u_2(I_1 - I_2) - \dot{u}_3I_3]\mathbf{n}_3 \end{aligned}$$

Consequently, the dynamical equations can be expressed as [see Equations (5.1)]

$$\begin{aligned} \dot{u}_1 &= \frac{u_2u_3(I_2 - I_3)}{I_1} \\ \dot{u}_2 &= \frac{u_3u_1(I_3 - I_1)}{I_2} \\ \dot{u}_3 &= \frac{u_1u_2(I_1 - I_2)}{I_3} \end{aligned} \tag{b}$$

To obtain a complete description of the motion of R, let Y_1, Y_2, Y_3 (see Figure 6.3) be a right-handed set of mutually perpendicular axes originating at A, with Y_2 parallel to AB and Y_3 directed vertically upward; and let q_1, q_2, q_3 be a set of "three-axis" Euler angles generated by aligning X_r with Y_r, for $r = 1, 2, 3$, and then performing successive right-handed rotations of amounts q_1 about X_1, q_2 about X_2, and q_3 about X_3. The angular velocity $\boldsymbol{\omega}$ can then be expressed as

$$\begin{aligned}\boldsymbol{\omega} \underset{(2.14)}{=} \Omega\mathbf{N}_3 &+ (c_2c_3\dot{\theta}_1 + s_3\dot{\theta}_2)\mathbf{n}_1 \\ &- (c_2s_3\dot{\theta}_1 - c_3\dot{\theta}_2)\mathbf{n}_2 \\ &+ (s_2\dot{\theta}_1 + \dot{\theta}_3)\mathbf{n}_3\end{aligned}$$

where $\mathbf{N}_3$ is a unit vector directed vertically upward, and s_i and c_i denote $\sin q_i$ and $\cos q_i$, respectively. Consequently, from Equations (a),

$$\begin{aligned}u_1 &= c_2c_3\dot{q}_1 + s_3\dot{q}_2 + \Omega(s_3s_1 - c_1s_2c_3) \\ u_2 &= c_3\dot{q}_2 - c_2s_3\dot{q}_1 + \Omega(c_3s_1 + c_1s_2s_3) \\ u_3 &= s_2\dot{q}_1 + \dot{q}_3 + \Omega c_1c_2\end{aligned} \tag{c}$$

and it follows that q_1, q_2, and q_3 must satisfy the differential equations

$$\begin{aligned}\dot{q}_1 &= \frac{u_1c_3 - u_2s_3 + \Omega c_1s_2}{c_2} \\ \dot{q}_2 &= u_2c_3 + u_1s_3 - \Omega s_1 \\ \dot{q}_3 &= u_3 - \frac{s_2(u_1c_3 - u_2s_3) + \Omega c_1}{c_2}\end{aligned} \tag{d}$$

which can be solved simultaneously with Equations (b). To use the energy integral as a check on such a solution, note that the kinetic energy K of R is given by [see Equations (4.14) and (4.16)]

$$K = \tfrac{1}{2}mc^2\Omega^2 + \tfrac{1}{2}\sum_{s=1}^{3} I_s u_s^2 \tag{e}$$

so that, from Equations (6.24), the generalized momenta p_1, p_2, p_3, are given by

$$p_r = \frac{\partial K}{\partial \dot{q}_r} = \sum_{s=1}^{3} I_s u_s \frac{\partial u_s}{\partial \dot{q}_r} \qquad r = 1, 2, 3$$

or, from Equations (c),

$$\begin{aligned}p_1 &= I_1u_1c_2c_3 - I_2u_2c_2s_3 + I_3u_3s_2 \\ p_2 &= I_1u_1s_3 + I_2u_2c_3 \\ p_3 &= I_3u_3\end{aligned} \tag{f}$$

The Hamiltonian H is given by

$$H \underset{(6.25)}{=} p_1\dot{q}_1 + p_2\dot{q}_2 + p_3\dot{q}_3 - K$$

and H can be expressed entirely in terms of u_r and q_r, with $r = 1, 2, 3$, by using Equations (d), (e), and (f). The energy integral

$$H = E$$

where E is a constant, then furnishes a relationship between the dependent variables in Equations (b) and (d), and this relationship must be satisfied for every value of the independent variable, t.

6.8 Momentum Integrals and Cyclic Coordinates. When the Hamiltonian of a system S (see Section 6.3) is expressed as a function H of $p_1, \cdots, p_n, q_1, \cdots, q_n$, and t, it may occur that some of $q_1, \cdots, q_n$, say the first k, do not appear, so that

$$\frac{\partial H}{\partial q_s} = 0 \qquad s = 1, \cdots, k \tag{6.41}$$

The canonic equations then possess the k integrals (see Section 6.4)

$$p_s = \alpha_s \qquad s = 1, \cdots, k \tag{6.42}$$

where $\alpha_1, \cdots, \alpha_k$ are constants. These integrals are known as *momentum integrals*, and a coordinate q_s that does not appear in H (or in the Lagrangian L) is called a *cyclic* or *ignorable* coordinate.

Proof: Substitution from Equations (6.41) into (6.33) gives

$$\sum_{r=1}^{n} \frac{\partial f}{\partial q_r} \frac{\partial H}{\partial p_r} - \sum_{r=k+1}^{n} \frac{\partial f}{\partial p_r} \frac{\partial H}{\partial q_r} + \frac{\partial f}{\partial t} = 0$$

and all of the partial derivatives of f that occur in this equation are equal to zero if $f = p_s$, with $s = 1, \cdots, k$. Consequently, f can be replaced with p_s, and α with α_s, in Equation (6.32).

▪ EXAMPLE

Equation (a) of the example in Section 6.4 shows that q_1 is a cyclic coordinate; and Equation (e) is the associated momentum integral.

6.9 Reduction from $2n$ to n First-Order Equations. When there exist n momentum integrals (see Section 6.8), or $n - 1$ momentum integrals and an energy integral (see Section 6.5) of the $2n$ canonic equations (6.26) and (6.27), the solution of the problem under consideration can be accomplished by solving a set of n differential equations of the form

$$\dot{q}_r = v_r(q_1, \cdots, q_n, \alpha_1, \cdots, \alpha_n) \tag{6.43}$$

where $\alpha_1, \cdots, \alpha_n$ are arbitrary constants and $v_1, \cdots, v_n$ are known functions of their arguments.

▪ EXAMPLE

Referring to the example in Section 4.10, suppose that $v(t) = 0$ throughout some time interval. Then, if A, B, and β are defined as

$$A = I_1 \qquad B = I_2 + M\left(\frac{3h}{4}\right)^2 \qquad \beta = \frac{3Mgh}{4}$$

the potential energy P and the kinetic energy K can be expressed as

$$P = \beta \cos \theta \tag{a}$$

and

$$K = \tfrac{1}{2}[A(\dot{\psi} + \dot{\phi} \cos \theta)^2 + B(\dot{\phi}^2 \sin^2 \theta + \dot{\theta}^2)] \tag{b}$$

The coordinates ϕ and ψ are cyclic (see Section 6.8). Hence the associated generalized momenta remain constant; that is,

$$p_\phi \underset{(6.24)}{=} \frac{\partial K}{\partial \dot{\phi}} = A(\dot{\psi} + \dot{\phi} \cos \theta) \cos \theta + B\dot{\phi} \sin^2 \theta = \alpha_1 \tag{c}$$

and

$$p_\psi = \frac{\partial K}{\partial \dot{\psi}} = A(\dot{\psi} + \dot{\phi} \cos \theta) = \alpha_2 \tag{d}$$

where α_1 and α_2 are arbitrary constants. The kinetic energy is a homogeneous function of the second degree in $\dot{\phi}$, $\dot{\psi}$, and $\dot{\theta}$. The Hamiltonian is, therefore, equal to the sum of K and P [see Equation (6.37)] and, since this sum does not involve the time t explicitly, there exists an energy integral that can be expressed as

$$K + P \underset{(a),(b)}{=} \tfrac{1}{2}[A(\dot{\psi} + \dot{\phi} \cos \theta)^2 + B(\dot{\phi}^2 \sin^2 \theta + \dot{\theta}^2)] + \beta \cos \theta \underset{(6.38)}{=} \alpha_3 \tag{e}$$

where α_3 is an arbitrary constant.

When Equations (c) and (d) are solved for $\dot{\phi}$ and $\dot{\psi}$, and the results are used to solve Equation (e) for $\dot{\theta}$, one obtains three expressions of the form of Equations (6.43):

$$\dot{\phi} = \frac{\alpha_1 - \alpha_2 \cos \theta}{B \sin^2 \theta} \tag{f}$$

$$\dot{\psi} = \frac{\alpha_2}{A} - \frac{\alpha_1 - \alpha_2 \cos \theta}{B \sin^2 \theta} \cos \theta \tag{g}$$

$$\dot{\theta} = \pm \left[\frac{2A\alpha_3 - \alpha_2{}^2}{AB} - \frac{2\beta \cos \theta}{B} - \frac{(\alpha_1 - \alpha_2 \cos \theta)^2}{B^2 \sin^2 \theta}\right]^{1/2} \tag{h}$$

Thus, when Equation (h) has been solved, so that θ is known explicitly as a function $\theta(t)$ of time t, then ϕ and ψ can be found by carrying out the indicated integrations in the equations

$$\phi \underset{(f)}{=} \phi_0 + \int_{t_0}^{t} \frac{\alpha_1 - \alpha_2 \cos \theta(\zeta)}{B \sin^2 \theta(\zeta)} \, d\zeta \tag{i}$$

and

$$\psi \underset{(g)}{=} \psi_0 + \frac{\alpha_2}{A}(t - t_0) - \int_{t_0}^{t} \frac{\alpha_1 - \alpha_2 \cos \theta(\zeta)}{B \sin^2 \theta(\zeta)} \cos \theta(\zeta) \, d\zeta \tag{j}$$

where t_0 is an arbitrary value of t and ϕ_0 and ψ_0 denote the values of ϕ and ψ at time t_0. [A detailed discussion of the solution of Equations (f) through (h) may be found in E. T. Whittaker, *A Treatise on the Analytical Dynamics of Particles and Rigid Bodies*, New York: Dover Publications, pp. 155–161.]

6.10 Canonic Variables and Canonic Transformations. The so-called variation of parameters method for solving differential equations can be cast into a form that is particularly convenient for dealing with the canonic equations (6.26) and (6.27). Even in this form, however, use of the method rarely simplifies the search for a complete solution of these equations, for it involves solving both a partial differential equation and two sets of ordinary differential equations which are similar in form to the canonic equations. The determination of *approximate* solutions, on the other hand, can be facilitated considerably by means of this technique. With this end in view, then, the method is set forth in Section 6.14. The present section and Sections 6.11 through 6.13 deal with necessary preliminaries.

When each of $2n$ quantities $P_1, \cdots, P_n, Q_1, \cdots, Q_n$ is defined as a function of the generalized momenta $p_1, \cdots, p_n$, the generalized coordinates $q_1, \cdots, q_n$, and the time t, these quantities satisfy differential equations of the form

$$\dot{P}_r = \hat{P}_r \qquad \dot{Q}_r = \hat{Q}_r \qquad r = 1, \cdots, n \tag{6.44}$$

where $\hat{P}_r$ and $\hat{Q}_r$ denote functions of $P_1, \cdots, P_n, Q_1, \cdots, Q_n$, and t. [This follows from Equations (6.26) and (6.27).] If there exists a function $\hat{H}$ of $P_1, \cdots, P_n, Q_1, \cdots, Q_n$, and t such that $\hat{P}_r$ and $\hat{Q}_r$ can be expressed as

$$\hat{P}_r = -\frac{\partial \hat{H}}{\partial Q_r} \qquad \hat{Q}_r = \frac{\partial \hat{H}}{\partial P_r} \qquad r = 1, \cdots, n \tag{6.45}$$

then the differential equations (6.44) assume the canonic form

$$\dot{P}_r = -\frac{\partial \hat{H}}{\partial Q_r} \qquad \dot{Q}_r = \frac{\partial \hat{H}}{\partial P_r} \qquad r = 1, \cdots, n \tag{6.46}$$

Under these circumstances, $P_1, \cdots, P_n, Q_1, \cdots, Q_n$ are called *canonic variables*, and the transformation from $p_1, \cdots, p_n, q_1, \cdots, q_n$ to $P_1, \cdots, P_n, Q_1, \cdots, Q_n$ is said to be *canonic with respect to the function* $\hat{H}$.

▪ EXAMPLE

Suppose $P_1, \cdots, P_n$ and $Q_1, \cdots, Q_n$ are defined as

$$P_r = -q_r \qquad Q_r = p_r \qquad r = 1, \cdots, n$$

or, in matrix form,

$$P = -q \qquad Q = p \tag{a}$$

Then

$$\dot{P} \underset{(a)}{=} -\dot{q} \underset{(6.31)}{=} -H_p{}^T \qquad \dot{Q} \underset{(a)}{=} \dot{p} \underset{(6.30)}{=} -H_q{}^T \tag{b}$$

and, if a function $\hat{H}$ of P, Q, and t is defined as

$$\hat{H} = H \tag{c}$$

then, in accordance with Equations (6.18) and (6.19),

$$H_p = \hat{H}_P P_p + \hat{H}_Q Q_p \underset{(a)}{=} 0 + \hat{H}_Q p_p = \hat{H}_Q$$

$$H_q = \hat{H}_P P_q + \hat{H}_Q Q_q \underset{(a)}{=} -\hat{H}_P q_q + 0 = -\hat{H}_P$$

and substitution into (b) gives

$$\dot{P} = -\hat{H}_Q{}^T \qquad \dot{Q} = \hat{H}_P{}^T$$

Consequently, the transformation given in Equation (a) is canonic with respect to $\hat{H}$ as given in (c)—that is, with respect to the Hamiltonian.

To see that the function $\hat{H}$ need not be equal to the Hamiltonian (as it was in this example) consider once again the particle moving in a vertical plane under the action of gravity, previously studied in the example in Section 6.4. The Hamiltonian was there shown to be

$$H = \frac{1}{2m}(p_1{}^2 + p_2{}^2) + mgq_2 \tag{d}$$

and the canonic equations are

$$\dot{p}_1 = 0 \qquad \dot{p}_2 = -mg \qquad \dot{q}_1 = \frac{p_1}{m} \qquad \dot{q}_2 = \frac{p_2}{m} \tag{e}$$

If P_1, P_2, Q_1, and Q_2 are defined as

$$P_1 = p_2 \qquad P_2 = p_1 \qquad Q_1 = Ct + q_2 \qquad Q_2 = q_1 \tag{f}$$

where C is any constant, then

$$\dot{P}_1 \underset{\text{(f)}}{=} \dot{p}_2 \underset{\text{(e)}}{=} -mg \tag{g}$$

$$\dot{P}_2 \underset{\text{(f)}}{=} \dot{p}_1 \underset{\text{(e)}}{=} 0 \tag{h}$$

$$\dot{Q}_1 \underset{\text{(f)}}{=} C + \dot{q}_2 \underset{\text{(e)}}{=} C + \frac{p_2}{m} \underset{\text{(f)}}{=} C + \frac{P_1}{m} \tag{i}$$

$$\dot{Q}_2 \underset{\text{(f)}}{=} \dot{q}_1 \underset{\text{(e)}}{=} \frac{p_1}{m} \underset{\text{(f)}}{=} \frac{P_2}{m} \tag{j}$$

while H can be expressed as a function $\bar{H}$ of P_1, P_2, Q_1, Q_2, and t by using Equation (f):

$$\bar{H} \underset{\text{(d),(f)}}{=} \frac{1}{2m}(P_1^2 + P_2^2) + mg(Q_1 - Ct) \tag{k}$$

The variables P_1, P_2, Q_1, Q_2 are not canonic with respect to $\bar{H}$, because

$$\frac{\partial \bar{H}}{\partial P_1} \underset{\text{(k)}}{=} \frac{P_1}{m} \tag{l}$$

so that, from Equation (i),

$$\dot{Q}_1 \neq \frac{\partial \bar{H}}{\partial P_1}$$

However, P_1, P_2, Q_1, Q_2 are canonic with respect to $\hat{H}$ defined as

$$\begin{aligned} \hat{H} &= CP_1 + \bar{H} \\ &\underset{\text{(k)}}{=} CP_1 + \frac{1}{2m}(P_1^2 + P_2^2) + mg(Q_1 - Ct) \end{aligned}$$

because

$$\begin{aligned} \frac{\partial \hat{H}}{\partial Q_1} &= mg \underset{\text{(g)}}{=} -\dot{P}_1 \\ \frac{\partial \hat{H}}{\partial Q_2} &= 0 \underset{\text{(h)}}{=} -\dot{P}_2 \\ \frac{\partial \hat{H}}{\partial P_1} &= C + \frac{P_1}{m} \underset{\text{(i)}}{=} \dot{Q}_1 \\ \frac{\partial \hat{H}}{\partial P_2} &= \frac{P_2}{m} \underset{\text{(j)}}{=} \dot{Q}_2 \end{aligned}$$

6.11 Generating Function for a Canonic Transformation. A transformation from generalized momenta $p_1, \cdots, p_n$, and generalized coordinates $q_1, \cdots, q_n$ that satisfy the canonic equations (6.26) and (6.27) to variables $P_1, \cdots, P_n, Q_1, \cdots, Q_n$ that are canonic (see Section 6.10) with respect to a function $\hat{H}$ of $P_1, \cdots, P_n, Q_1, \cdots, Q_n$, and t can be obtained by taking

$$p_r = \frac{\partial S}{\partial q_r} \qquad r = 1, \cdots, n \tag{6.47}$$

$$Q_r = \frac{\partial S}{\partial P_r} \qquad r = 1, \cdots, n \tag{6.48}$$

and

$$\hat{H} = H + \frac{\partial S}{\partial t} \tag{6.49}$$

where H is the Hamiltonian of the system and S is any function of $P_1, \cdots, P_n, q_1, \cdots, q_n$, and t such that Equation (6.47) can be solved for $P_1, \cdots, P_n$ in terms of $p_1, \cdots, p_n, q_1, \cdots, q_n$, and t. In other words, if $\hat{H}$ is formed in accordance with Equation (6.49), then $P_1, \cdots, P_n$ and $Q_1, \cdots, Q_n$ as given by (6.47) and (6.48) satisfy the equations

$$\dot{P}_r = -\frac{\partial \hat{H}}{\partial Q_r} \qquad \dot{Q}_r = \frac{\partial \hat{H}}{\partial P_r} \qquad r = 1, \cdots, n \tag{6.50}$$

The function S is called the *generating function* of the transformation. In matrix notation, Equations (6.47) through (6.50) are

$$p^T = S_q \tag{6.51}$$

$$Q^T = S_P \tag{6.52}$$

$$\hat{H} = H + S_t \tag{6.53}$$

and

$$\dot{P}^T = -\hat{H}_Q \qquad \dot{Q}^T = \hat{H}_P \tag{6.54}$$

Proof: Let p, q, P, and Q denote column matrices with elements p_r, q_r, P_r, and Q_r, with $r = 1, \cdots, n$. Then Equations (6.47) and (6.48) can be expressed as

$$p^T = S_q \tag{a}$$

$$Q^T = S_P \tag{b}$$

and, in accordance with (6.23), differentiation with respect to time t gives

$$\dot{p}^T \underset{(a)}{=} \dot{P}^T S_{qP} + \dot{q}^T S_{qq} + S_{qt} \tag{c}$$

$$\dot{Q}^T \underset{(b)}{=} \dot{P}^T S_{PP} + \dot{q}^T S_{Pq} + S_{Pt} \tag{d}$$

Consequently, multiplication of Equation (c) by $(S_{qP})^{-1}$, the inverse[2] of S_{qP}, leads to

$$\dot{P}^T \underset{(c)}{=} (\dot{p}^T - \dot{q}^T S_{qq} - S_{qt})(S_{qP})^{-1}$$

or, after elimination of $\dot{p}$ and $\dot{q}$ by means of Equations (6.30) and (6.31),

$$\dot{P}^T = -(H_q + H_p S_{qq} + S_{qt})(S_{qP})^{-1} \tag{e}$$

Substitution into Equation (d) now gives

$$\dot{Q}^T \underset{(d),(e)}{=} -(H_q + H_p S_{qq} + S_{qt})(S_{qP})^{-1} S_{PP} + H_p S_{Pq} + S_{Pt} \tag{f}$$

The partial derivatives of $\hat{H}$ with respect to P and Q, evaluated by using Equations (6.18) and (6.19), are

$$\begin{aligned} \hat{H}_P &\underset{(6.49)}{=} H_p p_P + H_q q_P + S_{tP} P_P + S_{tq} q_P \\ &\underset{(6.12)}{=} H_p p_P + (H_q + S_{tq}) q_P + S_{tP} \end{aligned} \tag{g}$$

and

$$\begin{aligned} \hat{H}_Q &= H_p p_Q + H_q q_Q + S_{tP} P_Q + S_{tq} q_Q \\ &= H_p p_Q + (H_q + S_{tq}) q_Q \end{aligned} \tag{h}$$

The quantities p_P, p_Q, q_P, and q_Q can be expressed in terms of S, as follows: Referring to Equation (6.22), and identifying G with Q^T, F with S_P, u with P, v with q, x with P, and y with Q, differentiate Equation (b) with respect to Q, obtaining

$$U = P_Q{}^T S_{PP} + q_Q{}^T S_{Pq} = q_Q{}^T S_{Pq}$$

or

$$q_Q = (S_{qP})^{-1} \tag{i}$$

Similarly, identification of F with S_q, and differentiation of Equation (a) with respect to Q gives

$$p_Q{}^T = P_Q{}^T S_{qP} + q_Q{}^T S_{qq} \underset{(i)}{=} (S_{Pq})^{-1} S_{qq}$$

so that

$$p_Q = [(S_{Pq})^{-1} S_{qq}]^T = S_{qq}{}^T [(S_{Pq})^{-1}]^T = S_{qq} (S_{qP})^{-1} \tag{j}$$

Next, Equation (b), differentiated with respect to P, yields

$$0 = P_P{}^T S_{PP} + q_P{}^T S_{Pq} = S_{PP} + q_P{}^T S_{Pq}$$

or

$$q_P = -(S_{qP})^{-1} S_{PP} \tag{k}$$

---

[2] The inverse of a square matrix A is a square matrix B such that $AB = BA = U$, the unit matrix [see Equation (6.12)].

and differentiation of Equation (a) with respect to P leads to

$$p_P = S_{Pq} - S_{qq}(S_{qP})^{-1}S_{PP} \tag{l}$$

$\hat{H}_P$ and $\hat{H}_Q$ can now be expressed as

$$\hat{H}_P \underset{(g)}{=} H_p[S_{Pq} - S_{qq}(S_{qP})^{-1}S_{PP}] - (H_q + S_{tq})(S_{qP})^{-1}S_{PP} + S_{tP} \tag{m}$$

and

$$\hat{H}_Q \underset{(h)}{=} H_p \underset{(j)}{S_{qq}(S_{qP})^{-1}} + (H_q + S_{tq})\underset{(i)}{(S_{qP})^{-1}} \tag{n}$$

and it follows from Equations (e) and (n) that

$$\dot{P}^T = -\hat{H}_Q$$

and Equations (f) and (m) give

$$\dot{Q}^T = \hat{H}_P$$

▪ EXAMPLE

As was shown in the example in Section 6.4, the Hamiltonian for a particle moving in a vertical plane under the action of gravity can be expressed as

$$H = \frac{1}{2m}(p_1^2 + p_2^2) + mgq_2 \tag{a}$$

The generating function given by

$$S = P_1q_2 + P_2q_1 + CP_1t \tag{b}$$

leads to the transformation [see Equations (6.47) and (6.48)]

$$p_1 = P_2 \qquad p_2 = P_1 \qquad Q_1 = q_2 + Ct \qquad Q_2 = q_1 \tag{c}$$

which is precisely the transformation introduced in Equation (f) of the example in Section 6.10, where it was shown also that this transformation is canonic with respect to a function $\hat{H}$ given by

$$\hat{H} = CP_1 + \frac{1}{2m}(P_1^2 + P_2^2) + mg(Q_1 - Ct) \tag{d}$$

As

$$CP_1 \underset{(b)}{=} \frac{\partial S}{\partial t}$$

and

$$\frac{1}{2m}(P_1^2 + P_2^2) + mg(Q_1 - Ct) \underset{(c)}{=} \frac{1}{2m}(p_1^2 + p_2^2) + mgq_2 \underset{(a)}{=} H$$

it appears that, in agreement with Equation (6.49),

$$\hat{H} \underset{(c)}{=} H + \frac{\partial S}{\partial t}$$

6.12 The Hamilton-Jacobi Equation. If the generating function S (see Section 6.11) is chosen in such a way that $\hat{H}$ vanishes, then it follows from Equations (6.50) that P_r and Q_r are constants, say α_r and β_r. The transformation equations (6.47) and (6.48) can then be expressed as

$$p_r = \frac{\partial S^*}{\partial q_r} \qquad r = 1, \cdots, n \tag{6.55}$$

$$\beta_r = \frac{\partial S^*}{\partial \alpha_r} \qquad r = 1, \cdots, n \tag{6.56}$$

where S^* denotes a function of $\alpha_1, \cdots, \alpha_n, q_1, \cdots, q_n$, and t that satisfies Equation (6.49) when $\hat{H}$ set equals to zero; that is, if $H(p_1, \cdots, p_n, q_1, \cdots, q_n, t)$ denotes the Hamiltonian, S^* satisfies the *Hamilton-Jacobi equation*

$$H\left(\frac{\partial S^*}{\partial q_1}, \cdots, \frac{\partial S^*}{\partial q_n}, q_1, \cdots, q_n, t\right) + \frac{\partial S^*}{\partial t} = 0 \tag{6.57}$$

Thus, if S^* is a complete solution of Equation (6.57)—that is, a solution containing n arbitrary constant $\alpha_1, \cdots, \alpha_n$ (none additive), then Equations (6.55) and (6.56) furnish a complete set of integrals of the canonic equations (6.26) and (6.27). The function S^* is called Hamilton's *principal function*, and the constants α_r and β_r, with $r = 1, n$, are known as *canonic constants*, β_r being called the *canonic conjugate* of α_r, and vice versa.

In matrix notation, Equations (6.55) through (6.57) are

$$p^T = S_q^* \tag{6.58}$$

$$\beta^T = S_\alpha^* \tag{6.59}$$

and

$$H(S_q^*, q, t) + S_t^* = 0 \tag{6.60}$$

▪ EXAMPLE

When the Hamiltonian for a particle moving in a vertical plane under the action of gravity is expressed as (see the example in Section 6.4)

$$H = \frac{1}{2m}(p_1^2 + p_2^2) + mgq_2 \tag{a}$$

$P_1 = \frac{\partial K}{\partial \dot{q}_1} = m\dot{q}_1$

then the Hamilton-Jacobi equation is

$$\frac{1}{2m}\left[\left(\frac{\partial S^*}{\partial q_1}\right)^2 + \left(\frac{\partial S^*}{\partial q_2}\right)^2\right] + mgq_2 + \frac{\partial S^*}{\partial t} = 0 \tag{b}$$

and it may be verified that this equation is satisfied by

$$S^* = -(\alpha_1 + \alpha_2)t + (2m\alpha_1)^{1/2}q_1 - \frac{1}{3m^2g}[2m(\alpha_2 - mgq_2)]^{3/2} \tag{c}$$

(The method used to find S^* is described in Section 6.13.)

The coordinates q_1 and q_2 can now be determined as follows:

$$\beta_1 \underset{(6.56)}{=} \frac{\partial S^*}{\partial \alpha_1} \underset{(c)}{=} -t + \left(\frac{2\alpha_1}{m}\right)^{-1/2} q_1 \tag{d}$$

Consequently,

$$q_1 = \left(\frac{2\alpha_1}{m}\right)^{1/2} (\beta_1 + t) \tag{e}$$

Similarly,

$$\beta_2 \underset{(6.56)}{=} \frac{\partial S^*}{\partial \alpha_2} \underset{(c)}{=} -t - \left[\frac{2(\alpha_2 - mgq_2)}{mg^2}\right]^{1/2} \tag{f}$$

so that

$$q_2 = \frac{\alpha_2}{mg} - \frac{g}{2}(\beta_1 + \beta_2 + t)^2 \tag{g}$$

Equations (e) and (g) have the same forms as Equations (h) and (i) of the example in Section 6.4. A striking difference between the two sets of equations is that the constants $c_1, \cdots, c_4$ in the example in Section 6.4 could be related easily to the initial values of the coordinates and their first time derivatives, whereas the canonic constants α_1, α_2, β_1, and β_2 are complicated functions of these initial values.

6.13 Separation of Variables. If S_r, $r = 1, \cdots, n$, and σ denote functions of q_r, $r = 1, \cdots, n$, and t, respectively, the substitution

$$S^* = \sigma + \sum_{r=1}^{n} S_r \tag{6.61}$$

may make it possible to bring the Hamilton-Jacobi equation (see Section

6.12) into the form

$$Y\left(\frac{d\sigma}{dt}, t\right) + \sum_{r=1}^{n} Z_r\left(\frac{dS_r}{dq_r}, q_r\right) = 0 \tag{6.62}$$

where Y and Z_r, $r = 1, \cdots, n$, denote functions of the arguments indicated. Under these circumstances, the Hamilton-Jacobi equation is said to be *separable*, and S^* can be determined by the method of *separation of variables*—that is, by solving the $n + 1$ ordinary differential equations

$$Z_r = \alpha_r \qquad r = 1, \cdots, n \tag{6.63}$$

and

$$Y = -\sum_{r=1}^{n} \alpha_r \tag{6.64}$$

where $\alpha_1, \cdots, \alpha_n$ are constants. Furthermore, without solving any differential equations, n integrals are obtained by substituting from Equations (6.55) and (6.61) into (6.63).

▪ EXAMPLE

As was shown in the example in Section 6.12, the Hamilton-Jacobi equation for a particle moving in a vertical plane under the action of gravity can be expressed as

$$\frac{1}{2m}\left[\left(\frac{\partial S^*}{\partial q_1}\right)^2 + \left(\frac{\partial S^*}{\partial q_2}\right)^2\right] + mgq_2 + \frac{\partial S^*}{\partial t} = 0 \tag{a}$$

If S_1, S_2, and σ denote functions of q_1, q_2, and t, respectively, then the substitution

$$S^* = \sigma + S_1 + S_2 \tag{b}$$

gives

$$\frac{1}{2m}\left[\left(\frac{dS_1}{dq_1}\right)^2 + \left(\frac{dS_2}{dq_2}\right)^2\right] + mgq_2 + \frac{d\sigma}{dt} \underset{\text{(a),(b)}}{=} 0$$

which is equivalent to

$$Y + Z_1 + Z_2 = 0$$

if Y, Z_1, and Z_2 are defined as

$$Y = \frac{d\sigma}{dt}$$

and

$$Z_1 = \frac{1}{2m}\left(\frac{dS_1}{dq_1}\right)^2 \qquad Z_2 = \frac{1}{2m}\left(\frac{dS_2}{dq_2}\right)^2 + mgq_2 \tag{c}$$

S_1, S_2, and σ are then found by solving the differential equations

$$\frac{1}{2m}\left(\frac{dS_1}{dq_1}\right)^2 \underset{(6.63)}{=} \alpha_1 \tag{d}$$

$$\frac{1}{2m}\left(\frac{dS_2}{dq_2}\right)^2 + mgq_2 \underset{(6.63)}{=} \alpha_2 \tag{e}$$

and

$$\frac{d\sigma}{dt} \underset{(6.64)}{=} -(\alpha_1 + \alpha_2) \tag{f}$$

Each of these can be reduced to the evaluation of an indefinite integral:

$$S_1 \underset{(d)}{=} \int (2m\alpha_1)^{1/2}\, dq_1 \tag{g}$$

$$S_2 \underset{(e)}{=} \int [2m(\alpha_2 - mgq_2)]^{1/2}\, dq_2 \tag{h}$$

$$\sigma \underset{(f)}{=} -\int (\alpha_1 + \alpha_2)\, dt \tag{i}$$

The omission of arbitrary constants in Equations (g) through (i), and the arbitrary choice of positive signs preceding the radicals in (g) and (h), are justified because neither prevents one from finding a complete solution of the Hamilton-Jacobi equation, which is all that matters. Carrying out the indicated integrations, one obtains

$$S_1 \underset{(g)}{=} (2m\alpha_1)^{1/2} q_1 \tag{j}$$

$$S_2 \underset{(h)}{=} -\frac{1}{3m^2 g}[2m(\alpha_2 - mgq_2)]^{3/2} \tag{k}$$

$$\sigma \underset{(i)}{=} -(\alpha_1 + \alpha_2)t \tag{l}$$

Consequently,

$$S^* \underset{(b)}{=} -(\alpha_1 + \alpha_2)t + (2m\alpha_1)^{1/2} q_1 - \frac{1}{3m^2 g}[2m(\alpha_2 - mgq_2)]^{3/2} \tag{m}$$

The two integrals obtained from Equation (6.63) are

$$\frac{1}{2m} p_1^2 \underset{(c),(d)}{=} \alpha_1 \qquad \frac{1}{2m} p_2^2 + mgq_2 \underset{(c),(e)}{=} \alpha_2$$

[The first of these is, essentially, the integral (e) of the example in Section 6.4; and the sum of the two integrals is the energy integral (see Section 6.5) previously encountered in Equation (j) of the example in Section 6.4.]

In the example in Section 6.12, q_1 and q_2 were found as functions of time by using Equation (6.56) with S^* as given in Equation (m). An alternative approach is to substitute from Equations (g) through (i) into (b), obtaining for S^* an expression involving indefinite integrals, and then to form Equations (6.56) by differentiating under the integral signs:

$$S^* \underset{(b)}{=} -\int (\alpha_1 + \alpha_2)\, dt + \int (2m\alpha_1)^{1/2}\, dq_1 + \int [2m(\alpha_2 - mgq_2)]^{1/2}\, dq_2 \tag{n}$$

$$\beta_1 = \frac{\partial S^*}{\partial \alpha_1} \underset{(n)}{=} -\int dt + \int \left(\frac{2\alpha_1}{m}\right)^{-1/2} dq_1$$

$$= -t + \left(\frac{2\alpha_1}{m}\right)^{-1/2} q_1 \tag{o}$$

$$\beta_2 = \frac{\partial S^*}{\partial \alpha_2} \underset{(n)}{=} -\int dt + \int m[2m(\alpha_2 - mgq_2)]^{-1/2}\, dq_2$$

$$= -t - \left[\frac{2(\alpha_2 - mgq_2)}{mg^2}\right]^{1/2} \tag{p}$$

Equations (o) and (p) are identical with (d) and (f) of the example in Section 6.12, respectively.

6.14 Variation of Parameters. The method of *variation of parameters* consists of using as a generating function (see Section 6.11) a complete solution of the Hamilton-Jacobi equation (see Section 6.12) associated with a *portion* of the Hamiltonian H for the problem under consideration. Specifically, suppose that $H^{(0)}$ and $H^{(1)}$ are two functions such that

$$H = H^{(0)}(p_1, \cdots, p_n, q_1, \cdots, q_n, t) + H^{(1)} \tag{6.65}$$

and let $S^{(0)}(\alpha_1, \cdots, \alpha_n, q_1, \cdots, q_n, t)$ denote a complete solution of the Hamilton-Jacobi equation associated with $H^{(0)}$—that is, of [see Equation (6.57)]

$$H^{(0)}\left(\frac{\partial S^{(0)}}{\partial q_1}, \cdots, \frac{\partial S^{(0)}}{\partial q_n}, q_1, \cdots, q_n, t\right) + \frac{\partial S^{(0)}}{\partial t} = 0 \tag{6.66}$$

If a function S of $P_1, \cdots, P_n, q_1, \cdots, q_n$, and t is now defined as

$$S = S^{(0)}(P_1, \cdots, P_n, q_1, \cdots, q_n, t) \tag{6.67}$$

then $p_1, \cdots, p_n$, and $q_1, \cdots, q_n$ are governed by Equations (6.47) and (6.48), that is, by

$$p_r = \frac{\partial S}{\partial q_r} \qquad r = 1, \cdots, n \tag{6.68}$$

$$Q_r = \frac{\partial S}{\partial P_r} \qquad r = 1, \cdots, n \tag{6.69}$$

where P_r and Q_r, $r = 1, \cdots, n$, are solutions of the so-called *variational equations*

$$\dot{P}_r = -\frac{\partial H^{(1)}}{\partial Q_r} \qquad r = 1, \cdots, n \tag{6.70}$$

$$\dot{Q}_r = \frac{\partial H^{(1)}}{\partial P_r} \qquad r = 1, \cdots, n \tag{6.71}$$

in which $H^{(1)}$ denotes a function of $P_1, \cdots, P_n, Q_1, \cdots, Q_n$, and t, obtained after using Equations (6.68) and (6.69) to express $p_1, \cdots, p_n$ and $q_1, \cdots, q_n$ in terms of $P_1, \cdots, P_n$ and $Q_1, \cdots, Q_n$. These equations follow from Equations (6.50) together with

$$\hat{H} \underset{(6.49)}{=} H^{(0)} \underset{(6.65)}{+} H^{(1)} + \underset{(6.67)}{\frac{\partial S^{(0)}}{\partial t}} \underset{(6.66)}{=} H^{(1)}$$

The method of variation of parameters can be applied repeatedly; that is, since Equations (6.70) and (6.71) are equations of the same form as (6.26) and (6.27), respectively, they can be solved by the method of variation of parameters, and this process can be continued ad infinitum.

In matrix notation, Equations (6.65) through (6.71) are

$$H = H^{(0)}(p, q, t) + H^{(1)} \tag{6.72}$$

$$H^{(0)}(S_q{}^{(0)}, q, t) + S_t{}^{(0)} = 0 \tag{6.73}$$

$$S = S^{(0)}(P, q, t) \tag{6.74}$$

$$p^T = S_q \tag{6.75}$$

$$Q^T = S_P \tag{6.76}$$

$$\dot{P}^T = -H_Q{}^{(1)} \tag{6.77}$$

$$\dot{Q}^T = H_P{}^{(1)} \tag{6.78}$$

▪ EXAMPLE

The Hamiltonian H for a particle moving in a vertical plane under the action of gravity can be expressed as

$$H = H^{(0)} + H^{(1)}$$

where (see the example in Section 6.4)

$$H^{(0)} = \frac{1}{2m}(p_1{}^2 + p_2{}^2)$$

and

$$H^{(1)} = mgq_2 \tag{a}$$

The Hamilton-Jacobi equation for $H^{(0)}$ is

$$\frac{1}{2m}\left[\left(\frac{\partial S^{(0)}}{\partial q_1}\right)^2 + \left(\frac{\partial S^{(0)}}{\partial q_2}\right)^2\right] + \frac{\partial S^{(0)}}{\partial t} \underset{(6.66)}{=} 0$$

and, using the method of separation of variables (see Section 6.13), one can express a complete solution of Equation (d) in the following form [see Equation (n) in the example in Section (6.13)]:

$$S^{(0)} = -\int (\alpha_1 + \alpha_2)\,dt + \int (2m\alpha_1)^{1/2}\,dq_1 + \int (2m\alpha_2)^{1/2}\,dq_2$$

Consequently,

$$S \underset{(6.67)}{=} -\int (P_1 + P_2)\,dt + \int (2mP_1)^{1/2}\,dq_1 + \int (2mP_2)^{1/2}\,dq_2$$

and, by Equations (6.69),

$$Q_1 = \frac{\partial S}{\partial P_1} = -\int dt + \int \left(\frac{2P_1}{m}\right)^{-1/2} dq_1$$

$$= -t + \left(\frac{2P_1}{m}\right)^{-1/2} q_1 \tag{b}$$

and

$$Q_2 = \frac{\partial S}{\partial P_2} = -\int dt + \int \left(\frac{2P_2}{m}\right)^{-1/2} dq_2$$

$$= -t + \left(\frac{2P_2}{m}\right)^{-1/2} q_2 \tag{c}$$

so that

$$q_1 \underset{(b)}{=} \left(\frac{2P_1}{m}\right)^{1/2} (Q_1 + t) \tag{d}$$

and

$$q_2 \underset{(c)}{=} \left(\frac{2P_2}{m}\right)^{1/2} (Q_2 + t) \tag{e}$$

$H^{(1)}$ can now be expressed as

$$H^{(1)} \underset{(a),(e)}{=} (2mg^2P_2)^{1/2}(Q_2 + t)$$

and the variational equations (6.70) and (6.71) are then

$$\dot{P}_1 = -\frac{\partial H^{(1)}}{\partial Q_1} = 0 \tag{f}$$

$$\dot{P}_2 = -\frac{\partial H^{(1)}}{\partial Q_2} = -(2mg^2P_2)^{1/2} \tag{g}$$

$$\dot{Q}_1 = \frac{\partial H^{(1)}}{\partial P_1} = 0 \tag{h}$$

$$\dot{Q}_2 = \frac{\partial H^{(1)}}{\partial P_2} = \left(\frac{2P_2}{mg^2}\right)^{-1/2} (Q_2 + t) \tag{i}$$

The general solutions of Equations (f) through (h) are

$$P_1 \underset{(f)}{=} a_1 \tag{j}$$

$$P_2 = \tfrac{1}{4}(a_2 - \sqrt{2m}\,gt)^2 \tag{k}$$

and

$$Q_1 = b_1 \tag{l}$$

respectively, where a_1, a_2, and b_1 are arbitrary constants; and Equation (i) thus becomes

$$\dot{Q}_2 \underset{(k)}{=} \sqrt{2m}\,g(Q_2 + t)(a_2 - \sqrt{2m}\,gt)^{-1}$$

which has the general solution

$$Q_2 = (b_2 + \sqrt{m/2}\,gt^2)(a_2 - \sqrt{2m}\,gt)^{-1} \tag{m}$$

The coordinates q_1 and q_2 can now be expressed as

$$q_1 \underset{(d),(j),(l)}{=} \left(\frac{2a_1}{m}\right)^{1/2} (b_1 + t) \tag{n}$$

and

$$q_2 \underset{(e),(k),(m)}{=} \frac{b_2 + a_2 t}{\sqrt{2m}} - \tfrac{1}{2}gt^2 \tag{o}$$

In Section 6.10 it was suggested that use of the method of variation of parameters may complicate, rather than simplify, the solution of a problem. This is certainly the case in the present example, where the second-order differential equations governing q_1 and q_2 are

$$\ddot{q}_1 = 0$$

and

$$\ddot{q}_2 = -g$$

so that repeated integration leads immediately to

$$q_1 = c_1 t + c_2$$

$$q_2 = c_3 + c_4 t - \frac{gt^2}{2}$$

where $c_1, \cdots, c_4$ are arbitrary constants; and this procedure is far simpler than the one used to obtain Equations (n) and (o).

6.15 Approximate Solutions. The method of variation of parameters (see Section 6.14) yields results only when the variational equations (6.70) and (6.71) have been solved. Frequently, useful information, if not an exact solution of the canonic equations, can be obtained by solving the variational equations approximately or by approximating $H^{(1)}$ before writing the variational equations.

▪ EXAMPLE

In the analysis of vibrations of elastic systems one sometimes encounters a differential equation of the form

$$\ddot{q} + \omega^2 q(1 + \epsilon^2 q^2) = 0 \tag{a}$$

where q represents a time-dependent quantity of interest, such as a displacement of one or more points of a system, and ω and ϵ are constants depending on physical properties. The solution to this equation can be expressed in "closed" form (in terms of elliptic functions), and it thus provides a good test for a method that leads to an approximate solution. Specifically, for example, the exact solution shows that q is a periodic function of t whose period τ and amplitude A are related as follows:

$$\tau = \frac{4K[\sqrt{B/2(1 + B)}]}{\omega\sqrt{1 + B}} \tag{b}$$

where B, defined as

$$B = \epsilon^2 A^2 \tag{c}$$

is an intrinsically positive quantity, and where $K[k]$ denotes the complete elliptic integral of the first kind, of modulus k. Equation (b) will be compared with an approximate relationship obtained by means of the method of variation of parameters.

In order to use the method of variation of parameters to solve Equation (a), one must first find a function H such that Equation (a) is equivalent to the canonic equations associated with H. This is the case if H is taken to be

$$H = \tfrac{1}{2}\left[p^2 + \omega^2 q^2\left(1 + \frac{\epsilon^2 q^2}{2}\right)\right]$$

For then

$$\frac{\partial H}{\partial q} = \omega^2 q(1 + \epsilon^2 q^2)$$

and

$$\frac{\partial H}{\partial p} = p$$

so that the canonic equations are

$$\dot{p} \underset{(6.26)}{=} -\omega^2 q(1 + \epsilon^2 q^2)$$

and

$$\dot{q} \underset{(6.27)}{=} p$$

from which Equation (a) follows immediately.

H can be expressed as

$$H = H^{(0)} + H^{(1)}$$

where

$$H^{(0)} = \tfrac{1}{2}(p^2 + \omega^2 q^2) \tag{d}$$

and

$$H^{(1)} = \frac{\epsilon^2 \omega^2 q^4}{4} \tag{e}$$

The Hamilton-Jacobi equation associated with $H^{(0)}$ is

$$\tfrac{1}{2}\left[\left(\frac{\partial S^{(0)}}{\partial q}\right)^2 + \omega^2 q^2\right] + \frac{\partial S^{(0)}}{\partial t} \underset{(6.66)}{=} 0 \tag{f}$$

and, when $S^{(0)}$ is expressed as

$$S^{(0)} \underset{(6.61)}{=} \sigma^{(0)}(t) + S_1^{(0)}(q) \tag{g}$$

then one can satisfy Equation (f) by taking

$$\tfrac{1}{2}\left[\left(\frac{dS_1^{(0)}}{dq}\right)^2 + \omega^2 q^2\right] \underset{(6.63)}{=} \alpha \tag{h}$$

and

$$\frac{d\sigma^{(0)}}{dt} \underset{(6.64)}{=} -\alpha \tag{i}$$

where α is an arbitrary constant. Equations (h) and (i) are, in turn, satisfied if

$$S_1^{(0)} \underset{(h)}{=} \int (2\alpha - \omega^2 q^2)^{1/2}\, dq \tag{j}$$

and

$$\sigma^{(0)} \underset{(i)}{=} -\alpha t \tag{k}$$

Consequently,

$$S^{(0)} \underset{(g),(j),(k)}{=} -\alpha t + \int (2\alpha - \omega^2 q^2)^{1/2}\, dq \tag{l}$$

and q is governed by

$$Q \underset{(6.69)}{=} \frac{\partial S}{\partial P} \tag{m}$$

where

$$S \underset{(6.67)}{=} -Pt + \int (2P - \omega^2 q^2)^{1/2}\, dq \tag{n}$$

It follows that

$$Q \underset{(m),(n)}{=} -t + \int (2P - \omega^2 q^2)^{-1/2}\, dq$$
$$= -t + \frac{1}{\omega} \text{ arc sin } \frac{\omega q}{\sqrt{2P}} \tag{o}$$

so that

$$q \underset{(o)}{=} \frac{\sqrt{2P}}{\omega} \sin \omega(Q + t) \tag{p}$$

The "perturbing" Hamiltonian $H^{(1)}$ can now be expressed as

$$H^{(1)} \underset{(e),(p)}{=} \frac{\epsilon^2 P^2}{\omega^2} \sin^4 \omega(Q + t)$$

and Equation (p) then furnishes the exact solution of (a) as soon as P and Q have been found as exact solutions of the variational equations

$$\dot{P} \underset{(6.70)}{=} -\frac{\partial H^{(1)}}{\partial Q} \underset{(n)}{=} -\frac{4\epsilon^2 P^2}{\omega} \sin^3 \omega(Q + t) \cos^3 \omega(Q + t)$$
$$\dot{Q} \underset{(6.71)}{=} \frac{\partial H^{(1)}}{\partial P} \underset{(n)}{=} \frac{2\epsilon^2 P}{\omega^2} \sin^4 \omega(Q + t)$$

If, on the other hand, approximate expressions for P and Q are used in Equation (p), then an approximate solution of (a) is obtained.

One way to approximate P and Q is to use the method of variation of parameters once again. To this end, note that

$$\sin^4 x = \tfrac{3}{8} - \tfrac{1}{2} \cos 2x + \tfrac{1}{8} \cos 4x$$

so that $H^{(1)}$ can be expressed as

$$H^{(1)} = \frac{\epsilon^2 P^2}{\omega^2} [\tfrac{3}{8} - \tfrac{1}{2} \cos 2\omega(Q + t) + \tfrac{1}{8} \cos 4\omega(Q + t)] \tag{q}$$

or as

$$H^{(1)} = \bar{H}^{(0)} + \bar{H}^{(1)}$$

where

$$\bar{H}^{(0)} = \frac{3\epsilon^2 P^2}{8\omega^2}$$

and

$$\bar{H}^{(1)} = \frac{\epsilon^2 P^2}{\omega^2} [-\tfrac{1}{2} \cos 2\omega(Q + t) + \tfrac{1}{8} \cos 4\omega(Q + t)] \tag{r}$$

If $\bar{S}^{(0)}$ denotes a function of $\bar{\alpha}$, Q, and t, where $\bar{\alpha}$ is an arbitrary constant, the Hamilton-Jacobi equation associated with $\bar{H}^{(0)}$ is

$$\frac{3\epsilon^2}{8\omega^2} \left[\frac{\partial \bar{S}^{(0)}}{\partial Q}\right]^2 + \frac{\partial \bar{S}^{(0)}}{\partial t} \underset{(6.66)}{=} 0 \tag{s}$$

and this is satisfied by

$$\bar{S}^{(0)} = -\bar{\alpha}t + \bar{S}_1^{(0)}$$

where $\bar{S}_1^{(0)}$ denotes a function of Q such that

$$\frac{3\epsilon^2}{8\omega^2}\left[\frac{d\bar{S}_1^{(0)}}{dQ}\right]^2 - \bar{\alpha} = 0$$

In other words

$$\bar{S}^{(0)} = -\bar{\alpha}t + \frac{\omega}{\epsilon}\sqrt{\frac{8\bar{\alpha}}{3}}\,Q \tag{t}$$

P and Q are now governed by

$$P \underset{(6.68)}{=} \frac{\partial \bar{S}}{\partial Q}$$

and

$$\bar{Q} \underset{(6.69)}{=} \frac{\partial \bar{S}}{\partial \bar{P}}$$

where $\bar{P}$ and $\bar{S}$ are new variables, and $\bar{S}$ is obtained by replacing $\bar{\alpha}$ with $\bar{P}$ in the expression for $\bar{S}^{(0)}$; that is,

$$\bar{S} \underset{(t)}{=} -\bar{P}t + \frac{\omega}{\epsilon}\sqrt{\frac{8\bar{P}}{3}}\,Q$$

Thus

$$P = \frac{\omega}{\epsilon}\sqrt{\frac{8\bar{P}}{3}} \tag{u}$$

and

$$\bar{Q} = -t + \frac{\omega}{\epsilon}\sqrt{\frac{2}{3\bar{P}}}\,Q$$

so that Q is given by

$$Q = \frac{\epsilon}{\omega}\sqrt{\frac{3\bar{P}}{2}}\,(\bar{Q} + t) \tag{v}$$

Furthermore, $\bar{P}$ and $\bar{Q}$ must satisfy the variational equations

$$\dot{\bar{P}} \underset{(6.70)}{=} -\frac{\partial \bar{H}^{(1)}}{\partial \bar{Q}} \qquad \dot{\bar{Q}} \underset{(6.71)}{=} \frac{\partial \bar{H}^{(1)}}{\partial \bar{P}} \tag{w}$$

where $\bar{H}^{(1)}$ is obtained by substituting from Equations (u) and (v) into (r).

If $\bar{P}$ and $\bar{Q}$ in Equations (u) and (v) are treated as constants, rather than as solutions of (w), then (u) and (v) furnish approximate expressions for P and Q. Substitution into Equation (p) then leads to an approximate solution of (a); namely,

$$q \approx A \sin\left[\omega\left(1 + \frac{3\epsilon^2 A^2}{8}\right)t + \phi\right] \tag{x}$$

where A and ϕ are arbitrary constants.

Before discussing this result, it is worth mentioning that relationships equivalent to Equations (u) and (v) when $\bar{P}$ and $\bar{Q}$ are treated as constants can be obtained most directly simply by taking

$$H^{(1)} \underset{(q)}{\approx} \frac{3\epsilon^2 P^2}{8\omega^2}$$

that is, by using only the so-called "slow" or zero-frequency term in Equation (q). Solution of the associated canonic equations

$$\dot{P} \approx -\frac{\partial H^{(1)}}{\partial Q} = 0$$

$$\dot{Q} \approx \frac{\partial H^{(1)}}{\partial P} = \frac{3\epsilon^2 P}{4\omega^2}$$

then yields the desired expressions for P and Q.

The approximate solution [Equation (x)] describes an oscillation of amplitude A and period $\tilde{\tau}$, with

$$\tilde{\tau} = \frac{2\pi/\omega}{1 + 3\epsilon^2 A^2/8}$$

Inasmuch as the ratio of $\tilde{\tau}$ to τ [see Equation (b)] depends solely on $\epsilon^2 A^2$, τ and $\tilde{\tau}$ can be compared by evaluating this ratio for various values of $\epsilon^2 A^2$. Figure 6.4, which shows the results of such computations, suggests

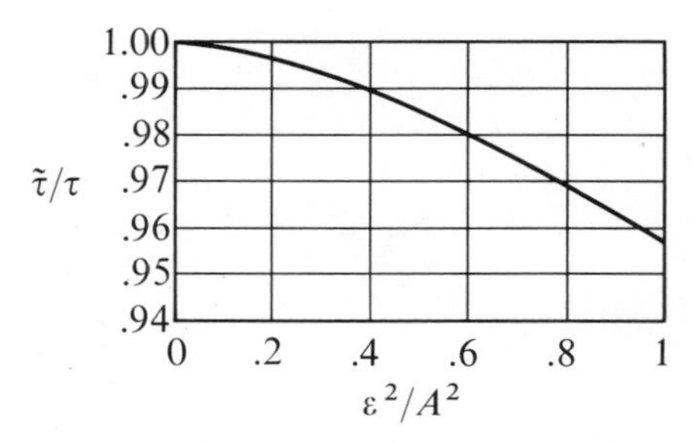

Figure 6.4

that the method of variation of parameters can, indeed, lead to useful approximations. This becomes evident when it is noted that comparison of $\epsilon^2 A^2$ with unity furnishes a measure of the relative importance of the nonlinear term in Equation (a): Even when $\epsilon^2 A^2 = 1$, the approximate value of the period is seen to differ from the exact value by only slightly more than 4 percent.

Problems

▪ PROBLEM SET 11

(Sections 5.1–5.8)

11(a) Figure 11(a) is a schematic representation of a gyroscopic compass. Differential equations governing the operation of this device are to be derived on the basis of the following assumption: The earth E is a sphere of radius R that rotates in an inertial reference frame R^* with a constant angular speed Ω about a line NS passing through the center of the sphere and fixed both in E and in R^*.

The gyroscope consists of a rotor A, an inner gimbal ring B, a pendulous mass P, an outer gimbal ring C, and a frame F. The rotor A is driven by a motor that forms an integral part of the inner gimbal ring B, and the pendulous mass P is rigidly attached to B.

A general configuration of the system may be described as follows.

Let x_1, x_2, and x_3 be mutually perpendicular axes fixed in E, intersecting at a point O on the surface of E, and directed vertically upward, toward the east, and northward, respectively, as shown. Regard x_1, x_2, and x_3 as fixed also in F, and let c_1, c_2, and c_3 be mutually perpendicular lines fixed in C, taking c_1 to be coincident with x_1, and letting θ measure the angle between x_2 and c_2 (and hence between x_3 and c_3). Next, let b_1, b_2, and b_3 be mutually perpendicular lines fixed in B, b_2 coinciding with c_2, and let ϕ measure the angle between c_3 and b_3. Finally, let a_1, a_2, and a_3 be mutually perpendicular lines fixed in A, a_3 coinciding with b_3, and let ψ measure the angle between b_1 and a_1.

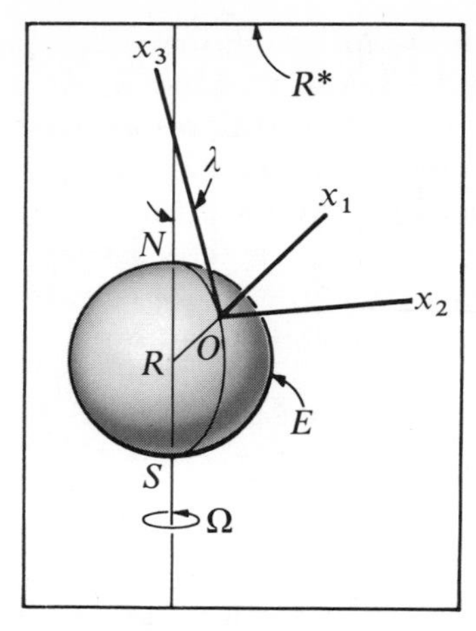

Figure 11(a)-1

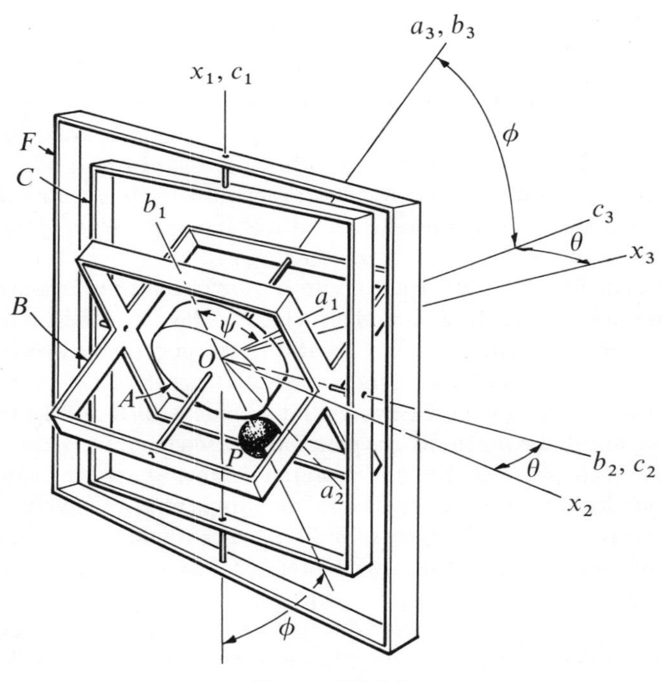

Figure 11(a)-2

A, B, C, and P have the following inertia properties: Point O is the mass center of each of A, B, and C. Lines a_i, b_i, and c_i, $i = 1, 2, 3$, are principal axes of inertia of A, B, and C, respectively, for point O, and the associated principal moments of inertia are A_i, B_i, C_i, $i = 1, 2, 3$, with $A_2 = A_1$. P has a mass m, and the distance between O and P is L.

When A is driven in such a way that $\dot{\psi} = \omega$, a constant, and if all second- or higher-degree terms in θ and ϕ are dropped, the differential equations governing θ and ϕ assume the form

$$r_1\ddot{\theta} + r_2\theta + r_3\dot{\phi} = 0$$
$$s_1\ddot{\phi} + s_2\phi + s_3\dot{\theta} = s_4$$

where r_1, r_2, r_3 and s_1, $\cdots$, s_4 are constants. Determine these constants, assuming that the gravitational force exerted on P by E has the magnitude mg, where g is the gravitational acceleration at point O, and the line of action of this force is parallel to x_1.

Results:

$$r_1 = A_1 + B_1 + C_1$$
$$r_2 = \omega\Omega A_3 \cos\lambda + \Omega^2(A_3 - A_2 + B_3 - B_2 + C_3 - C_2)\cos^2\lambda$$
$$r_3 = \omega A_3 - \Omega(A_1 + A_2 - A_3 + B_1 + B_2 - B_3)\cos\lambda$$
$$s_1 = A_2 + B_2 + mL^2$$
$$s_2 = mgL + \omega\Omega A_3\cos\lambda + \Omega^2[(A_3 - A_1 + B_3 - B_1)(\cos^2\lambda - \sin^2\lambda) - mL^2 + mL(2L - R)\cos^2\lambda]$$
$$s_3 = -\omega A_3 + \Omega(A_1 + A_2 - A_3 + B_1 + B_2 - B_3)\cos\lambda$$
$$s_4 = \Omega(\omega A_3 - mLR\Omega\cos\lambda)\sin\lambda + \Omega^2(A_3 - A_1 + B_3 - B_1 + mL^2)\sin\lambda\cos\lambda$$

11(b) Regarding the earth E, moon M, and sun S each as a particle, let Q be the mass center of E and S, and let R_E, R_M, R_S, and R_Q designate reference frames in which E, M, S, and Q, respectively, are fixed and whose relative orientations do not change with time. Furthermore, assume that R_S can be chosen in such a way that E moves in R_S on a circle with a radius of 93 million miles, traversing this orbit once per year.

Assuming that R_Q is an inertial reference frame, assess the advisability of regarding R_E, R_M, and R_S as inertial reference frames for the purposes of analyzing motions of E, M, S, e, and m, where e and m designate a low-altitude satellite of the earth and a low-altitude satellite of the moon, respectively.

Results: The choices that can be expected to yield useful results are E: R_S; M: R_S; e: R_S, R_E; m: R_S, R_M.

11(c) Referring to Problems 5(e) and 8(d), and taking $\phi = 0$ and $\dot{q}_1 = -\dot{q}_2 = \Omega$ at time $t = 0$, show that the midpoint of S moves on a curve having the form indicated in Figure 11(c). Determine the quantity λ.

Result: $(w + W)L^2 \sin \beta/(J_1 + J_2)\Omega^2$.

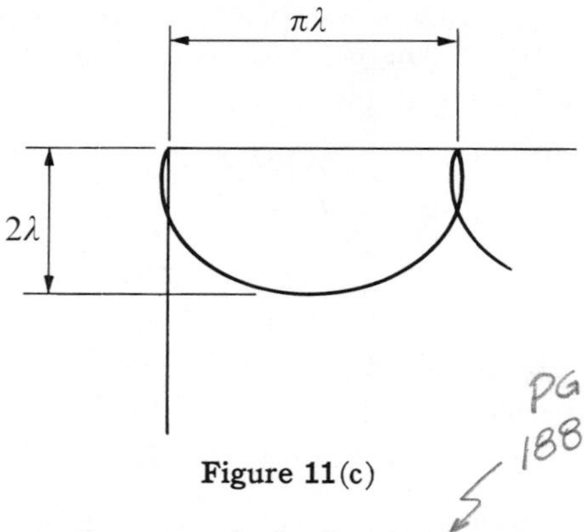

Figure 11(c)

11(d) Referring to the example in Section 5.4, suppose that a set of friction forces exerted on S by T is equivalent to a couple of torque $-cu_1\mathbf{n}_1$, where c is a constant; and let v_2 and v_3 be scalar functions of time such that the velocity $\mathbf{v}$ of S^* is given by

$$\mathbf{v} = v_2\mathbf{n}_2 + v_3\mathbf{n}_3$$

Assuming that the axis of T is vertical, and taking $v_2 = 0$, $v_3 = 4c(R - r)/mr^2$, and $u_1 = 0$ at $t = 0$, show that v_2 approaches a limiting value $v_2{}^*$ as t approaches infinity, and plot $v_2/v_2{}^*$ versus ct/mr^2.

Result: See Figure 11(d).

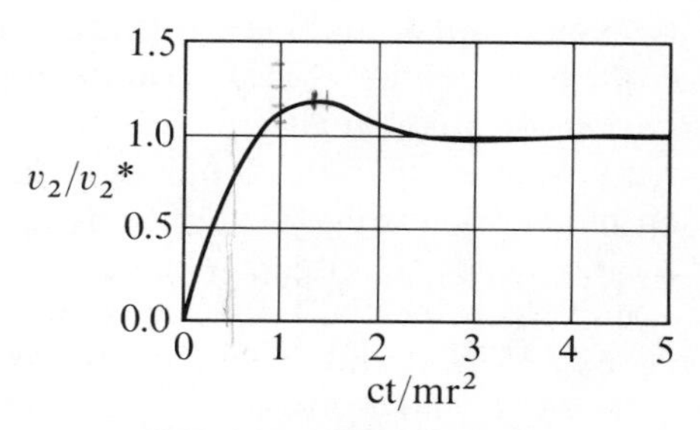

Figure 11(d)

11(e) When a rigid body moves in an inertial reference frame R^* under the action of a system of body and contact forces equivalent to a couple of torque $\mathbf{T}$ together with a force $\mathbf{F}$ applied at the mass center B^* of B, the vectors **T** and **F** can always be expressed as

$$\mathbf{T} = \sum_{i=1}^{3} T_i \mathbf{n}_i \qquad \mathbf{F} = \sum_{i=1}^{3} F_i \mathbf{N}_i$$

where $\mathbf{n}_1$, $\mathbf{n}_2$, and $\mathbf{n}_3$ comprise a right-handed set of mutually perpendicular unit vectors respectively parallel to principal axes of inertia of B for B^* (but not necessarily fixed in B) and $\mathbf{N}_1$, $\mathbf{N}_2$, and $\mathbf{N}_3$ are any three non-coplanar unit vectors fixed in R^*; and the velocity $\mathbf{v}$ of B^* and angular velocity $\boldsymbol{\omega}$ of B in R^* can be expressed as

$$\mathbf{v} = \sum_{i=1}^{3} v_i \mathbf{N}_i \qquad \boldsymbol{\omega} = \sum_{i=1}^{3} \omega_i \mathbf{n}_i$$

Letting

$$u_r = \begin{cases} v_r & r = 1, 2, 3 \\ \omega_{r-3} & r = 4, 5, 6 \end{cases}$$

and, referring to Section 5.4 of the text, show that

$$F_1 = m\dot{v}_1 \qquad F_2 = m\dot{v}_2 \qquad F_3 = m\dot{v}_3$$
$$I_1\alpha_1 - (I_2 - I_3)\omega_2\omega_3 = T_1$$
$$I_2\alpha_2 - (I_3 - I_1)\omega_3\omega_1 = T_2$$
$$I_3\alpha_3 - (I_1 - I_2)\omega_1\omega_2 = T_3$$

where m is the mass of B, I_i is the moment of inertia of B about a line passing through B^* and parallel to $\mathbf{n}_i$, and $\alpha_i = \boldsymbol{\alpha} \cdot \mathbf{n}_i$, $\boldsymbol{\alpha}$ being the angular acceleration of B in R^*. (The first three equations express *Newton's second law* of motion, and the last three are known as *Euler's dynamical equations.*)

11(f) Assuming that the only body and contact forces acting on the system described in Problem 8(g) are gravitational forces, show that $u_1, \cdots, u_4$, and θ are governed by the differential equations

$$\dot{x}_i = f_i(x_1, \cdots, x_5) \qquad i = 1, \cdots, 5$$

where $x_i = u_i$ for $i = 1, \cdots, 4$, and $x_5 = \theta$. Dropping all terms of second or higher degree in x_4 and/or x_5, determine $f_1, \cdots, f_5$.

Result: $-x_2x_3$, $-x_3(x_1 + 2x_4)$, x_1x_2, $2x_2x_3 + (x_3^2 - x_2^2)x_5$, x_4.

11(g) Four spheres, A, B, C, and D, each of radius R, are supported in hemispherical sockets, as shown in Figure 11(g); and a heavy object E (not shown) with a planar base is supported by the spheres. If E moves in such a way that no slip occurs at the point of contact between E and any sphere, the system comprised of $A, \cdots, E$ possesses seven degrees

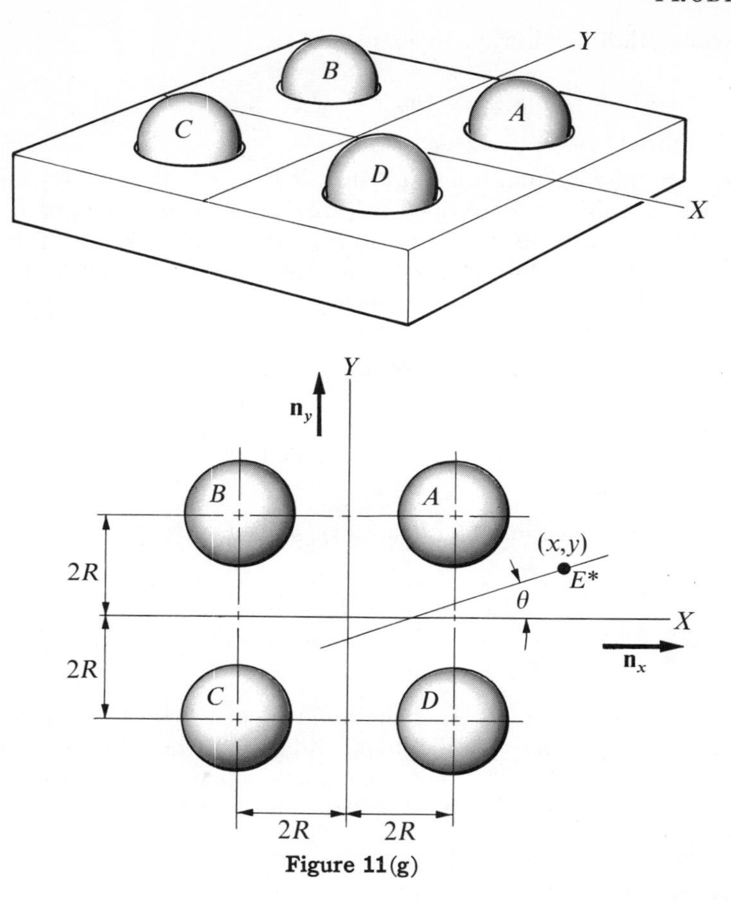

Figure 11(g)

of freedom. Consequently, seven equations of the form of Equations (5.3) govern quantities $u_1, \cdots, u_7$ defined as

$$u_1 = \frac{\dot{x}}{R} \qquad u_2 = \frac{\dot{y}}{R} \qquad u_3 = \dot{\theta}$$

$$u_4 = \boldsymbol{\omega}^A \cdot \mathbf{n}_z, \quad \cdots \quad, u_7 = \boldsymbol{\omega}^D \cdot \mathbf{n}_z$$

where x and y are coordinates of a point E^* [see Figure 11(g)] fixed in the base of E; θ is the angle between X and a line fixed in the base of E; $\boldsymbol{\omega}^A, \cdots, \boldsymbol{\omega}^D$ denote the angular velocities of $A, \cdots, D$; and $\mathbf{n}_z = \mathbf{n}_x \times \mathbf{n}_y$.

Determine the number of quantities whose values must be specified for time $t = 0$ in order to obtain a complete description of the state of the system for $t > 0$; and, supposing that the system is set into motion in such a way that, initially,

$$x = y = \theta = 0$$
$$\boldsymbol{\omega}^A = \Omega \mathbf{n}_x \qquad \boldsymbol{\omega}^B = 2\Omega \mathbf{n}_x$$

find the initial values of u_1, u_2, and u_3.

Results: 22; $\Omega/2$, $-3\Omega/2$, $\Omega/4$.

11(h) Referring to Problem 2(g) [see also Problem 3(c)], and assuming that the support S is horizontal, use the principle of virtual work to derive three second-order differential equations governing θ, ϕ, and ψ. Next, referring to Section 5.4, introduce u_1, u_2, and u_3 as

$$u_1 = -\dot{\theta} \qquad u_2 = \dot{\phi} \cos \theta \qquad u_3 = \dot{\psi} + \dot{\phi} \sin \theta$$

and derive three first-order equations governing u_1, u_2, and u_3. Comment briefly on the difference between these two methods for deriving dynamical equations.

Results:

$$\ddot{\phi} \cos \theta + 2\dot{\theta}\dot{\psi} = 0$$
$$5\ddot{\theta} - 6\dot{\phi}\dot{\psi} \cos \theta - 5\dot{\phi}^2 \sin \theta \cos \theta - \frac{4g \sin \theta}{r} = 0$$
$$3(\ddot{\psi} + \ddot{\phi} \sin \theta) + 5\dot{\phi}\dot{\theta} \cos \theta = 0$$
$$5\dot{u}_1 + 6u_2u_3 - u_2^2 \tan \theta + \frac{4g \sin \theta}{r} = 0$$
$$\dot{u}_2 - 2u_1u_3 + u_1u_2 \tan \theta = 0$$
$$3\dot{u}_3 - 2u_1u_2 = 0$$

11(i) Considering the systems described in Problems 5(a), 5(b), 5(d), 5(e), 5(h), 8(a), 8(g), 9(b), 9(g), 11(a), and 11(g), assess in each case the relative advantages of using Equations (5.1), (5.5), and (5.7) to derive dynamical equations.

▪ PROBLEM SET 12

(Sections 5.9–5.17)

12(a) Referring to Problem 5(b), and letting Q, R, and S have the values $62W$, $24W$, and $68W$, respectively, determine the equilibrium values of q_1, q_2, and q_3.

Results: 45, 63.4, 71.6 degrees.

12(b) Four bars, each of length L, are connected by hinges and linear springs, as shown in Figure 12(b). Each of the springs has a natural length $L/2$, and the spring constants have the values k_1 and k_2, respectively.

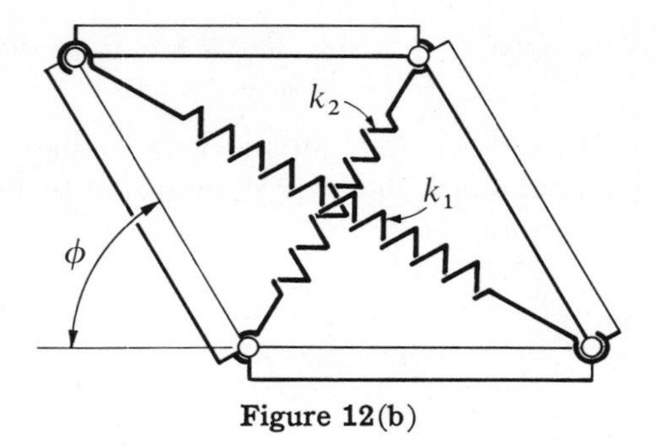

Figure 12(b)

Determine the value of k_1/k_2 such that this system is in equilibrium when $\phi = 60$ degrees.

Result: 0.7.

12(c) Assuming that the system described in Problem 5(g) is in equilibrium, determine the ratio T_b/T_e.

Result: -244.

12(d) Referring to Problem 2(d), and letting $T\mathbf{N}$, $T'\mathbf{N}$, and $t\mathbf{n}$ be the torques of couples applied to A, A', and D, respectively, show that the system is in equilibrium only if T and T' are each equal to $-ta/2b$.

12(e) Two particles of weights W_1 and W_2 are attached to each other by two light strings, one of these having a length l while the other has a length $L(L > l)$. The longer string is supported by a cylindrical roller whose axis is mounted in smooth bearings, as indicated in Figure 12(e).

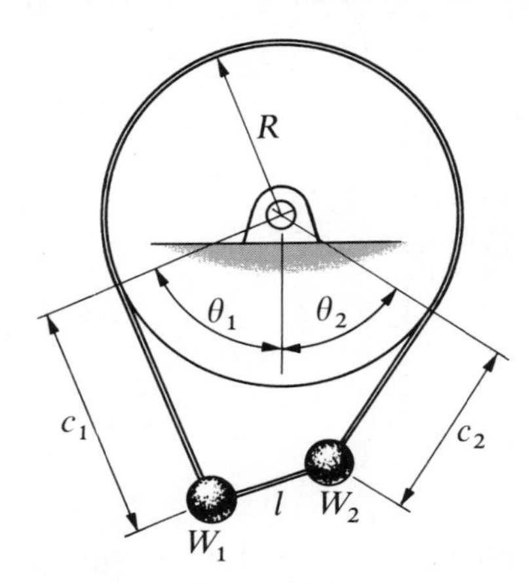

Figure 12(e)

Taking $l/R = 1$, $L/R = 8$, and $W_1/W_2 = 0.5$, determine θ_1 for equilibrium.

SUGGESTION: Introduce the auxiliary variables $\xi_1 = c_1/R$ and $\xi_2 = c_2/R$, where c_1 and c_2 are the lengths indicated in Figure 12(e), and show that, for equilibrium,

$$\theta_1 = \theta_2 = \theta$$

$$\tan\theta = 2\xi_2 - \xi_1 \qquad \xi_1 + \xi_2 - 2\theta = 8 - 2\pi$$

and

$$[(\xi_1 - \xi_2)\sin\theta]^2 + [2\sin\theta - (\xi_1 + \xi_2)\cos\theta]^2 = 1$$

Result: 1.28 radians.

12(f) The system described in Problem 5(b) [see also Problems 9(d) and 12(a)] is in equilibrium, with $q_1 = q_2 = q_3 = 45$ degrees. Letting B designate the third bar from the left in the middle row of bars, determine the magnitude of the reaction of B on the pin supporting the upper end of B; and evaluate Q, R, and S.

Results: $5.3W; 27W, 8W, 49W$.

12(g) Figure 12(g) shows a linkage used in a circuit breaker. The

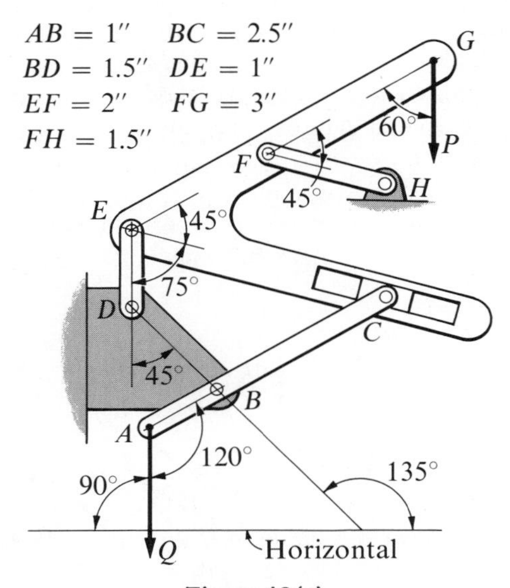

Figure 12(g)

system is in equilibrium when forces of magnitude P and Q are applied as indicated in the sketch.

Assuming that all contact surfaces are smooth and that weight forces are negligible, determine the ratio P/Q.

Result: 0.43.

12(h) In Figure 12(h), $A, \cdots, E$ represent identical uniform square

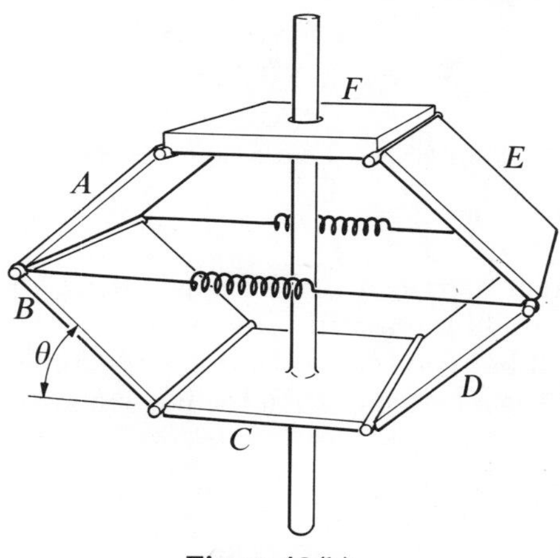

Figure 12(h)

plates, each of mass m and side L. These are attached to each other and to a uniform square plate F of mass $5m$, side L, by means of smooth hinges. A vertical shaft, to which C is rigidly attached, passes through an opening in F, thus leaving F free to slide on the shaft. Finally, two light, linear springs, each of natural length L and modulus k, connect the plates as shown.

One possible motion of this system is described as follows: The shaft is made to revolve with a constant angular speed Ω, and $\theta = \bar{\theta}$, a constant. Determine Ω^2.

Result:

$$\frac{6 \cot \bar{\theta}}{3 + 4 \cos \bar{\theta}} \left(\frac{4k \sin \bar{\theta}}{m} - \frac{7g}{L} \right)$$

12(i) In Figure 12(i), X and Y designate a fixed horizontal line and a fixed vertical line, respectively; L is a line normal to a thin circular disk C; M is a line fixed in the plane of C; and N is the common perpendicular to Y and L (and hence horizontal).

Show that C can move in such a way that $q_1 = \pi/3$ radians, $\dot{q}_2 = 2\pi$ rad/sec, and $\dot{q}_3 = s$, a constant; and determine s.

Result: $-\pi/2$ rad/sec.

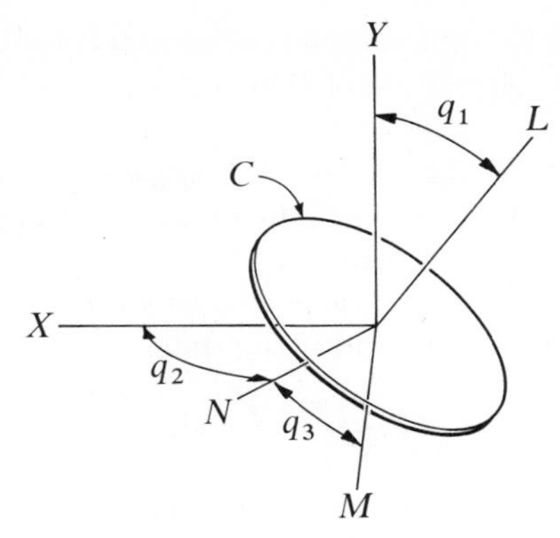

Figure 12(i)

12(j) The rod AB of Problem 5(a) [see also Problems 8(c), 9(e), and 10(b)] is moving in such a way that θ remains constant (and not equal to zero). During this motion, the system of forces exerted by one half of the rod on the other half can be replaced with a couple together with a force applied at the contact point between the two halves. Resolving the torque of the couple into two components, one normal to the plane of the wire, the other parallel to this plane, determine the magnitude of the former component.

Result:

$$\frac{WL}{4}\left[A + \frac{(R^2 - L^2)^{1/2}\Omega^2(1 - A^2)}{g}\right]$$

where

$$A = \frac{3g(R^2 - L^2)^{1/2}}{\Omega^2(3R^2 - 4L^2)}$$

▪ PROBLEM SET 13

(Sections 5.18–5.21)

13(a) A rigid body B forms part of a simple nonholonomic system S possessing $n - m$ degrees of freedom in an inertial reference frame R. Letting $\mathbf{L}$ and $\mathbf{H}$ denote, respectively, the linear momentum of B in R and the angular momentum of B in R relative to the mass center B^* of B, show that the contribution of B to the generalized momenta of S can be expressed as

$$(p_r)_B = \tilde{\mathbf{v}}_{u_r} \cdot \mathbf{L} + \tilde{\boldsymbol{\omega}}_{u_r} \cdot \mathbf{H} \qquad r = 1, \cdots, n - m$$

where $\tilde{\mathbf{v}}_{u_r}$ and $\tilde{\boldsymbol{\omega}}_{u_r}$ are nonholonomic partial rates of change of the position of B^* in R and of the orientation of B in R, respectively.

13(b) A rigid body B forms part of a simple nonholonomic system S possessing $n - m$ degrees of freedom in an inertial reference frame R, and forces $\mathbf{F}_i$ are applied at points Q_i, with $i = 1, \cdots, \beta$.

Let $\mathbf{I}_i$, $i = 1, \cdots, \beta$, be a set of bound vectors applied at Q_i, $i = 1, \cdots, \beta$, respectively, and defined as

$$\mathbf{I}_i = \int_{t_1}^{t_2} \mathbf{F}_i \, dt \qquad i = 1, \cdots, \beta$$

Next, replace this set of vectors with a bound vector $\mathbf{I}$, applied at a point Q of B, together with a couple of torque $\mathbf{J}$, and show that the contribution of $\mathbf{F}_1, \cdots, \mathbf{F}_\beta$ to the generalized impulse can be expressed as

$$(I_r)_B = \tilde{\mathbf{v}}_{u_r}{}^Q \cdot \mathbf{I} + \tilde{\boldsymbol{\omega}}_{u_r} \cdot \mathbf{J}$$

where $\tilde{\mathbf{v}}_{u_r}{}^Q$ and $\tilde{\boldsymbol{\omega}}_{u_r}$ are values at time t_1 of a nonholonomic partial rate of change of the position of Q in R and of the orientation of B in R, respectively.

13(c) Referring to Problem 5(e), suppose that $\beta = 0$ and that the system is set into motion by being struck at a point Q on the shaft S, the distance between Q and the center of disk D_1 being equal to s. Show that the midpoint of S moves on a circular path, and determine the radius of the circle.

Result:

$$\frac{4W + 3w(6 + r^2/L^2)}{6(2W + 3w)(1 - s/L)}$$

13(d) Assuming that R and S in Problem 5(b) are equal to zero, but that the force of magnitude Q is applied suddenly at an instant when the system is at rest, with $q_1 = q_2 = q_3 = 0$, determine the value of $\dot{q}_2/\dot{q}_1$ immediately subsequent to the application of this load.

Result: 10.

13(e) The uniform block shown in Figure 13(e) is projected in such a way that, at $t = 0$, the velocity of point A has a magnitude of 10 inches per second and is directed from A toward B, the angular velocity of the block has the direction CB and a magnitude of 5 radians per second, and the edge BD is vertical.

There exists an instant t^* at which the block can be brought to rest by a single blow. Determine t^*.

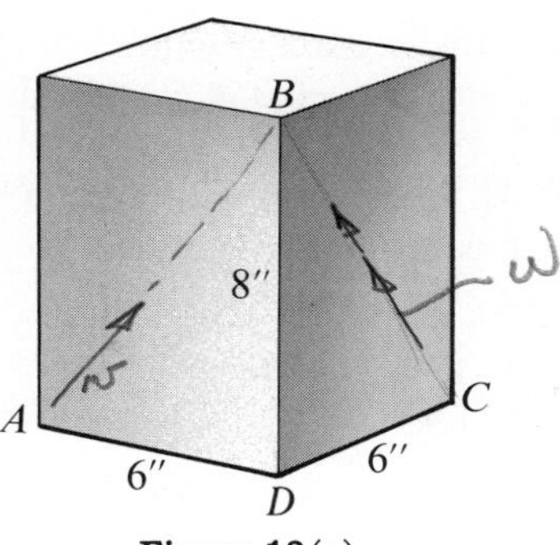

Figure 13(e)

SUGGESTION: Show that $\mathbf{H} \cdot \mathbf{v} = 0$ where $\mathbf{H}$ is the angular momentum of the block with respect to the mass center of the block and $\mathbf{v}$ is the velocity of the mass center, both evaluated for time t^*.

Result: 0.0116 sec.

13(**f**) A rigid body B is attached to a shaft that is free to rotate about its axis. Show that if the axis of rotation is a principal axis of inertia of B for one point of this axis, and a blow applied to B is perpendicular to the plane containing the mass center of B and the axis of rotation, then the shaft is not subjected to impulsive reactions when B is struck.

13(**g**) A uniform, thin, circular disk D of mass M and radius R is rolling in a straight line with a speed V on a horizontal support. A particle P of mass $M/10$, moving horizontally with the same speed, collides with D, striking the center of D at an instant at which the velocity vector of P is parallel to the axis of D. Thereafter, P adheres to D.

Assuming that D does not slip on its support, determine the magnitude of the angular velocity of D immediately subsequent to the collision.

Result: $0.94V/R$.

13(**h**) Two uniform spheres, S and S, collide at an instant at which the line joining their centers is parallel to the unit vector $\mathbf{n}_1$ of a right-handed set of fixed, mutually perpendicular unit vectors $\mathbf{n}_1$, $\mathbf{n}_2$, and $\mathbf{n}_3$. Immediately prior to the collision, the velocities $\mathbf{v}$ and $\bar{\mathbf{v}}$ of the centers of S and $\bar{S}$, and the angular velocities $\boldsymbol{\omega}$ and $\bar{\boldsymbol{\omega}}$ of S and $\bar{S}$, are given by

$$
\begin{aligned}
\mathbf{v} &= 0.04\mathbf{n}_1 + 0.02\mathbf{n}_2 + 0.03\mathbf{n}_3 \text{ ft/sec} \\
\bar{\mathbf{v}} &= -0.04\mathbf{n}_1 - 0.02\mathbf{n}_2 - 0.03\mathbf{n}_3 \text{ ft/sec} \\
\boldsymbol{\omega} &= 0.01\mathbf{n}_1 + 0.02\mathbf{n}_2 + 0.03\mathbf{n}_3 \text{ rad/sec} \\
\bar{\boldsymbol{\omega}} &= 0
\end{aligned}
$$

The radii R and $\bar{R}$, and the masses m and $\bar{m}$, of S and $\bar{S}$ have the values

$$R = 0.25 \text{ ft} \qquad \bar{R} = 0.50 \text{ ft}$$
$$m = 0.20 \text{ slug} \qquad \bar{m} = 1.60 \text{ slugs}$$

Taking the coefficient of restitution equal to 0, 0.5, and 1.0, determine, in each case, (a) the value of the coefficient of friction necessary to prevent slip, and (b) the percentage loss of kinetic energy of the system during the collision, assuming that slip does not occur.

Results: (a) 0.26, 0.17, 0.13; (b) 27, 21, 5.

13(i) A billiard ball moves on a curved (parabolic) path if it is struck in such a way that the initial velocity of the point of the ball that is in contact with the support is neither equal to zero nor parallel to the initial velocity of the center of the ball. For example, this happens when the cue is used as indicated in Figure 13(i), where $\mathbf{n}_1$, $\mathbf{n}_2$, and $\mathbf{n}_3$ are mutually

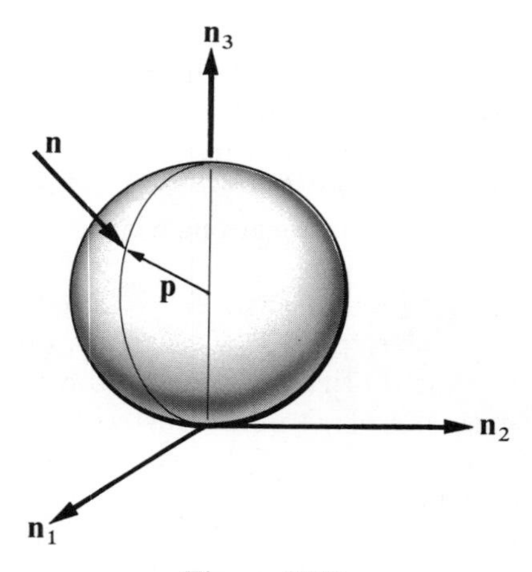

Figure 13(i)

perpendicular unit vectors, with $\mathbf{n}_3$ vertical; $\mathbf{p} = (1/\sqrt{2})(\mathbf{n}_1 + \mathbf{n}_3)$ is the position vector, relative to the center of the ball, of the point at which the ball is struck; and $\mathbf{n} = (1/\sqrt{2})(\mathbf{n}_2 - \mathbf{n}_3)$ is a unit vector parallel to the cue.

Assuming that the force applied to the ball by the cue is at all times parallel to $\mathbf{n}$, and taking the coefficient of friction for the ball and table equal to 0.1, determine the cosine of the angle between the initial velocity of the center and the initial velocity of the contact point.

Result: -0.477.

▪ PROBLEM SET 14

(Sections 6.1–6.9)

14(a) Letting p_1 and p_2 denote the generalized momenta associated with the coordinates q_1 and q_2 of Problem 8(a) [see also Problem 10(c)], show that

$$\dot{q}_1 = \frac{p_1 m_{22} - p_2 m_{12}}{m_{11} m_{22} - m_{12}^2}$$

and

$$\dot{q}_2 = \frac{p_2 m_{11} - p_1 m_{12}}{m_{11} m_{22} - m_{12}^2}$$

where

$$m_{11} = \frac{m}{6}(3b^2 + a^2 \sin^2 q_2)$$

$$m_{12} = \frac{mab}{4} \cos q_2$$

$$m_{22} = \frac{ma^2}{6}$$

14(b) Letting f designate a continuously differentiable function of the $2n + 1$ variables $x_1, \cdots, x_n, y_1, \cdots, y_n$, and t, and defining X, Y, and T as

$$X = f_x \qquad Y = f_y \qquad T = f_t$$

show that

$$T_x = X_t$$

and

$$X_y = Y_x^T$$

where Y_x^T denotes the transpose of Y_x.

14(c) Letting v denote an n-dimensional column matrix and U the matrix defined in Equation (6.12), verify that

$$(Uv)^T = v^T U = v$$

and that

$$U^T = U$$

14(d) Letting u and v denote n-dimensional column matrices and w an $n \times n$ matrix, verify that

$$(u^T v)^T = v^T u$$

and

$$(wv)^T = v^T w^T$$

14(e) Referring to Problem 14(b), express the partial derivative of $f_x f_y^T$ with respect to t in terms of X, Y, X_t, and Y_t.

Result: $XY_t^T + YX_t^T$.

14(f) Letting H be the Hamiltonian of the rod whose motion is described in Problem 5(a) [see also Problems 9(e) and 10(b)], evaluate $H - (K + P)$, where K and P are the kinetic and potential energy of the rod.

Result: $-(W\Omega^2/3g)[3(R^2 - L^2)\sin^2\theta + L^2\cos^2\theta]$.

14(g) Form the Hamiltonian H for the system described in Problem 8(a) [see also Problems 10(c) and 14(a)] and show that $H = K + P$, where K and P are the kinetic and potential energy of the plate.

Result:

$$H = \frac{m_{22}p_1^2 - 2m_{12}p_1p_2 + m_{11}p_2^2}{2(m_{11}m_{22} - m_{12}^2)} - (mga/3)\cos q_2$$

14(h) Letting p_1 denote the generalized momentum associated with the coordinate q_1 in Problem 8(a) [see also Problems 14(a) and 14(g)], show that

$$p_1 = C, \text{ a constant}$$

is an integral of the equations of motion of the plate. Next, use this fact together with the energy integral to determine the magnitude of the angular velocity of the plate for an instant at which $q_2 = 0$, assuming that the plate is released from rest when $q_2 = \pi/3$ radians.

Result:

$$\frac{2}{b}\left[2ga\left(\frac{1}{4} + \frac{b^2}{a^2}\right)\right]^{1/2} = \left[2g\frac{a}{b^2} + 8g\frac{1}{a}\right]^{1/2}$$

14(i) Taking $R = 2L$ in Problem 5(a) [see also Problem 14(f)], determine Ω^2 such that $\dot{\theta} = 0$ whenever $\theta = \pi/4$ or $\theta = \pi/2$.

Result:

$$\frac{3\sqrt{3}}{2\sqrt{2}}\frac{g}{L}$$

14(j) Figure 14(j) shows a gyroscopic device consisting of a frame F, spring assembly S, gimbal ring G, and rotor R. These parts have the following inertia properties:

The point of intersection of the spin axis and the output axis is the

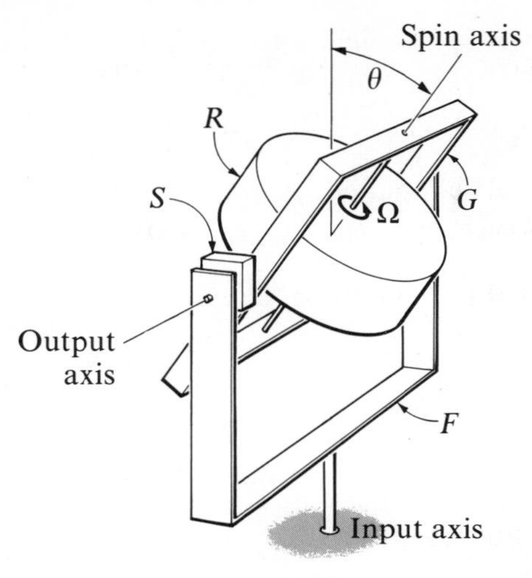

Figure 14(j)

mass center of R, and R has a moment of inertia J about the spin axis, and a moment of inertia I about any line passing through the mass center of R and perpendicular to the spin axis. The mass center of G coincides with that of R, and the spin axis, the output axis, and a line perpendicular to both of these and passing through their point of intersection are all principal axes of inertia of G, the corresponding moments of inertia being A, B, and C. Finally, the mass center of F lies on the input axis, and F has a moment of inertia K about this axis.

Consider the following class of motions of this system: F is free to rotate about the input axis, which is fixed; G can rotate about the output axis, but a resisting torque of magnitude $k\theta$ is associated with such rotations; and R is made to rotate with constant angular speed Ω in G (by means of a motor that forms an integral part of G).

Assuming that at time $t = 0$ the frame F is at rest, $\theta = \pi/2$ radians, and $\dot{\theta} = 0$, determine the value of the spring constant k such that $\dot{\theta}$ vanishes for $t > 0$ when $\theta = \pi/4$ radians.

Result:

$$\frac{(4J\Omega/\pi)^2}{3(A + C + I + J + 2K)}$$

14(k) If L is the kinetic potential of a system S whose first k generalized coordinates are ignorable, and if $\alpha_1, \cdots, \alpha_k$ are the constant values of the associated generalized momenta, then the quantity R

defined as

$$R = L - \sum_{s=1}^{k} \alpha_s \dot{q}_s$$

may be regarded as a function of $\rho_1, \cdots, \rho_{n-k}$ and $\dot{\rho}_1, \cdots, \dot{\rho}_{n-k}$, where $\rho_r = q_{k+r}$, with $r = 1, \cdots, n - k$. (R is called the "Routhian" of S.)

Show that R satisfies the differential equations

$$\frac{d}{dt}\frac{\partial R}{\partial \dot{\rho}_r} - \frac{\partial R}{\partial \rho_r} = 0 \qquad r = 1, \cdots, n - k$$

▪ PROBLEM SET 15

(Sections 6.10–6.11)

15(a) A transformation from p_1, p_2, q_1, q_2 to P_1, P_2, Q_1, Q_2 is performed by setting

$$q_1 = a_1 Q_1 + a_2 Q_2 \qquad q_2 = a_3 Q_1 + a_4 Q_2$$
$$p_1 = -2P_1 + P_2 \qquad p_2 = 1.5P_1 - 0.5P_2$$

Determine the values of $a_1, \cdots, a_4$ such that this transformation is canonic with respect to the Hamiltonian $H(p_1,p_2, q_1,q_2, t)$.

Results: 1, 3, 2, 4.

15(b) Letting p and q be n-dimensional column matrices representing generalized momenta and generalized coordinates, respectively, show that a transformation defined by the equations

$$P = Ap \qquad Q = Bq$$

where P and Q are n-dimensional column matrices and A and B are constant $n \times n$ matrices, is canonic with respect to the Hamiltonian $H(p, q, t)$ if and only if $AB^T = U$, where U denotes the $n \times n$ unit matrix.

15(c) A transformation is defined by the equations

$$Q_1 = q_1 \cos \omega t - q_2 \sin \omega t$$
$$Q_2 = q_1 \sin \omega t + q_2 \cos \omega t$$
$$P_1 = p_1 \cos \omega t - p_2 \sin \omega t$$
$$P_2 = p_1 \sin \omega t + p_2 \cos \omega t$$

Assuming that p_r and q_r, with $r = 1, 2$, satisfy Hamilton's canonic equations, show that this transformation is not canonic with respect to the Hamiltonian.

15(d) A transformation from generalized coordinates $q_1, \cdots, q_n$ and generalized momenta $p_1, \cdots, p_n$ to new variables $Q_1, \cdots, Q_n$ and $P_1, \cdots, P_n$ is called a *point transformation* if Q_r is a function of the q's and t while P_r is a function of the p's, q's, and t; that is,

$$Q_r = F_r(q_1, \cdots, q_n, t) \qquad r = 1, \cdots, n$$

and

$$P_r = G_r(p_1, \cdots, p_n, q_1, \cdots, q_n, t) \qquad r = 1, \cdots, n$$

Use the generating function S defined as

$$S = \sum_{r=1}^{n} F_r P_r$$

to show that a point transformation is canonic if

$$G = (F_q{}^T)^{-1} p$$

where p, q, F, and G are the column matrices having p_r, q_r, F_r, and G_r as the elements in the rth row, respectively; and find a quantity H' such that the transformation is canonic with respect to $H + H'$, where H is the Hamiltonian.

Result: $H' = F_t{}^T P$.

15(e) Use the results of Problem 15(d) to show that the transformation defined in Problem 15(c) is canonic with respect to $H + \omega(Q_1 P_2 - Q_2 P_1)$.

15(f) Show that the generating function S defined as

$$S = \sum_{r=1}^{n} P_r q_r$$

can be used to obtain the *identity transformation;* that is, $P_r = p_r$ and $Q_r = q_r$, with $r = 1, \cdots, n$.

15(g) The Hamiltonian H of a certain system is of the form

$$H = F(p_1, p_2, q_1, q_2 + ct)$$

where c is a constant. Show that the generating function S defined as

$$S = P_1 q_1 + P_2(q_2 + ct)$$

can be used to discover a set of canonic variables P_1, P_2, Q_1, Q_2 such that the associated Hamiltonian $\hat{H}$ does not involve the time t explicitly. Determine $\hat{H}$.

Result: $F(P_1, P_2, Q_1, Q_2) + cP_2$.

15(h) The quantities p_1, p_2, q_1, and q_2 satisfy the equations

$$\dot{p}_r = -\frac{\partial H}{\partial q_r}$$

$$\dot{q}_r = \frac{\partial H}{\partial p_r}$$

where

$$H = p_1 p_2 \sin(\epsilon t + \omega_1 q_1 - \omega_2 q_2)$$

and ω_1, ω_2, and ϵ are constants. Express p_r and q_r in terms of t and new variables P_r and Q_r, $r = 1, 2$, in such a way that these satisfy the equations

$$\dot{P}_r = -\frac{\partial \hat{H}}{\partial Q_r}$$

$$\dot{Q}_r = \frac{\partial \hat{H}}{\partial P_r}$$

where $\hat{H}$ is a function of P_1, P_2, Q_1, Q_2 that does not contain t explicitly; and determine $\hat{H}$.

Results:

$$p_1 = P_1 \qquad p_2 = P_2 \qquad q_1 = Q_1 \qquad q_2 = Q_2 + \frac{\epsilon t}{\omega_2}$$

$$\hat{H} = P_1 P_2 \sin(\omega_1 Q_1 - \omega_2 Q_2) - \frac{\epsilon P_2}{\omega_2}$$

15(i) Letting p_r and q_r, with $r = 1, \cdots, n$, denote generalized momenta and generalized coordinates that satisfy Hamilton's canonic equations, show that a transformation to variables P_r and Q_r that are canonic with respect to a function $\hat{H}$ of P_r and Q_r, $r = 1, \cdots, n$, can be obtained by letting

$$\hat{H} = H + \frac{\partial S}{\partial t}$$

where H is the Hamiltonian, and then proceeding in any one of the following three ways:

(a) Let

$$p_r = \frac{\partial S}{\partial q_r}$$

$$P_r = -\frac{\partial S}{\partial Q_r}$$

where S is a function of $q_1, \cdots, q_n, Q_1, \cdots, Q_n$, and t;

(b) Let

$$q_r = -\frac{\partial S}{\partial p_r}$$

$$P_r = \frac{\partial S}{\partial Q_r}$$

where S is a function of $p_1, \cdots, p_n, Q_1, \cdots, Q_n$, and t;

(c) Let

$$q_r = -\frac{\partial S}{\partial p_r}$$

$$Q_r = \frac{\partial S}{\partial P_r}$$

where S is a function of $p_1, \cdots, p_n, P_1, \cdots, P_n$, and t.

15(j) When the Hamiltonian H of a single-degree-of-freedom system can be expressed as

$$H = \tfrac{1}{2}(p^2 + \omega^2 q^2)$$

where ω^2 is a constant, the system is called a *harmonic oscillator* because the Hamiltonian of a particle of mass m that is attached to a linear spring of modulus k and is constrained to move on a horizontal straight line assumes precisely this form if q is defined as $x\sqrt{m}$, where x is the displacement of the particle from its equilibrium position, and ω is taken equal to $\sqrt{k/m}$.

Show that the generating function

$$S = \frac{\omega}{2}\left[q\left(\frac{2P}{\omega^2} - q^2\right)^{1/2} + \frac{2P}{\omega^2}\arcsin\frac{\omega q}{\sqrt{2P}}\right] - Pt$$

can be used to find canonic variables P and Q such that the associated Hamiltonian $\hat{H}(P, Q, t)$ is identically equal to zero. (It follows that P and Q are constants.)

15(k) A transformation from the variables p and q of a harmonic oscillator [see Problem 15(j)] to canonic variables P and Q is obtained by using the generating function

$$S = \tfrac{1}{2}\omega q^2 \cot Q$$

Find the differential equations governing P and Q.

Results: $\dot{P} = 0,\ \dot{Q} = \omega.$

15(l) The Hamiltonian of a certain system is given by

$$H = \omega^2 p(q + t)^2$$

where ω is a constant. Determine q as a function of time (a) by making use of the generating function

$$S = P(q + t)$$

and solving the associated canonic equations, and (b) by direct integration of Hamilton's canonic equations.

Result: $q = -t + \dfrac{1}{\omega} \tan \omega(t - t_0).$

▪ PROBLEM SET 16

(Sections 6.12–6.15)

16(a) Derive the Hamilton-Jacobi equation for a harmonic oscillator [see Problem 15(j)]; next, after replacing P with α, a constant, show by direct substitution that the function S given in Problem 15(j) is a solution of this equation.

16(b) A particle of mass m moves under the action of the gravitational force exerted on it by a particle of mass M that is fixed in an inertial reference frame.

Defining generalized coordinates q_1, q_2, and q_3 as

$$q_1 = r\sqrt{m} \qquad q_2 = \theta \qquad q_3 = \phi$$

where r is the distance between the particles and θ and ϕ measure angles as indicated in Figure 16(b), show that the Hamilton-Jacobi equation for this system can be expressed as

$$\tfrac{1}{2}\left[\left(\frac{\partial S^*}{\partial q_1}\right)^2 + \frac{1}{q_1{}^2}\left(\frac{\partial S^*}{\partial q_2}\right)^2 + \frac{1}{q_1{}^2 \cos^2 q_2}\left(\frac{\partial S^*}{\partial q_3}\right)^2\right] + \frac{\partial S^*}{\partial t} - \frac{\mu}{q_1} = 0$$

where $\mu = GMm^{3/2}$, G being the universal gravitational constant.

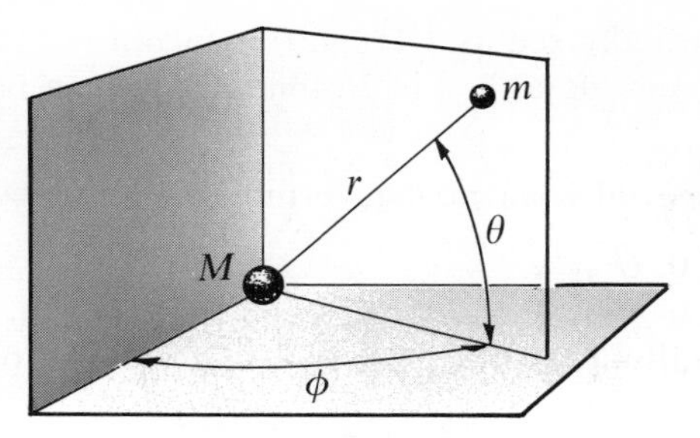

Figure 16(b)

16(c) Use the method of separation of variables to solve the Hamilton-Jacobi equation for a harmonic oscillator [see Problem 16(a)], and thus show that the relationship between q and t can be expressed as

$$\omega(\beta + t) = \pm \text{ arc sin } \frac{q\omega}{\sqrt{2\alpha}}$$

while p is given by

$$p = \pm(2\alpha - \omega^2 q^2)^{1/2}$$

16(d) Letting $H(p_1, \cdots, p_{n-1}, p_n, q_1, \cdots, q_{n-1})$ denote the Hamiltonian of a certain system (note the absence of q_n and t), show that a complete set of integrals of the canonic equations for the system is given by the equations

$$p_r = \frac{\partial W}{\partial q_r} \qquad r = 1, \cdots, n-1$$

$$p_n = \alpha_n$$

$$\beta_1 = -t + \frac{\partial W}{\partial \alpha_1}$$

$$\beta_r = q_n + \frac{\partial W}{\partial \alpha_r} \qquad r = 2, \cdots, n-1$$

$$\beta_n = q_n + \frac{\partial W}{\partial \alpha_n}$$

where α_1, α_n and $\beta_1, \cdots, \beta_n$ are constants; W is a function of $q_1, \cdots, q_{n-1}$ and the n constants $\alpha_1, \cdots, \alpha_n$ (none additive); and W satisfies the *reduced Hamilton-Jacobi equation*

$$H\left(\frac{\partial W}{\partial q_1}, \cdots, \frac{\partial W}{\partial q_{n-1}}, \alpha_n, q_1, \cdots, q_{n-1}\right) = \alpha_1$$

16(e) Show that the reduced Hamilton-Jacobi equation [see Problem 16(d)] for the system described in Problem 16(c) can be expressed as

$$q_1^2\left[\left(\frac{\partial W}{\partial q_1}\right)^2 - 2\alpha_1\right] - 2\mu q_1 + \left(\frac{\partial W}{\partial q_2}\right)^2 + (\alpha_3 \sec q_2)^2 = 0$$

16(f) Letting W_r, with $r = 1, \cdots, n-1$, denote functions of $q_1, \cdots, q_{n-1}$, respectively, and supposing that the substitution

$$W = \sum_{r=1}^{n-1} W_r$$

makes it possible to bring the reduced Hamilton-Jacobi equation [see Problem 16(d)] into the form

$$\sum_{r=1}^{n-1} Z_r\left(\frac{dW_r}{dq_r}, q_r, \alpha_1, \alpha_n\right) = 0$$

where Z_r denotes a function of the arguments indicated, note that W can then be found by solving the $n-1$ ordinary differential equations

$$Z_r = \alpha_r \qquad r = 2, \cdots, n-1$$

$$Z_1 = -\sum_{r=2}^{n-1} \alpha_r$$

Use this fact to verify that the relationship between the time t and the coordinates q_1, q_2, and q_3 in Problem 16(b) [see also Problems 16(e) and 16(d)] can be expressed as

$$\beta_1 = -t \pm \int \left(2\alpha_1 + \frac{2\mu}{q_1} - \frac{\alpha_2}{q_1^2}\right)^{-1/2} dq_1$$

$$\beta_2 = \pm\tfrac{1}{2} \int (\alpha_2 - \alpha_3^2 \sec^2 q_2)^{-1/2}\, dq_2$$

$$\mp \tfrac{1}{2} \int q_1^{-2}\left(2\alpha_1 + \frac{2\mu}{q_1} - \frac{\alpha_2}{q_1^2}\right)^{-1/2} dq_1$$

$$\beta_3 = q_3 \mp \int \alpha_3 \sec^2 q_2 (\alpha_2 - \alpha_3^2 \sec^2 q_2)^{-1/2}\, dq_3$$

✓ **16(g)** Taking

$$H^{(0)} = \tfrac{1}{2}p^2 \qquad H^{(1)} = \tfrac{1}{2}\omega^2 q^2$$

use the method of variation of parameters to find q as a function of t for a harmonic oscillator. [To solve the variational equations, refer to Problem 15(l).]

16(h) Figure 16(f) shows a system comprised of a light, linear spring of modulus k and a particle of mass m. The spring is free to rotate about a horizontal axis, and the length of the spring is such that L, the static equilibrium distance between the particle and the point of support is given by

$$L = \frac{N^2 mg}{k}$$

where g denotes the acceleration of gravity and N is a constant.

If $N = 2$ and the particle is released from rest with $x \ll L$ and $\theta \approx 0$ [see Figure 16(f) for x and θ], it appears, at first, that x has a periodic

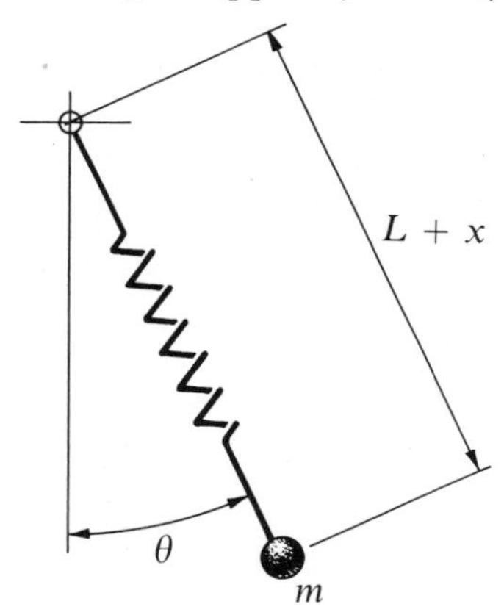

Figure 16(f)

character and that θ remains very small; later, it can be observed that θ has an oscillatory character; and the amplitude of the x oscillations can be seen to decrease as that of the θ oscillations increases. After a certain time, the θ oscillations disappear again, and the process then repeats itself.

To obtain an analytical description of this phenomenon, proceed as follows:

1. Defining two generalized coordinates, q_1 and q_2, as

$$q_1 = x\sqrt{m} \qquad q_2 = \theta L \sqrt{m}$$

and two "natural" frequencies, ω_1 and ω_2, as

$$\omega_1 = \left(\frac{k}{m}\right)^{1/2} \qquad \omega_2 = \left(\frac{g}{L}\right)^{1/2}$$

show that the Hamiltonian H of the system can be expressed as

$$H = H^{(0)} + H^{(1)} + f + \text{const}$$

where

$$H^{(0)} = \tfrac{1}{2}(p_1^2 + \omega_1^2 q_1^2) + \tfrac{1}{2}(p_2^2 + \omega_2^2 q_2^2)$$
$$H^{(1)} = \frac{q_1}{L\sqrt{m}}\left(\frac{\omega_2^2}{2} q_2^2 - p_2^2\right)$$

and f denotes a function of fourth or higher degree in q_1, q_2, p_1, and p_2.

2. Treating the function f as negligible, obtain the following approximate expressions for q_1 and q_2:

$$q_1\omega_1 \approx (2P_1)^{1/2} \sin [\omega_1(Q_1 + t)]$$
$$q_2\omega_2 \approx (2P_2)^{1/2} \sin [\omega_2(Q_2 + t)]$$

where P_1, P_2, Q_1, and Q_2 are certain functions of time t.

3. Verify that $H^{(1)}$ contains a "zero-frequency" term if $\omega_2 = \omega_1/2$ (or, equivalently, $N = 2$); and, using this value of ω_2 and approximating $H^{(1)}$ with the zero-frequency term, show that

$$H^{(1)} \approx -AP_1^{1/2}P_2 \sin (\omega_1 Q)$$

where

$$A = \frac{3\sqrt{2}}{4\omega_1 L\sqrt{m}}$$

and

$$Q = Q_1 - Q_2$$

4. Verify that P_1, P_2, Q_1, and Q_2 must now satisfy the differential equations

$$\dot{P}_1 = A\omega_1 P_1^{1/2}P_2 \cos \omega_1 Q$$
$$\dot{P}_2 = -A\omega_1 P_1^{1/2}P_2 \cos \omega_1 Q$$
$$\dot{Q}_1 = -\frac{AP_1^{-1/2}P_2 \sin \omega_1 Q}{2}$$
$$\dot{Q}_2 = -AP_1^{1/2} \sin \omega_1 Q$$

5. To find P_2, note that $H^{(1)}$ must remain constant [because $H^{(1)}$ does not contain t explicitly], so that

$$-AP_1^{1/2}P_2 \sin \omega_1 Q = C_1$$

where C_1 is a constant; add the first two equations in Step 4, which yields

$$P_1 + P_2 = C_2$$

where C_2 is a constant; and use these two equations to eliminate P_1 and Q from the second of the equations in Step 4, obtaining

$$\dot{P}_2 = \pm\omega_1[A^2(C_2 - P_2)P_2^2 - C_1^2]^{1/2}$$

6. Letting x, θ, $\dot{x}$, and $\dot{\theta}$ have the initial values

$$x(0) = x_0 \qquad \theta(0) = \theta_0 \qquad \dot{x}(0) = \dot{\theta}(0) = 0$$

verify that the following initial values of P_1, P_2, Q_1, and Q_2 are compatible with the given values:

$$P_1(0) = \frac{kx_0^2}{2} \qquad P_2(0) = \frac{kL^2\theta_0^2}{8}$$

$$Q_1(0) = \frac{\pi}{2\omega_1} \qquad Q_2(0) = \frac{\pi}{\omega_1}$$

Furthermore, verify that C_1 and C_2 (see Step 5) can be expressed as

$$C_1 = \frac{3}{32}\,kL^2\left(\frac{x_0}{L}\right)\theta_0^2$$

$$C_2 = \frac{1}{2}\,kL^2\left[\left(\frac{x_0}{L}\right)^2 + \left(\frac{\theta_0}{2}\right)^2\right]$$

7. To bring the differential equation governing P_2 into a convenient, dimensionless form, introduce a new dependent variable y and a new independent variable z by letting

$$y = \frac{P_2}{kL^2}$$

$$z = \omega_1 t$$

Use the results of Steps 5 and 6 to show that y must then satisfy the initial condition

$$y(0) = \frac{\theta_0^2}{8}$$

and the differential equation

$$\frac{dy}{dz} = \pm\,\frac{3}{32}\,[128y^2(a - y) - b]^{1/2}$$

where

$$a = \tfrac{1}{2}\left[\left(\frac{x_0}{L}\right)^2 + \left(\frac{\theta_0}{2}\right)^2\right] \qquad b = \left[\left(\frac{x_0}{L}\right)\theta_0^2\right]^2$$

Finally, letting $\tilde{\theta}$ denote the "amplitude" of the θ oscillations, show that

$$\tilde{\theta} = 2\sqrt{2y}$$

8. The differential equation found in Step 7 can be expressed as

$$\frac{dy}{dz} = \pm\,\frac{3}{32}\sqrt{Y}$$

where Y denotes a third-degree polynomial in y. Show that Y vanishes for three distinct, real values of y, and let y_1, y_2, and y_3 denote these values, taking

$$y_1 < y_2 < y_3$$

9. Under the conditions described in Step 8, y is a periodic function of z, the period Z being given by

$$Z = \frac{8\sqrt{2}}{3}(y_3 - y_1)^{-1/2}K(\mu)$$

where $K(\mu)$ denotes the complete elliptic integral of the first kind of modulus μ, and

$$\mu = \left(\frac{y_3 - y_2}{y_3 - y_1}\right)^{1/2}$$

Letting T_1 denote the period of pure x oscillations, and noting that $\tilde{\theta}$ (see Step 7) is also a periodic function, show that the period $\tilde{T}$ of $\tilde{\theta}$ is related to T_1 as follows:

$$\frac{\tilde{T}}{T_1} = \frac{4\sqrt{2}}{3\pi}(y_3 - y_1)^{-1/2}K(\mu)$$

Taking $x_0/L = 0.1$ and $\theta_0 = 0.01$ radian, evaluate $\tilde{T}/T_1$.

Result: 34.2.

Index